1996

WRITING BY DESIGN
A Handbook for Technical Professionals

Michael Greene
Wentworth Institute of Technology

Jonathan G. Ripley
Wentworth Institute of Technology

REGENTS/PRENTICE HALL
Englewood Cliffs, New Jersey 07632

Library of Congress Cataloging-in-Publication Data

Greene, Michael, 1943–
Writing by design: a handbook for technical professionals/
Michael Greene, Jonathan G. Ripley.
p. cm.
Includes index.
ISBN 0-13-969346-7
1. Technical writing—Handbooks, manuals, etc. I. Ripley, Jonathan G., 1949– . II. Title.
T11.G682 1992
808'.0666—dc20 91-31611
CIP

Editorial/production supervision: ***Anthony Calcara***
Cover design: ***Rich Dombrowski***
Acquisition editor: ***Maureen Hull***
Prepress buyer: ***Ilene Levy***
Manufacturing buyer: ***Ed O'Dougherty***
Marketing manager: ***Tina Culman***
Supplements editor: ***Cindy Harford***
Editoral assistant: ***Marianne Bernotsky***

Dedicated

Elizabeth Nicole Greene
1967–1992

A Division of Simon & Schuster
Englewood Cliffs, New Jersey 07632

Printed in the United States of America

10 9 8 7 6 5 4 3 2 1

ISBN 0-13-969346-7

Prentice-Hall International (UK) Limited, *London*
Prentice-Hall of Australia Pty, Limited, *Sydney*
Prentice-Hall Canada Inc., *Toronto*
Prentice-Hall Hispanoamericana, S.A., *Mexico*
Prentice-Hall of India Private Limited, *New Delhi*
Prentice-Hall of Japan, Inc., *Tokyo*
Simon & Schuster Asia Pte. Ltd., *Singapore*
Editora Prentice-Hall do Brasil, Ltda., *Rio de Janeiro*

Contents

Preface

This handbook is designed for technical students and technical professionals who need to communicate their ideas to others. It moves step by step through the process of designing a document or presentation effectively, from initial conception to final production. Individual task chapters cover the communication tasks essential to success in technical fields.

Task-focused, *Writing By Design* provides practical advice and answers for persons who need to communicate in writing, at a meeting, or in front of a group of people. Persons who use this handbook can begin directly with the task they need to accomplish or they can approach their communication needs indirectly, working through the Document Design Process and the process chapters. In either case, the book is designed for use, concentrating on common technical communication tasks, from letters and memos to technical directions and proposals. Students and professionals can go directly to what they need to find, working out the design process as they move forward.

Special chapters focus on electronic tools for communication, on a simulated consulting firm to serve as the basis for exercises and projects, and on issues affecting communication across cultures. The book provides full coverage of electronic tools: word processors, page layout software, graphics packages, and other computer-based writing and presentation devices and programs. In addition the text covers a wide variety of technical communication skills, including communicating by phone, participating in meetings, and reading technical material. We provide a human context for learning about the world of work. A fictitious technical consulting company provides an organizational context for writing and speaking and this case study approach is incorporated into the exercises which follow each chapter. Ethical considerations boxes are frequent throughout the text, providing an ethical framework as a basis for discussion and practice. A chapter on communicating across cultures focuses students' attention on the problems and opportunities of international communication.

ACKNOWLEDGMENTS

The journey towards the completion of this manuscript has been eased by a great many people. The following list of individuals provided us with technical information, helpful advice, and enduring support along the way.

Mark Fontanella
Kathy Spirer
Dick Gates
Sandy Pascal
Rich Benedetto
Gary Ham
Barbara Karanian
Angela Times
Michael and Linda Mass
Pat Moore
Russ Bramhall
Ken Gordon
Bob Gundersen
Kathy Kirk
David Chadbourne
Sandy Pearlman
John Crocker
Mike Feller
Joanne Tuck
Jessica Romin
Pam Walsh
Paul Lazarovich
Amos St. Germain

We want to give special thanks to the staff at the Wentworth Memorial Library, including its Director, Anne Montgomery Smith, and Rosemary Walker, Priscilla Biondi, Barbara Coffey, Mike Logan, Betsy Holmes Murray, Dennis Berthiaume, Mary Ellen Flaherty, Sara Niesobecki, and Chris Abraham. We received valuable advice from Robert Villanucci, Fred Driscoll, and David Stevens. Special thanks for help with the Instructor's Manual goes to Jessica Romin.

A number of our students helped us to formulate the ideas included in this text, including but not limited to the following:

Richard Mansfield
Patrick Hafford
Andy Boyce
Kevin Stockwood
Brian Vanlaarhoven
Bill Chadwick
Andy Croce
Justin Cincotti
Jose Rondon
Don Fraser
Mike Tavares
Brian Babineau
Reggie Pagan
June Joseph
David Gerns
John Donnellan
Stuart Grabar
Scott Morris
Lou Bedard
Stan Black
Mark Richmond
Hugh Tripp
Denis Foley
Brendan Farrell
Francis Frey
Donald Welch
Salvatore Purpura
Thomas Armstrong
Charles Gachanja
Parviz Sadri
Dale Anderson
Richard Holland
William Elliott
Donovan Malvy
David Landry
Robert Gerami
Tom Talbot
Gerry Scerra
John Nardone
Barry Sullivan
Anita Bota
Timothy Micklich
Scott Ayles
Richard Smith
Richard Fay
Joseph Laccone
Bruce Yogel
Duc Tran
Glenn Savoy
Robert Gay
Jose Martinez

Suzanne Pierre
Joel Fahy
Robert Jones
Joseph Mercurio
Patrick Hetherton
Rajiv Patel
Thomas Silvestro
Richard Gadbois
David Pinney
Patrick Dumont
Wesley Thomas
Mary Donnellan
Michael Gouchie
Janice Graves
John Buettner
Rhonda Tilmans

We would like to thank all of the students we have had the pleasure of teaching in our Technical Writing, Technical Communications, and Professional Communications courses throughout the years; we have tested our output with them.

We would also like to thank the individuals who reviewed our manuscript for their support and helpful suggestions: Maris Roze of the DeVry Institute of Evanston, Illinois; Leanne D. Murray; William N. McGaw of the Texas A&M University System; Chris W. Grevesen of DeVry Technical Institute in Woodbridge, New Jersey; Linda Dobbs of DeVry Technical Institute in Irving, Texas; Joel P. Bowman of Western Michigan University; Constance Y. Dees of Alabama Agricultural and Mechanical University; Jolene D. Scriven of Northern Illinois University; Margaret Baker Graham of Iowa State University; and Rebecca Limback of Central Missouri State University.

We would especially like to acknowledge Hillary for she gave the time and the support to complete this manuscript. Our children—Kristin, Elizabeth, and Jennifer, and Joshua, Jared, and Rebecca—gave us a reason for the journey.

Michael Greene
Jon Ripley

1

Readme.1st

1.1 INTRODUCTION

This chapter is a step-by-step set of directions for using this handbook. This book has been designed as a desktop reference, a tool to help you communicate clearly and effectively. The main reason for you to be using this tool is that you have a job to do: you have something to write, or you have a meeting to attend, or you have something to present at a meeting. In other words, you have a task that requires you to communicate in writing, at a meeting, or in front of a group of people. If you have a task to do, you can find information about your task in this text.

To help you better understand this tool, we have included several features to enable you to access information in different ways. Since not everyone will use this document in the same way, we have included multiple entry points into the text.

1.2 ENTRY POINTS

We have designed this book so that you can easily enter it from many different locations. Here is a list of the various ways you can enter the book.

- The Table of Contents presented on the back cover.
- The Table of Contents located after the title page.
- The Index at the rear of the book.
- The chapter outlines preceding each chapter.
- The chapter titles, the section headings, and the sub-headings.
- Quick Reference Boxes designed to present capsule summaries of important material. These boxes are shaded so they can be located quickly.
(This is an example of a Quick Reference Box.)

When you are reading material in a chapter, you may be directed to another chapter that will contain related information that may prove useful. For example, you will see *See Chapter 9, Designing the Memo* in italics. Obviously, you may ignore this direction and continue with your reading, or you may go to Chapter 9, gain the necessary information, and return to your initial position.

We have designed a book in which you can quickly and easily locate the material that you need. Ideally, if you are a student taking a course and this is a required text, you will find this book useful enough so that you will take it with you to your places of employment.

Similarly, if you are currently an engineer or a technical professional, we hope you employ this book frequently to assist you with the many communication tasks required of you in your field. In the industrial world, you may only use very small portions that apply to specific communication tasks. When you are designing the layout of a document, for example, you might consult *Chapter 7, Selecting a Format.* If you have been assigned the task of writing a set of procedures for using a device, you might review *Chapter 17, Giving Technical Directions.*

The chapter you are now reading contains an overview of the Document Design Process that you can use to design any document. You may want to review this chapter whenever you have a large or important writing task ahead of you.

To locate the specific material you need, you can turn to the back cover of this book. From here, you can proceed to the outline that precedes each of the chapters in the text. The outline will direct you to the specific section you need. See Fig. 1-1 for an illustration of how to move from the Table of Contents on the back cover to the outline at the beginning of a chapter and then to a specific section.

A detailed Contents can also be found at the front of the handbook.

Now that you are aware of the various ways of approaching the material in this book, we would like you to know what it can do for you. This tool helps you to:

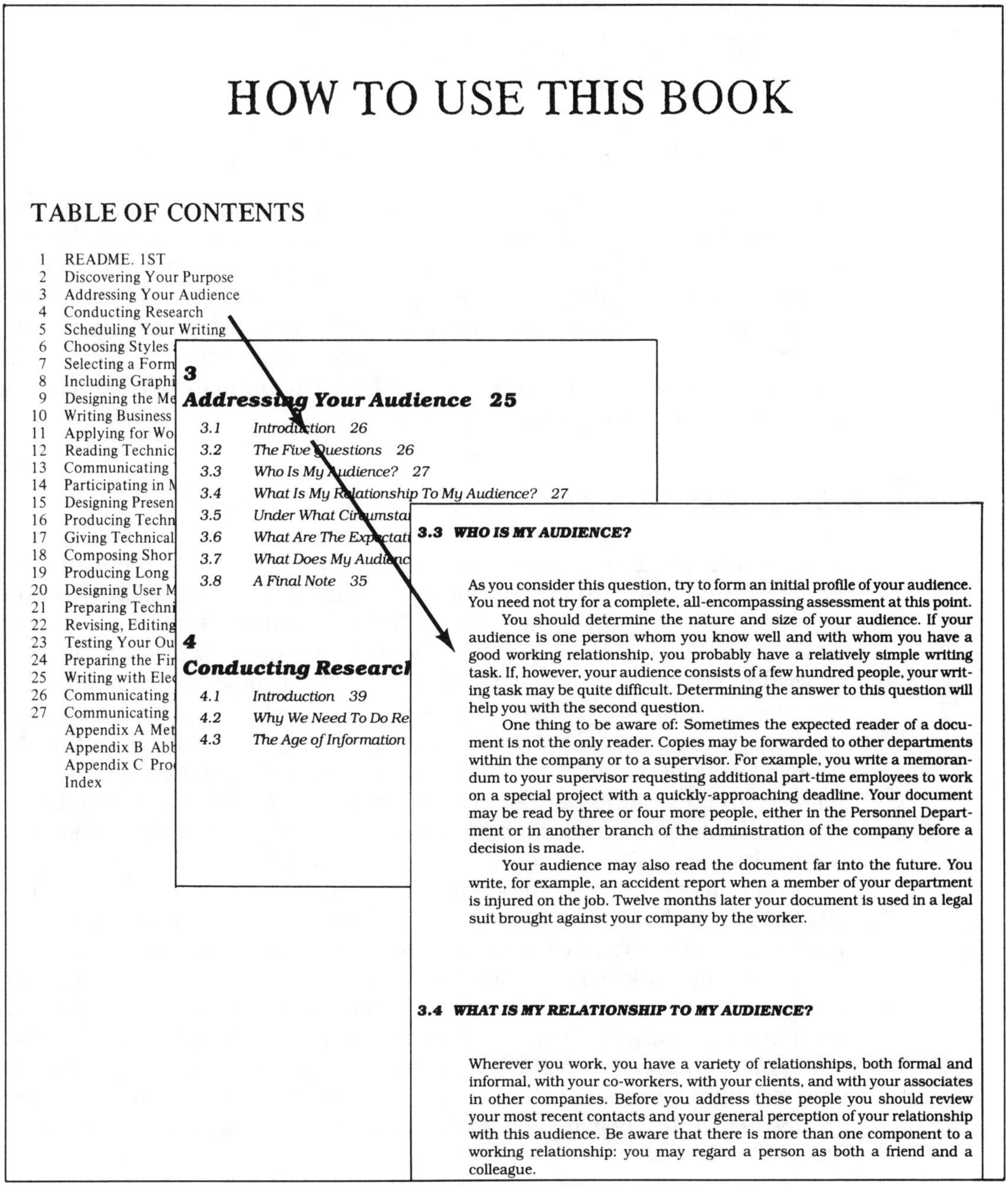

HOW TO USE THIS BOOK

TABLE OF CONTENTS

3

Addressing Your Audience 25

4

Conducting Researc

3.3 WHO IS MY AUDIENCE?

As you consider this question, try to form an initial profile of your audience. You need not try for a complete, all-encompassing assessment at this point.

You should determine the nature and size of your audience. If your audience is one person whom you know well and with whom you have a good working relationship, you probably have a relatively simple writing task. If, however, your audience consists of a few hundred people, your writing task may be quite difficult. Determining the answer to this question will help you with the second question.

One thing to be aware of: Sometimes the expected reader of a document is not the only reader. Copies may be forwarded to other departments within the company or to a supervisor. For example, you write a memorandum to your supervisor requesting additional part-time employees to work on a special project with a quickly-approaching deadline. Your document may be read by three or four more people, either in the Personnel Department or in another branch of the administration of the company before a decision is made.

Your audience may also read the document far into the future. You write, for example, an accident report when a member of your department is injured on the job. Twelve months later your document is used in a legal suit brought against your company by the worker.

3.4 WHAT IS MY RELATIONSHIP TO MY AUDIENCE?

Wherever you work, you have a variety of relationships, both formal and informal, with your co-workers, with your clients, and with your associates in other companies. Before you address these people you should review your most recent contacts and your general perception of your relationship with this audience. Be aware that there is more than one component to a working relationship: you may regard a person as both a friend and a colleague.

Figure 1-1 How to locate specific information from the Table of Contents.

DESIGN A VARIETY OF DOCUMENTS
PREPARE FOR PRESENTATIONS
PARTICIPATE EFFECTIVELY IN MEETINGS
INITIATE CORRESPONDENCE
CREATE REPORTS, MANUALS AND PROPOSALS

We also cover reading technical documents, using the telephone, and writing a number of specific technical documents. The book is not designed to help you with college compositions, essays, poetry or creative fiction. For these tasks you need other resources and tools. This handbook is for technical professionals who need to communicate effectively in a work setting.

Since communicators must also weigh ethical issues in their writing and speaking, we have included boxes of information labelled Ethical Considerations at various points in the text.

1.3 PROCESS AND TASK CHAPTERS

Two basic kinds of chapters in this textbook are Process chapters and Task chapters. Process chapters contain information on ways to improve all your communication, and view written and oral communication as a process to be followed. You will learn guidelines and techniques that apply to the kinds of writing and speaking tasks you need to do throughout your adult life. *Chapter 3, Addressing Your Audience* is a process chapter, for example, because you should consider your audience every time you communicate.

Task chapters, on the other hand, offer specific guidelines for producing the different kinds of documents and presentations you must do in technical careers. Tasks will be treated individually. Thus, you will find an entire chapter on *Preparing Technical Proposals* (Chapter 21).

Technical students often believe that all of their careers will be spent "hands-on" and that they will rarely need to write or make presentations. In truth, you will be expected to communicate frequently and well in any technical place of work. Your promotions and salary increases will frequently reflect how well you communicate, and generally, the higher you go in the company the more you will be expected to communicate formally. As you move into supervisory and decision-making positions, more and more of your time will be spent at a desk and at meetings. You will communicate to groups of people who will be judging you and your ideas by three things:

- the way you use words
- the way you organize your ideas
- the attitude they perceive in you

You can use this text to improve in each of these areas.

1.4 THE DOCUMENT DESIGN PROCESS

Whenever you are designing a document, you can follow one basic, underlying process: whatever you are composing, you can use the Document Design Process. In order to emphasize the importance of this process, we have organized this book to correspond to the stages we recommend. See Figure 1-2 for an overview of the Document Design Process.

1.4.1 The Answer Stage

The first stage in Document Design is a four-step process that involves a series of steps similar to the familiar Who, What, When, Where and Why of the newspaper business. The steps are, however, in a different order.

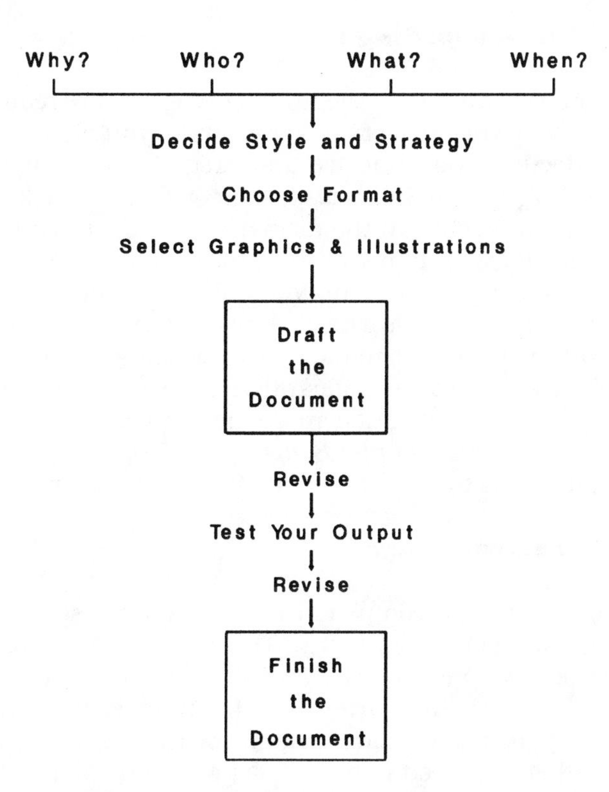

Figure 1-2 The Document Design Process.

WHY?	Why are you designing this document? What do you want to happen? What is it you are trying to accomplish?
WHO?	To whom are you sending your document? What do they know? What don't they know?
WHAT?	What is your message? Do you have the information you need? How can you get this information?
WHEN?	When is your deadline? How much time do you need to prepare?

Chapters 2 through 5 explore each of these questions in turn. There are other questions you may want to ask when you are preparing a specific document which will be discussed in detail in other chapters, but the questions above will help you successfully compose any document.

1.4.2 The Design Stage

Once you have answered these questions, you are ready to design your document. Using the answers as your guide, you choose an appropriate style for the document and devise a strategy to get your message across. *See Chapter 6, Choosing Styles and Strategy* for a more detailed explanation.

Now you decide on the format for the document. As you begin to put your ideas down on paper or in front of you on the computer screen, try to visualize how the document will look. *See Chapter 7, Selecting a Format.* Keep in mind that this is a draft version. If you have allowed yourself enough time, you will be able to make many revisions in your document.

Illustrations can significantly enhance your document. Some readers can grasp a concept much more clearly when they see a representation rather than text. *In Chapter 8, Including Graphics and Illustrations,* we offer basic guidelines to help you create illustrations for your readers.

1.4.3 The Draft Stage

Now that you have completed the pre-writing stages of the process, you should draft your document. The particular document you write will depend on your responses to the design stage of the process. Chapters 9 through 21 deal with a variety of different tasks you may find yourself doing as part of your career. Each communication task is given a separate chapter. Whether you need to write a set of technical directions for a manufacturing process, prepare a user manual for a product your company distributes, or complete a technical proposal for a government contract, you can employ the Document Design Process.

While most of the chapters involve the preparation of a written document, some involve communication activities where no written document is produced. Preparing a presentation, for example, involves the same underlying process as designing a document. You start by answering the four questions presented above: WHO? WHY? WHAT? and WHEN? Notice that the order has been changed to emphasize the importance of analyzing your audience.

If you take the time to write down your responses to these questions, you will find that you have a clearer sense of what you need to do when you get in front of your audience and you will most likely feel more confident about what you say.

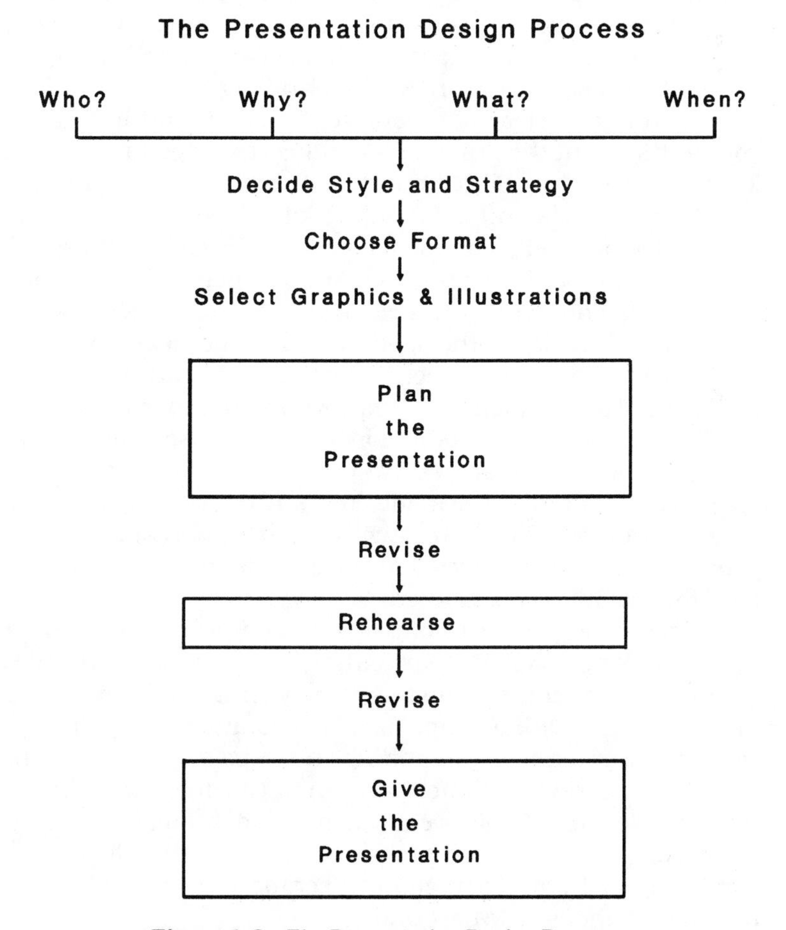

Figure 1-3 The Presentation Design Process.

Now that you have answered the questions, you can follow through with the Design Stage. Developing a strategy, planning the format of your presentation, and including graphics and illustrations are equally as important in preparing a spoken presentation. Since audio and visual aids are especially worthwhile as part of an oral presentation, we strongly recommend that you look at *Chapter 8, Including Graphics and Illustrations* when you are planning a presentation.

1.4.4 The Edit Stage

Once you have a draft, or in the case of an oral presentation, an outline and some visual aids, you have something you can work with. You can begin the Edit Stage by proofreading your work. If you are working with one of the most common word processing packages, you have access to a spelling checker which can find many spelling errors. Read the document out loud to hear how it sounds. Work on the phrasing of your sentences to make sure that each one is clear and concise. You may find it difficult to pick apart your writing, but the practice of making revisions until you get it right is the key to good writing. *See Chapter 22, Revising, Editing, and Proofreading.*

You may also find it difficult to let someone else read your work, but a trusted friend, colleague, or fellow student can be an excellent sounding board for your work-in-progress. Another person can frequently find errors that you may not be able to see. As writers, we see what we wanted to write, not what is actually on the page. What may be clear to you as the writer may be unclear to your audience. By testing your output before you make your final copy of a document, you improve your chances of success. *See Chapter 23, Testing Your Output* for a more complete discussion of this topic.

Now you proofread your work again, taking care to double-check any recent revisions you have made. For particularly important documents, you may want to test your output with a variety of readers. Then, once you are satisfied with the changes you have made, you prepare your final version for submission to your audience.

One of the pleasures of using a word processor when you write is the ease of making revisions, particularly when you need to cut-and-paste sections of your document. Instead of rewriting the entire document, you can move passages with a simple series of commands. Within seconds you can see how your changes will appear on the page, and just as quickly you can move the section to another position in the document. In *Chapter 25, Writing With Electronic Tools,* we discuss the advantages of word processing and other computer software programs. One major advantage is this: if you employ a word processor, you can compare potential formats for your document and choose the best one.

1.4.5 The Final Stage

If you have followed the steps in the Document Design Process, you are ready to complete your document with the added confidence that comes from having carefully designed the final version. You have greatly increased your chances of success.

We treat this stage in *Chapter 24, Preparing the Finished Product.* At the point when your document is done, you print the final copy, sign your name, if appropriate, and distribute it to the individuals who need to see your document.

Make sure you allow yourself enough time to accomplish these tasks. No matter how much time and effort you put into the four previous stages, you must follow through to the end. If a bid on a U. S. Government contract is late, it is not even considered, no matter how good the document is.

1.5 CONTEXTS FOR COMMUNICATION

Since most of the communication we do in our careers involves organizations, whether private or public military, educational, industrial or governmental institutions, we have included a chapter at the end of this textbook that details an organizational environment. *Chapter 26, Communicating in an Organizational Environment* describes Gates and Associates, a consulting firm based in Washington, D. C., and places you within this organization. We believe this is a beneficial way to provide you with a context or framework for your writing. Many of the exercises at the end of the chapters in this book involve Gates and Associates. We recommend that you read this chapter before you do the exercises.

We have also provided a chapter on communication across cultures. We live in a global environment, and most companies now do business on an international basis. *Chapter 27, Communicating Across Cultures* addresses the difficulties of writing and speaking to people who come from different cultural backgrounds, and offers strategies to pursue when you need to get your message across.

1.6 A FINAL WORD

Now that you have completed this first chapter, we would like you to look again at the Table of Contents on the back cover and at the beginning of this textbook. Notice that the organization of the book mirrors the stages in the document design process we recommend. See Figure 1-4 for an overview of how the Table of Contents coincides with the Document Design Process.

Whenever you have a task to do, you can move through the design process and get help quickly on any stage you desire. As you become experienced with the process, it will become part of the way you write.

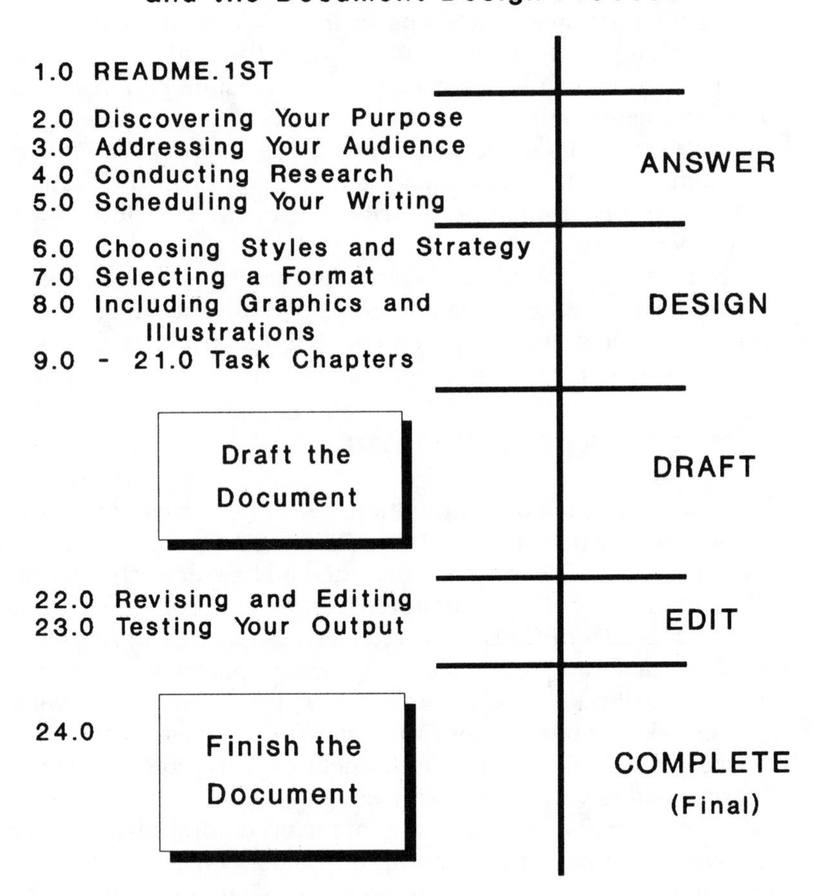

Figure 1-4 Notice how the chapters are organized to take you through the stages of the Document Design Process.

README.1ST EXERCISES

1. A logo is a symbol designed to represent a company or organization. Some companies and organizations spend thousands of dollars on the design of the logo since members of the general public often identify the organization through the logo. Examine the logos below. You can probably readily identify them, and as you do, a number of associations may spring to mind.

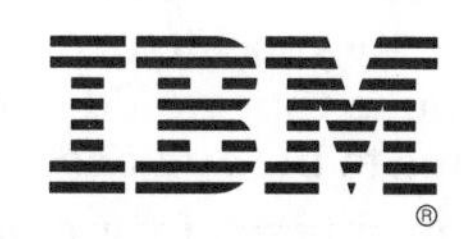

Reprinted with the permission of Shell Oil Co., IBM, and Eastman Kodak Company.

(a) Write down the associations you have with the logos presented above.
(b) Take one of the logos, and answer the following questions:
What is the purpose of this logo?
Who is the intended audience?
Does the design reflect the purpose?
Does the design show that the designers understood their audience? Why or why not?

2. Using the company *(Your Last Name) and Associates,* design a company logo or insignia to be used on all office stationary and all of the business cards of your company.

3. Using the logo you designed in Exercise 2, design a company business letter (8 and 1/2″ by 11″) to be used by all employees of ____________ and Associates. **Please note:** before you design the business letter, answer the questions in exercise 4.

4. Answer the following questions about the company business letter you are preparing to design in exercise 3:

Why are you designing it?
Who is your audience?
What is your message?
When is your deadline?
Now, as well as you can, describe the **style** of your document.
What **strategy** have you employed to get your message across?
What is the **format** for your document?
What **graphics** and **illustrations** have you included?

Now return to exercise 3 and draft your document. Use your own or an imaginary address and phone number in the stationery.

5. Locate a short article in a technical magazine. Examine the way the article has been designed on the page. Write down your observations on the design, and answer the following questions about the design of the article.
- What is the purpose of the article?
- Who is the intended audience?
- Do you think that the design is appropriate?
- Can you think of any ways to improve the design?

6. Examine the front page of a local newspaper. What illustrations appear on the front page? Where are they located? Are the illustrations appropriate? Can you think of any illustrations that would enhance the design of the front page and make it more appealing to the audience?
7. Interview one of the members of your class to find out what process he or she uses to write a long paper. Focus on the individual's most recent long paper. Investigate how the writer planned, researched, drafted, and revised the document before submitting it. Write a memo to your instructor in which you share your results.

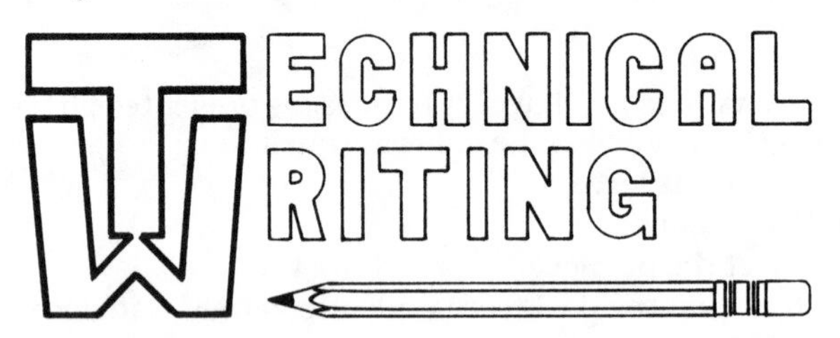

Reprinted with the permission of Michael Gouchie.

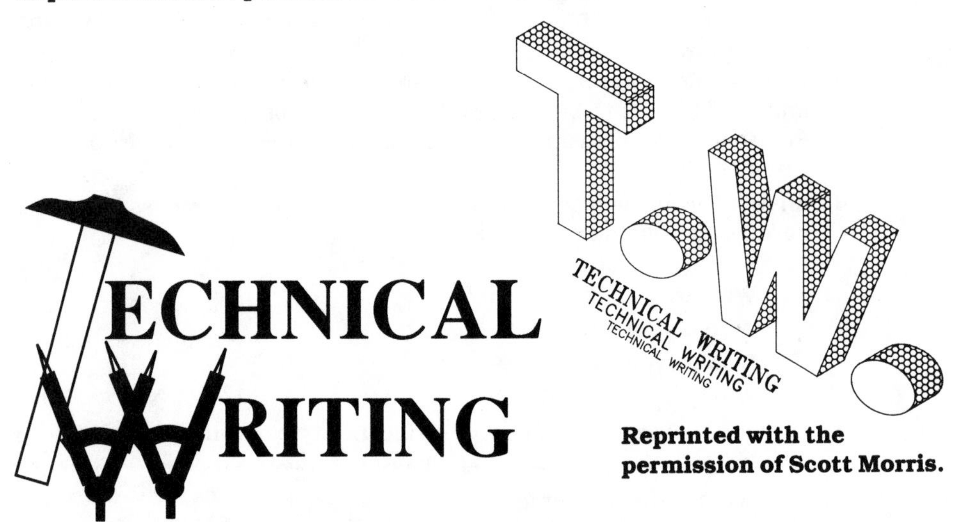

Reprinted with the permission of Scott Morris.

Reprinted with the permission of David Pinney.

Reprinted with the permission of Andy Boyce.

2

Discovering Your Purpose

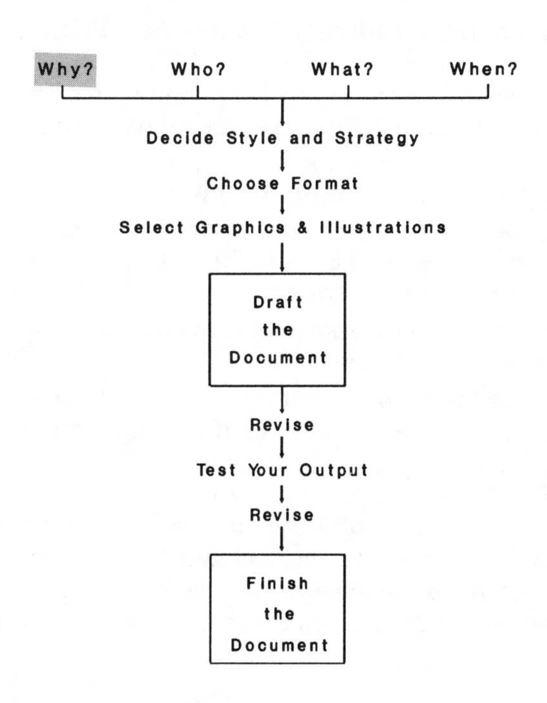

2.1 INTRODUCTION

Before you begin writing, you should decide on your purpose. In a sense, *the underlying purpose of all technical communication is to convey technical information clearly and accurately.* Technical writing is technical because of its subject matter, and many of the documents you write on the job require the inclusion of technical information. It is vital that your readers be able to comprehend this information and that the information is accurate. If it is not, a product may fail to work, a bridge may collapse, a rocket may explode, or a patient may die.

There may, of course, be times when you write a document, as part of your job, that contains no technical information. Some of the tasks performed by people in a technical field are not technical in nature. For example, if you write a memo to announce the hiring of a new secretary in your department, you may not include any technical information. The memo could be classified as business writing. The concepts we discuss in this chapter, however, will apply equally to technical and business communication. Our primary focus will be technical communication, but there will be times when we use examples that are not technical in nature.

This chapter will present a variety of ways of discovering your purpose. We begin with a number of ways of asking questions to identify and limit your purpose. You will be encouraged to write down your purpose. We will also discuss some of the benefits of knowing your purpose.

2.2 ASK QUESTIONS TO BEGIN YOUR DISCOVERY

Discovering your purpose involves asking questions to find out why you are doing a particular task. You ask questions such as the following:

What message do I want to convey?

Why do I need a (any) document? For example, would it be better to phone this person? Should I hold a meeting where all of the parties are present to go over strategy?

Why do I need this particular document? Could this be presented better in another format?

What is the point I want my audience to remember? What action do I want my audience to take? If this document is successful, what will happen?

Notice how many of these questions require considering your readers. In this way, purpose is very related to audience, which is discussed in Chapter 3. One of the reasons why we listed the four questions on the same line in the Document Design Process is that the four questions are so closely

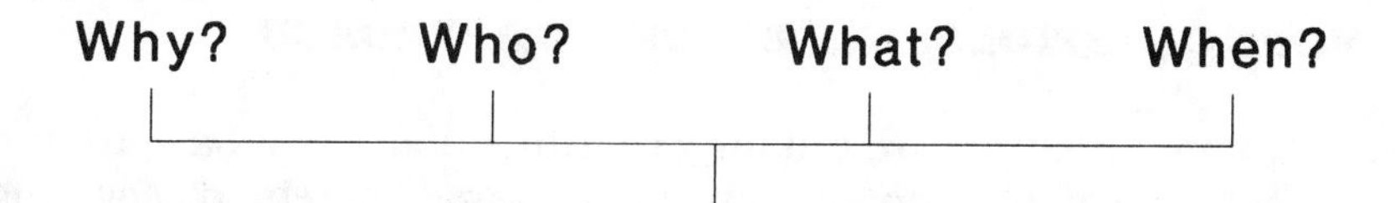

Figure 2-1 The top line from the Document Design Process.

related. You need to think about why?, who?, what?, and when? in conjunction with each other.

Different technical tasks have different purposes. You may want to **propose** a course of action or behavior for the future, **report** on behavior that occurred in the past, **document** or record information for future reference, **persuade** other people concerning your ideas, qualifications, and products, or **instruct** them on how to follow certain procedures or processes. The following lists some of the potential purposes of the technical writing tasks you may undertake in your career.

propose	technical proposals feasibility studies memos (for example, agendas for meetings) business letters
report	trip reports progress reports short and long reports
document	lab manuals technical descriptions, definitions, and specifications procedures minutes of a meeting work, service, and repair reports contracts and other legal agreements
persuade	resumes and cover letters proposals presentations sales brochures, bulletins, and catalogues
instruct	technical directions user manuals

This is not an absolute list. A memo is varied to the extent that it could be used for all of the purposes listed above. In all cases, however, you are seeking to inform your audience. You assume that they need the technical information you have for them.

2.3 MORE QUESTIONS: THE TAGMEMIC APPROACH

To know what we are talking or writing about, we have to understand it. Another way of defining your subject, and thereby discovering your purpose, is the Tagmemic Approach, an exercise described in *Writing for the 21st Century: Computers and Research Writing* by William Wresch, Donald Pattow, and James Gifford (McGraw-Hill, 1988). The exercise is designed to help you understand what you are writing about by asking questions to limit your subject.

> What makes this item unique?
> What possible forms can it take?
> How does it fit into the larger world?

What you are looking for is a way of making your subject distinct, of isolating it from everything that surrounds it. For example, if you are asked to design a prototype for a new product in your company, you could start by answering the three questions in the box above. These questions can be broken down further:

What makes this item special?
How is this item important?
What are the major features of this item?
What are some of the forms it can take?
What is the most unique form of this item?
Is it different from what it used to be?
Will this item change in the future? How?
What makes this item different from other items?
What else relies on this item?
What causes changes in this item?

When you have finished answering the questions, you should have a much clearer idea of your subject matter, and about its uniqueness, its range of forms, and its relationships with the larger world. Together with the questions listed in Section 2.2, the Tagmemic Approach questions should provide you with a clear and limited purpose.

What we mean by limited is this: when you are deciding on purpose, you should also develop a clear understanding of what you are not attempting to do. When you are designing your resume, for example, it is vital that you realize that the purpose is **not** to get you a job, but to get you an interview. You have accomplished your goal if your resume enables you to gain interviews. If you are not finding a job, but you are getting interviews, you do not need to revise your resume; you need to work on your interviewing skills.

2.4 *THE DIFFERENCE BETWEEN GOALS AND OBJECTIVES*

We can break down the question of purpose in another way. We can distinguish between **goals** and **objectives.**

"What are my basic goals?"
"What are my basic objectives?"

There is confusion in the minds of many people about the difference between goals and objectives. We can understand the difference in terms of being specific: Objectives are specific ways of accomplishing your general goal.

For example, let's say you are the manager of a Quality Assurance Department. You have noticed that the twelve employees within your department do not get along well. There is constant bickering that sometimes flares into angry arguments, and some members of the department hardly talk to each other. They can be working on a project together without speaking. It is clear to you that something must be done. Your goal is to improve relationships within the department. (In this sense, goal and purpose are synonymous.) Therefore, you call a meeting of the department at which you state that all of the members need to be more friendly, and to accomplish this, there will be regular luncheons every Friday at which department members will be encouraged to interact. Furthermore, each member of the department is assigned a specific day to be responsible for, such as bringing doughnuts for the morning coffee break. Attending the luncheons and bringing the doughnuts become the objectives of the people in your department, specific ways of reaching the goal of better office harmony.

You need to realize that simply having a goal will not get it done. You must work out ways to accomplish your goals, and this means specifying your objectives. When you are preparing to write, you need to know your purpose, **and** you need to consider ways of accomplishing your purpose.

2.5 *MULTIPLE PURPOSES*

With each specific writing task you do, you will have one or more purposes. Sometimes you have a combination of purposes. It's important to recognize this before you begin writing.

For example, if you are completing a laboratory experiment to test new equipment for your company, you may have a number of purposes in mind when you sit down to write up your report in your lab manual. You want to **inform** your readers. The company engineers may need the results to verify their assumptions about the design and performance of the new equipment. Your supervisor will also look at the work you do, and so a secondary purpose may be to impress your supervisor, to **persuade** your supervisor that you are doing a competent job. It is worth mentioning a third purpose in this example. By completing your report, you **document** the results, insuring that your company has a written history of the procedures you followed and the results you found. This may be very valuable information in the future: you may no longer be with the company or your memory of the experiment may have faded.

Whenever you have a writing task, no matter how small or insignificant it may appear at first, you should be aware that you may have more than a single purpose. Make sure that your final product accomplishes all of your purposes.

2.6 *WRITING DOWN YOUR PURPOSE*

To insure that you have a clear understanding of your purposes, we recommend that you complete the exercise below whenever you have a writing task. You will find that it clarifies your thinking and enables you to begin your communication task efficiently.

> **Exercise:** Complete the following.
> The purpose of this document is to ______________________________
> my audience about ______________________________
> ____
> ____.
> My secondary purpose is to ______________________________
> ____
> ____.
> When my audience finishes reading this document, I would like the audience to ______________________________
> ____
> ____.

Don't do this in your head. Write it down. The process of writing something down forces us to shape our meaning into a clear statement. **You are not ready to begin writing until you know what your purpose is.**

There will be occasions when you do not want your readers to know your secondary purposes. While we recommend that you state your purpose, there will be times when we would advise you not to state your secondary purposes. For example, you have been asked to write an objective report on the costs and recent sales of five products manufactured by your company. Division managers will meet to review the products and make decisions about the future direction of the products. You may have a strong feeling about one of the products; although you may write the report in such a way as to influence the managers, your approach is to remain impartial. Your main purpose is to inform, and although you may have a secondary purpose, you would not state it.

Another example of an instance where you would not want to state your secondary purpose is when you write a paper for a grade. You would not state that you are writing the paper to receive a high grade in the course.

ETHICAL CONSIDERATIONS

Is there ever a situation where it is ethically permissible to hide your primary purpose from your audience? Since much of what you will write depends on the particular context of a specific communication situation, this is a difficult question to answer. Much depends on your role within an organization, who your audience is, the history of events between you and your audience, whether you are intentionally misleading (or misinforming by omission) your audience, and your personal ethical stance.

We recommend that you go through the entire Document Design Process before you make a decision to hide your primary purpose. You need to consider all of the factors involved and have a sound reason before you decide to hide your purpose. You should not make this decision lightly, whatever the writing situation. As a communicator, you have an obligation to be honest to your audience.

2.7 THE BENEFITS OF KNOWING YOUR PURPOSE

A number of good things result from knowing your purpose.

One benefit of making the effort of writing down your purpose is this: you may be able to use your written statement in your document. Knowing the purpose of a document is very important for your readers also. In most cases, we highly recommend when you are writing a memo that you inform your audience of your purpose at the beginning. The audience now has a purpose to focus its attention on.

A second benefit of this effort is that you focus on your message very early in the planning process. If you need to gather information (*See Chapter 4*), you will not spend time searching for data you do not need. If you are unclear about your purpose, you do not yet know what you want, and you may spend valuable hours hunting for information that you will eventually discard.

A third benefit involves scheduling your writing. Your purpose may determine how long you have to get the document done. For example, if you discover that your purpose is to inform each member of your department about a change in company policy before they attend a Division meeting on Friday, you can schedule your writing to ensure that everyone receives the message on time.

A related benefit is this: once you know your purpose, the organizational pattern of the document should begin to take shape. You have many patterns to choose from. Let the purpose help determine which one you select.

First event → last event Use this approach when you want to follow a process in chronological order (organized according to time). Step-by-step instructions within technical directions are written in chronological order, as are progress reports and lab manuals. You would also use this pattern to explain the history of a process.

Part → part This pattern divides a mechanism into parts, and describes each in turn. When you are writing a technical description of a mechanism or a product, you can use this pattern. Sales catalogues are also organized in this way.

Left → right Use this pattern when you are describing a physical or material item, such as you would in a technical description.

Function → function When your focus is on what happens as a result of a process, you may want to employ this pattern. For example, if you want to explain what each system of an engine does, use this pattern.

Cause → effect When you are looking at a process across time, particularly when you are looking into the future and attempting to predict possible outcomes, use this pattern. Since so much of what is technical in nature can be viewed in terms of cause and effect, of events occurring as a result of a previous event, this pattern is frequently used in scientific and technological documents.

Problem → solution If your purpose is to solve a problem, you might first state the problem, then discuss how to solve it. A long report investigating how to correct a failure might be organized in this way.

Theory → plan of action Similar to the problem → solution pattern, this pattern begins with an abstract concept and then illustrates how to put this theory into operation. When engineers conceptualize a prototype and then begin to build it, they are following this pattern.

Simple → complex This pattern is valuable when you are dealing with a subject that is very difficult to understand. By starting with the simplest concepts, you prepare the audience for understanding you as you begin to get more and more intricate. You could employ this pattern in a section of a user manual, for example, so that your audience could learn the most basic features first. The more complicated procedures would thus be presented after your audience had a basic understanding.

Most important → least important Use this pattern to describe items in terms of their priority. You could use this pattern to organize the agenda for a meeting.

Particular → general Also referred to as induction, this pattern takes specific individual instances and attempts to build a general rule or application. Use this pattern when you want to persuade an audience. For example, you want to persuade your manager to allow flexible working hours instead of the rigid 9-to-5 schedule now in force; you could show several instances where company productivity would have improved if the flexible schedules had been in place as a way of building toward the general principle you are attempting to establish.

General → particular Use this pattern when you want to look at the overall picture, and then break this down into specific instances. Also referred to as deduction, this is a particularly appropriate pattern for a long report or a presentation of recommendations. When you place an overview at the beginning of a report, and then use the rest of the document to explain in detail, you are using this pattern.

The pattern or patterns you select depend on your purpose. This is not to say that there is one correct pattern for each communication task. The same subject matter can be organized in many different ways, and as you proceed further in the Document Design Process you may decide to change the pattern. Having a sense of organization at the beginning of a project is helpful, however, and we encourage you to develop one. Let your discovery of your purpose help you decide which organizational pattern you will initially follow.

2.8 *A FINAL NOTE*

If you do not know why you are writing something, do not write it. It may be the case that you should not send that memo. Spend time considering your purpose and asking questions of yourself and other people. Then sit down and try again to write down your purpose. When you are satisfied that you know your purpose, then proceed to analyze your audience.

PURPOSE EXERCISES

1. For each of the writing tasks and communication activities listed on the next page, write down two possible purposes. For some of the tasks, you may have to specify the exact nature of the communication. For example, you may need to

specify what kind of memo you are considering. If you are unable to think of two purposes, try to analyze why. What is it about this particular form of communication? (In parentheses is the number of the chapter in this book where the task can be found.)

Document	Purposes
Memo (9) **Letter (10)** **Resume and Cover Letter (11)** **Technical Description (16)** **Technical Directions (17)** **Short Report (18)** **Long Report (19)** **User Manual (20)** **Technical Proposal (21)**	
Activity	**Purposes**
Reading a Technical Document (12) **Conversing on the Telephone (13)** **Leading a Meeting (14)** **Attending a Meeting (14)** **Presenting Orally (15)**	

2. This exercise employs an assignment you have for another course you are taking. It can be a homework exercise, a paper, or a lab report. Describe the assignment. Speculate about why the instructor assigned this particular task; in other words, what were the intentions of the instructor? What is your purpose in completing the assignment? If you have multiple purposes, describe these.
3. You have been asked to introduce yourself to the class. Go through the list of questions presented in Section 2.2 and write down your responses to the questions.
4. You have been asked to write a 300-word article about a technical subject. Your purpose is to inform your instructor and your classmates. Decide on your subject and then use the Tagmemic Approach to help you understand your subject more clearly.
5. Smith and Jones Company has experienced four straight quarters of losses. A meeting of all the major decision-makers within the company has been called for Friday. The goal of the meeting is to develop a two-year plan to make the company profitable. You have been asked to lead the meeting. List some of your objectives, specific ways of helping the meeting to run smoothly and to accomplish its goal.

6. Complete the following for the next writing assignment you need to do:

The purpose of this document is to ____________________
my audience about ____________________
____
____________________.

My secondary purpose is to ____________________
____
____________________.

When my audience finishes reading this document, I would like the audience to ____________________
____
____________________.

7. Examine the following memo. What are the basic organizational patterns employed in the memo? Is there one pattern that provides the framework for the memo?

June 20, 1992

To: Sheena Ferguson, Division Manager
From: Debra Margoles, Head of Research
Re: Performance Review of Charles Constance

On Tuesday, Charles Constance arrived at work 35 minutes late. On Wednesday, he called in sick. On Friday, he arrived two hours late. When I asked for the Kestral Report on Friday, he replied that he had been unable to complete it.

In the last month, Mr. Constance has been out of work a total of 8 days and has been late on 7 occasions. He has been in charge of two major reports, both of which he has failed to complete on time. I have issued a verbal warning and a written warning, neither of which has had an impact on his behavior.

In sum, the current situation can not continue. With your permission, I intend to begin a formal performance review.

cc: Charles Constance
Personnel

8. Examine the following memo. What are the basic organizational patterns used in the memo? (There may be a combination of patterns.)

May 23, 1992

To: Chemical Engineering Team
From: Harriet Hamilton
Re: This week's schedule

As you know, Nancy Brown of Cook, Markham and Brown will meet with the Chemical Engineering and the Civil Engineering Teams on Thursday to establish the timetable for the first phase of the Princeton Project.

To prepare for this meeting, I have asked Joseph Obolowicz to brief us on the work being done at the Ferber Petroleum Plant. He will arrive on Tuesday morning and will meet individually with the members of the team during the day. I will inform you of the meeting times. On Wednesday, we will meet together with Mr. Obolowicz at 11:00 A.M. in the Board Room. If necessary, we can meet again at 3:30 P.M. to iron out any problem areas.

The meeting on Thursday begins sharply at 9:00 A.M. Please be on time.

3

Addressing the Audience

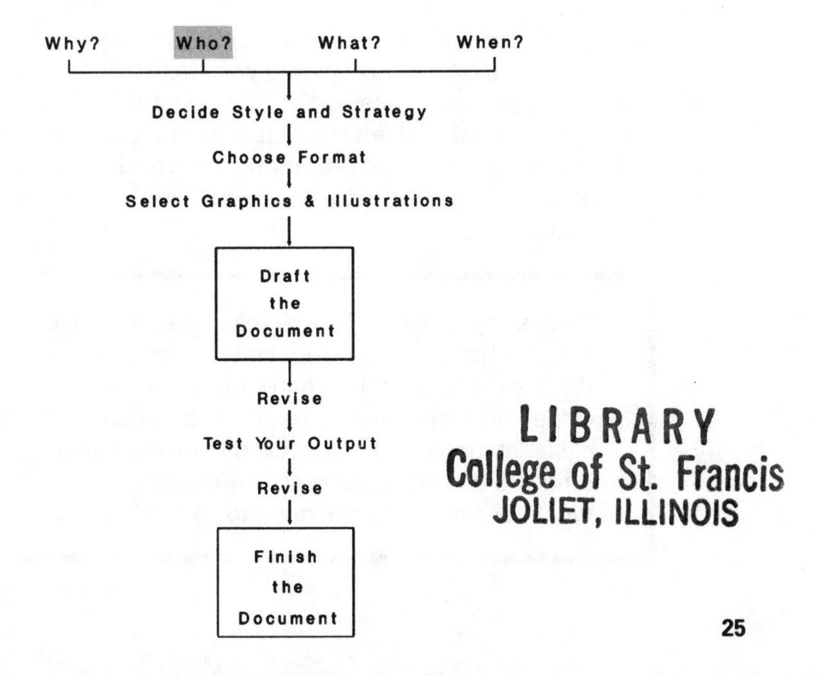

3.1 INTRODUCTION

Communication never occurs in a vacuum; it is an interaction between the presenter and the audience. Whenever you have a document to write or a presentation to make, you need to fit your material to your audience. The better you are able to gauge your audience, the more successful you will be as a writer and as a speaker.

If you have never had a job that required you to write memos, letters or reports, then most, if not all, of your required writing has been for a specific, limited audience: your teacher. Your goals have been to learn certain skills and to receive a good grade, or at the least a passing grade. As you progressed through a semester, you learned more and more about the teacher and what he or she expected of you.

If you were a particularly careful observer, sometimes it was possible to do quite well in a course even if you were not a great writer. The same principle holds true in the business world: *if you can give your audience the material they need to know without committing glaring errors, your written work will be received well.*

In a work setting you may not have the luxury of getting to know your audience and receiving feedback from this audience across a long period of time. You may be communicating to a large, mixed audience, or to someone you have never met, or to someone you have only spoken to once on the phone. Whatever the situation, however, there is a process you can follow that will help you assess the situation, make an informed judgment about your audience, and tailor your writing to this audience.

3.2 THE FIVE QUESTIONS

As we recommended when we wrote about purpose, we would like you to ask a series of questions to help you analyze your audience. Although there are many potential questions that you could ask yourself as you plan a document or an oral presentation, it is simpler and more efficient to get into the habit of asking a few questions before you begin. These questions will help acquaint you with your audience and enable you to gear your presentation to this audience.

- Who is my audience? (How many people will read or listen to this message? What do I know about my audience?)
- What is my relationship to this person or these people?
- Under what circumstances will this document be read? Or under what circumstances will I make this oral presentation?
- What are the expectations of my audience?
- What does my audience need to know?

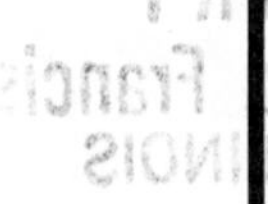

Each of these questions will be explored in turn. You should try to get into the habit of asking these questions each time you have a writing or speaking task. With a little practice, you will be able to go through these questions without referring to the above list.

3.3 *WHO IS MY AUDIENCE?*

As you consider this question, try to form an initial profile of your audience. You need not try for a complete, all-encompassing assessment at this point.

You should determine the nature and size of your audience. If your audience is one person whom you know well and with whom you have a good working relationship, you probably have a relatively simple writing task. If, however, your audience consists of a few hundred people, your writing task may be quite difficult. Determining the answer to this question will help you with the second question.

One thing to be aware of: Sometimes the expected reader of a document is not the only reader. Copies may be forwarded to other departments within the company or to a supervisor. For example, you write a memorandum to your supervisor requesting additional part-time employees to work on a special project with a quickly-approaching deadline. Your document may be read by three or four more people, either in the Personnel Department or in another branch of the administration of the company before a decision is made.

Your audience may also read the document far into the future. You write, for example, an accident report when a member of your department is injured on the job. Twelve months later your document is used in a legal suit brought against your company by the worker.

3.4 *WHAT IS MY RELATIONSHIP TO MY AUDIENCE?*

Wherever you work, you have a variety of relationships, both formal and informal, with your co-workers, with your clients, and with your associates in other companies. Before you address these people you should review your most recent contacts and your general perception of your relationship with this audience. Be aware that there is more than one component to a working relationship: you may regard a person as both a friend and a colleague.

Perhaps the most important consideration when answering this question about your audience is:

What is my working relationship with this audience?

When you enter a company as an employee, you should attempt to make a quick evaluation of the organizational structure of the company and your position within this structure. Almost all companies in the United States and Europe are organized according to a hierarchy. As you move upwards, each level of the company represents a higher degree of power, authority in decision-making, and income; the top level would represent the company President or Chief Executive Officer, or in some cases the Board of Directors. Hierarchies are most often organized in a pyramid-like structure, with more and more employees represented at each level as you go down the pyramid.

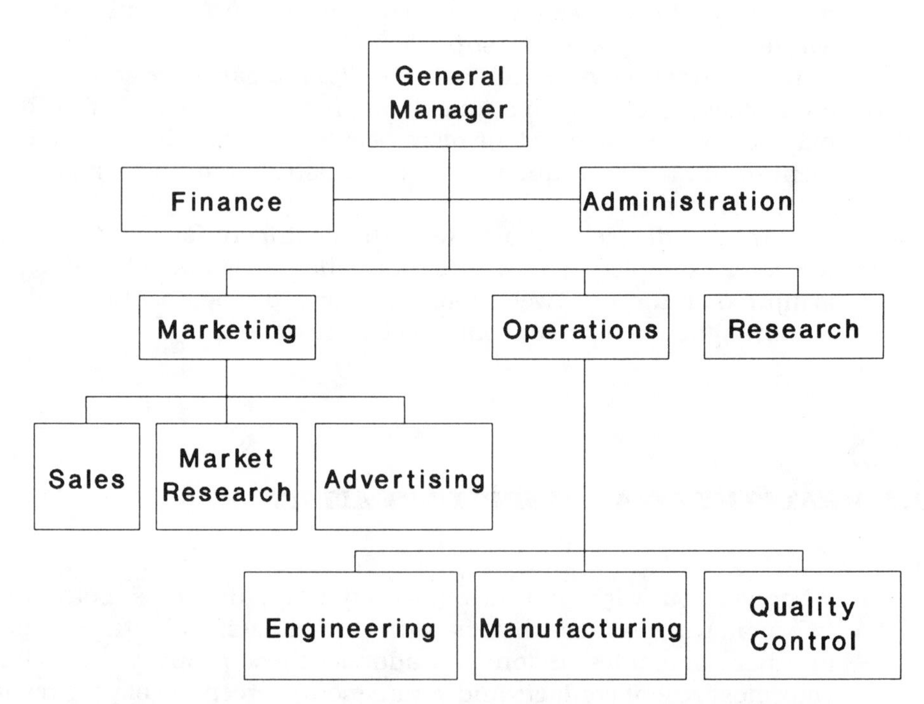

Figure 3-1 The hierarchal structure of a large research and development company.

Your standing within this company structure will influence the words you choose and in some cases the amount of information you divulge to your audience. Given the corporate hierarchy described above, you will usually be on a higher level than your audience, at the same level, or on a lower rung on the ladder. One other possibility is that you will be communicating to all three levels. Notice the importance of considering the hierarchical level of your audience as we move through the following examples.

EXAMPLE

As Head of the Electronic Engineering Department, you need to inform all of your department members that they will be required to submit weekly progress reports. They have not had to do this before. You decide to send a memo addressed to all department members stating the new policy.

ANALYSIS Obviously, you out-rank all of the people in your department. They may not be happy with the new policy, but they must adhere to the new policy. As department head, you write regular evaluations of employee performance and you have direct impact on raises and promotions. Your audience in this example must respond to your memo. (This does not imply that the employees will be happy to receive this memo nor that you should write in a disrespectful tone to your audience.) As long as you state your message clearly and directly, you can be certain that your audience will hear your message.

EXAMPLE

You are the Head of the Electrical Engineering Department. Your secretary received a call while you were out of the office. The Vice-President of Administration and Personnel wants a listing of all of the major purchases made by your department in the last three months. She needs this information by Friday, three days ahead. You gather all the necessary information and decide to write a memo to the Vice-President.

ANALYSIS Your audience is the Vice President of the company. She needs information in a hurry. Your task is to research the required information and present it in a clear, organized fashion. You may want to briefly present some of the reasons for some of the purchases on the list. This one particular task may not be crucial to your standing within the company, but if you frequently fail to submit accurate and clear responses to requests for information from those higher within the company it may affect your standing.

You are not always communicating with people in your own company, and you are not always able to know exactly the status of an employee of another company. There will be times when the only information you have about your audience is a job title, but this information is an important clue for you. The job title should give you a general notion of position within a company, of power and authority, and of status relative to you. By making these assumptions, you will have a better chance of communicating successfully.

We are certainly not recommending that you act humbly to the company president or that you treat employees under you on the corporate ladder with disdain and anger. What we do recommend is that you adjust your communication to your audience's position. Use whatever knowledge you have about your audience to your advantage.

3.5 UNDER WHAT CIRCUMSTANCES WILL THE AUDIENCE RECEIVE MY MESSAGE?

The answer to this question may save you quite a lot of time and potential embarrassment. Try to imagine the audience as they experience your message. If you can accustom yourself to doing this, you can almost guarantee a better presentation of information.

Let's look at some situations. If you are writing a memo to a busy executive who reads all of her mail standing over a wastebasket, it would be foolish to send a ten-page document. Present the facts quickly and clearly, and then end the message. If you are writing a set of instructions for the assembly of a chaise lounge, realize that your reader will be reading your instructions while doing the assembly. (As you write, picture your reader actually putting together the product.) *See Chapter 17, Giving Technical Directions.*

If you are presenting an hour-long demonstration of a product for another company, make sure that you plan enough material to completely fill an hour. (The most common error that inexperienced presenters make is that they do not prepare enough material.) You should also prepare visual aids for your audience. Imagine the questions you may be asked and prepare responses. *See Chapter 15, Designing Presentations.*

3.6 WHAT ARE THE EXPECTATIONS OF MY AUDIENCE?

All audiences expect communication that they can understand and put to use. No matter how brilliant your ideas are, you have failed as a communicator if your audience can not understand your message. The following is an example of a good idea that is poorly communicated:

EXAMPLE

August 4, 1991

Dr. William Novack
General Dynamo
77 Sunset Lane
Smithtown, New York 11787

Dear Dr. Novack:

Thank you very much for being our guest speaker at our annual Appreciation Day ceremony. I'd like to apologize for the poor attendance by our employees to your speech, but there was a softball-game-and-keg-party going on at the same time. Everyone who bothered to attend found your talk on "Applying Quality Control Standards to the Entire Workplace" an interesting subject.

I didn't want to bring this up in front of the entire audience, but why didn't you use more humor? Your topic could have used some levity.

Anyway, thanks again.

Sincerely,

Bob Curtain

Bob Curtain

ANALYSIS The letter is unintentionally insulting. The writer was not aware that the way he phrased his message would insult his audience, but this is exactly what he does.

Bob Curtain was angry with his employees for not showing up, but the message says something quite different. It focuses on the poor attendance for the speech. The phrase "who bothered to attend" implies that employees felt annoyed about attending.

There are other errors in judgment. The phrase "interesting subject" implies that the presentation wasn't very good. The criticism in paragraph two is uncalled for and not very constructive. "Thanks again" is tossed off without thought because it is preceded by "Anyway".

It would have been better if this letter had not been sent.

3.6.1 *Developing a Profile of the Audience*

The more contact you have with a potential audience, the more information you have about the expectations of your audience. For example, as your relationship with your supervisor at work develops across time, you build a storehouse of knowledge about who this person is and what your supervisor finds important. Since this person plays such a significant role in your career, you pay careful attention to the messages you receive from this person. Soon you develop a profile that you utilize each time you communicate.

Each individual is different, and your writing should reflect these differences. Sometimes it is necessary to write two different documents to two people even though your message is the same. Because your audience is different, you phrase your message differently, based on your working history and current relationship, and your knowledge of their expectations, including their personal likes and dislikes.

You would, however, only send different messages if the situation warranted this strategy. In most situations, you will send a copy of your document to each member of your audience since you will have taken all of them into consideration in the design and content of the document.

Usually, when someone gives you a writing task, they provide you with some guidelines for the project. When you are taking a course in school, your teacher will give you guidelines and rules for each assignment. He may say, "Papers with more than seven spelling errors will be returned." Or he may return a paper with a low grade and with comments that explain why you received this grade.

The instructor will also give you hints and clues as to what he expects. You may be able to tell from what he stresses in lectures or discussions with the class what he feels is important, both in terms of content and style. The more perceptive you are about the teacher and the teacher's expectations, the more successful you will be in the course. (Good students can sometimes predict the questions that will appear on an exam simply by using their knowledge of an instructor.) The principle is the same in a work situation: **use what you know of your audience.**

Let's look at an example of a case where a writer uses all of his resources to find out about his audience. Imagine the following. Paul Fields has been working as a Civil Engineering Technician for Disposition Inc. Paul has spent the last two years working on waste disposal projects. His supervisor, Shirley Miller, asks him to write a section of a proposal that will be submitted to the County of Lynville. Lynville will award a contract of $500,000.00 to dispose of the nuclear wastes from the Central Valley Power Plant. Paul's specific writing task is to describe how the waste material will be contained while it is being transferred to the disposal site.

Paul has worked on similar projects in the past but this is the first time he has had to write a technical proposal.

In this case Paul's audience is the County of Lynville, specifically the Board of Managers that will review the proposals and assign the contract. The county has published a document that describes the specific format for each section. The audience **expects** that the proposal will be clear, accurate, without error, and professionally written. The document will also be judged on appearance.

Disposition Inc. has previously submitted and won bids on projects with the County of Lynville. These bids are available in the company files. After Paul has reviewed the details of the writing task with his supervisor, he should immediately consult the company files. Here he can learn what style of writing was successful in the past. He may be able to use these proposals as models for his task.

If Paul has specific questions about this project, he can telephone the County Board and ask his questions directly. Paul can also ask his fellow employees who have had experience with the Board of Managers for information.

By the time Paul ends his research and begins designing his part of the document, he should have a good idea of his audience even though he may have had no idea at all about the County's political structure before he was assigned the project. When he does begin the design stage, he can feel confident that he knows the expectations of his audience.

Paul has a second audience for his document: his supervisor. Shirley Miller will review his writing before the entire proposal is assembled and submitted.

3.6.2 *Three Suggestions*

- If you write a document for a person and receive feedback, try to come to some conclusions about your audience's standards for a written document.
- If you have an important document to write and you do not know what your audience expects, call the person and ask. You will produce a better document and your audience will appreciate your effort.
- Many writers ignore what they know about a person and **write** messages that they would never **say** to a person. Read your document aloud. Hear the way it sounds. Imagine your audience reading your message.

3.7 *WHAT DOES MY AUDIENCE NEED TO KNOW?*

Once you have determined the answers to the above questions, you can proceed to the final question. Deciding how much information your audience requires is a skill that is developed through experience. You may, for example, research a topic for a full week in order to write a three-sentence memorandum. You may be able to dash off a two-sentence reply in response to a request for information, or you may need ten pages to explain in full the same information to a different audience.

Again, the answer to this question frequently depends on the history of contacts you have had with your audience. As a communicator, your success is connected to how well you can adjust your message to fit your audience, and this depends on your judgment of what your audience needs to know.

3.7.1 *The Three Levels of Audience*

One approach to this difficult judgment is to assume that there are three levels of audience based on knowledge and experience:

You can assume that a **general** audience is intelligent, possesses common sense and has some high school education, but has little or no experience or training in the subject matter about which you are communicating.

You can assume that an **informed** audience is intelligent, possesses common sense and has a high school education, and has had some training and experience applying their knowledge about your subject matter.

You can assume that an **expert** audience is intelligent, possesses common sense and has a high school or higher education, and has spent considerable time learning a field and applying knowledge to a variety of situations. The expert may have more knowledge about your subject matter than you do.

In most situations there is a larger general audience than an expert audience. You need to be sure which audience you are addressing.

The document you compose—and particularly the language you employ—will be determined by the level of your audience. Experts often speak in a kind of shorthand: abbreviated sentences filled with technical terms. Two carpenters, for example, might discuss dado bits, routers, risers, waney lumber, gussets, newel posts, mullions, and sidelights in a dialogue that lasts less than ten minutes. An untrained bystander, overhearing the conversation, would be able to follow very little of the talk.

As a general rule, the longer you are in a field, the more your speech patterns and your writing style change to include the vocabulary of that field. Sometimes you may forget that your audience has not had your experience or your training, so check your documents carefully to ensure that you have considered the level of your audience.

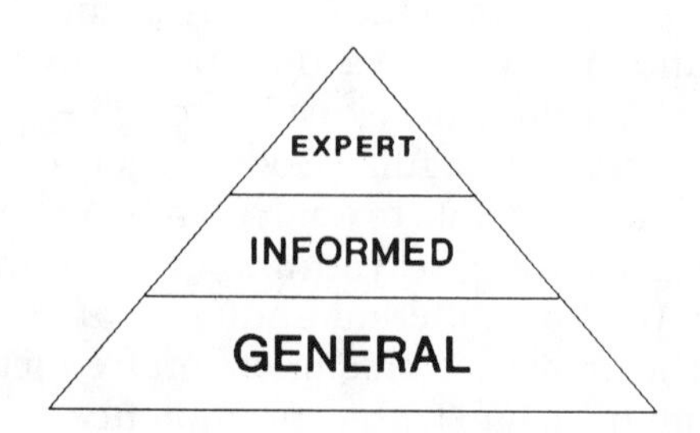

Figure 3-2 The three levels of audience

3.7.2 A Comparison of Documents

When you are writing to experts, you do not need to provide as much background information and detail. You can assume that they will understand the terms you use and will be familiar with the situations you describe. Here is a message revised for an expert, an informed and a general audience.

DOCUMENT 1 **expert audience**

The VAX 780 has twenty-seven p c boards.

DOCUMENT 2 **informed audience**

The VAX 11/780 is a 32-bit supermini computer; the CPU is composed of twenty-seven printed-circuit boards.

DOCUMENT 3 **general audience**

One type of computer is the supermini, a very fast machine that is smaller and cheaper than a mainframe. One 32-bit supermini is the VAX 11/780 made by Digital Equipment Company, a U. S. company that originated in Massachusetts.

A bit is a single high or low voltage signal. By putting many bits in a row, the number of pieces of information that can be represented increases exponentially. A 32-bit machine can send information to 4.3 billion storage compartments.

The Central Processing Unit, the heart of the machine, on the VAX is composed of twenty-seven printed circuit boards. Printed-circuit boards contain the intricate electronic circuits that carry information.

3.8 A FINAL NOTE

Until you feel yourself to be an experienced writer or when you have a difficult document to write or presentation to make, write down the answers to the five questions about audience presented in Section 3.2 and discussed throughout the chapter. The answers will help you get started with your task, and they will encourage you to *write for your readers.*

AUDIENCE EXERCISES

1. Write three definitions of a technical term, one to an expert, one to an informed audience and one to a general audience. See the example in Section 3.7.2. A good source for this technical term is the vocabulary of your major field of study.
2. You want to design a form that will be used by all the employees of Gates and Associates when they schedule their vacations for this calendar year. You need to know the following information from each employee:

    Length of vacation time allotted? (You will have a checklist on each employee, but you want to doublecheck the information on your sheet.)

    When? (You should ask for a number of preferences.)

    Employee's name?

    Department?

    Telephone extension?

    You need to remind all employees that they may not receive their first choice for a vacation time. Design the form and have everyone return it to your office in the Personnel Department of Gates and Associates.
3. You are writing a memo to one individual, Bob Fraser, someone you have worked with for ten years. You have developed a close personal friendship, to the point where you occasionally go on family outings with Bob's wife and two children, Lisa and Paul, aged 10 and 7.

    Bob is one of the accountants with Gates and Associates. He handles the company's retirement plan as one of his duties. Company employees can have contributions to the plan deducted from their weekly paychecks on a regular basis. You want to inform Bob that you would like to have $15.00 a week deducted from your pay and placed in the Employee Retirement Plan. Your account number is 1122768-04431.
4. You need to write a memo to F. Robert Gates, the Chief Executive Officer of Gates and Associates, for whom you work. You are to report the preliminary results of your investigation into the situation described in Section 26.7. To find out more about F. Robert Gates, consult Sections 26.3 and 26.4. To find out more about your role within the company, read Sections 26.2 and 26.6.
5. You need to write the same information as in exercise 4, but this time your audience is your supervisor, Sheila O'Brien, who is an "overview" person. She wants a short, crisp summary that tells her the essentials. She won't read more than one paragraph. To find out more about Sheila O'Brien, consult Section 26.5.
6. Locate a technical article in a magazine or journal. Read through it and analyze the assumptions made by the author about the level of expertise of the audience. Point out one paragraph that illustrates your analysis of the writer's assumptions about the audience.
7. Terence Cooper left Gates and Associates under "less-than-happy" circumstances two weeks ago. As a matter of fact, Terence was fired over a dispute con-

cerning expense account vouchers. Rumors circulating through the company grapevine suggest that he had been regularly treating his friends to meals at company expense.

Terence has secured employment in another consulting firm, The Outward Group, but is reportedly still upset with the circumstances concerning his dismissal.

When he left Gates and Associates, he may have taken with him a very important file (Number 47-3816) concerning a building project where a wall had collapsed in Atlanta, Georgia. After Cooper left the firm, the case was turned over to you. You have located all of the other documentation, but you were unable to find this particular file.

You need this file, and you need to contact Terence Cooper. (You knew Cooper casually, enough to have discussed current company business, but you never socialized with him. You got along well with him.)

Write down your strategy for dealing with this audience. How would you proceed? If you were able to reach Terence Cooper by phone, what would you say? Assuming that you were unable to speak with him, how would you phrase your letter?

8. Gates and Associates has dealt with the Wilson Travel Agency for the past thirteen years. The agency has provided prompt, reliable service until recently. Five months ago, Harvey Wilson, the founder of the agency, retired and left the business to his son, Brad.

Under Brad, the service has become sloppy. Three times in the last four months you have been given incorrect, out-of-date departure times. Then, on Tuesday of last week, you were to make a connecting flight at Chicago's O'Hare Airport. Your plane arrived on time at 11:52 a. m. The plane you were to take to Madison, Wisconsin left on time at 11:59 a. m. The travel package supplied by the Wilson Travel Agency listed the departure time on the Wisconsin flight as 12:59 p. m. Because of the mix-up, you were late for an important meeting with clients at the University of Wisconsin.

You have tried to reach Brad Wilson by phone, but each of the five times you have called you were told that he was out of the office. He has not returned any of the calls.

Now write a letter to the travel agency to address this situation. (Here is some additional information: F. Robert Gates knows what happened and knows you are writing this letter. He is displeased with the service of Wilson Travel Agency, but does not want to switch to another agency at this time because he has a sense of loyalty to Harvey Wilson.)

Wilson Travel Agency
1234 Pennsylvania Ave.
Washington, D.C. 55555

For more information about F. Robert Gates, consult Sections 26.3 and 26.4.

4

Conducting Research

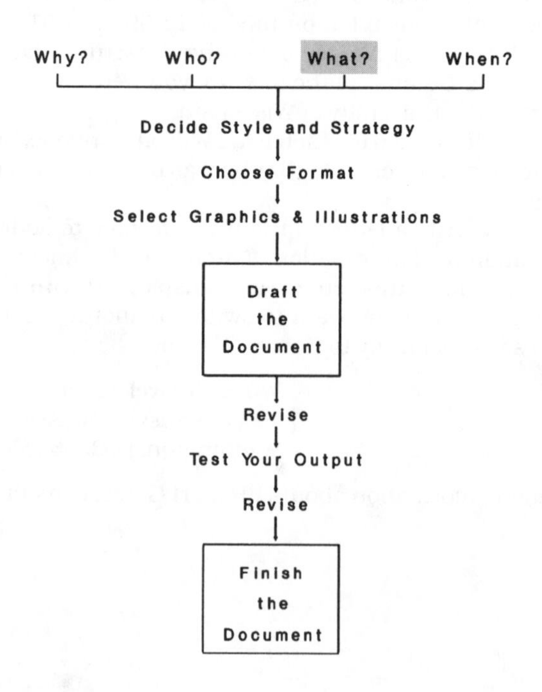

4.1 INTRODUCTION

Once you have determined why you are communicating (your purpose) and to whom you are communicating (your audience), you need to decide on **what** you are communicating. When you approach any communication task, you either:

- know all you intend to communicate, or
- need to find out something to include.

Often you will know the content of your message before you begin. If so, you can proceed to the next step in the Document Design Process, answering the question "When?". (*See Chapter 5, Scheduling Your Writing.*) Many times, however, you will conduct research to find out information.

Research does **not** necessarily mean going to the library to search through the books and magazines preserved there. It can also mean picking up the phone to call a colleague to get the exact date, time, and location for a meeting. It can mean walking down the corridor and asking someone for the exact amount paid for a shipment of replacement parts. It can mean accessing a computer file for the specifications of a product your company manufactures. In this sense, we are conducting research every day of our lives. When we need to know something, we find it out.

If you are like many people, the word *research* may cause you to feel anxiety or boredom. You perceive research to be a turn-off, a long and tedious process that should be avoided whenever possible. Research can be time-consuming and it can be tedious, but it can also be exciting and rewarding.

Research is discovery. Consider the remarks of Roald Hoffman in the shaded box below.

> The writing of a research paper to me is in no way an activity divorced from the process of discovery itself. I have inklings of ideas, half-baked stories, a hint that an observation is relevant. But almost never do I get to a satisfactory explanation until I have to, which is when I write a paper. Then things come together, or maybe I make them come together.
>
> Source: Roald Hoffman, Nobel Prize winning chemist, in an article on research and teaching in the *Boston Sunday Globe.* (November 5, 1989)

The process of writing down our ideas forces us to check our statements, and by so doing sharpens our thinking. There are magical moments when you can lose your sense of time in the quest for information. And, the more research you do and the more familiar you become with your sources of information, the more exciting the search becomes.

4.2 WHY WE NEED TO DO RESEARCH

Technical writing must be clear and accurate. We must ensure that our writing is accurate by verifying our information. Expert opinion is crucial to the process of writing or speaking about technical matters.

4.3 THE AGE OF INFORMATION

We live in a time when we have more research options than ever before. Some of the ways we conduct research to get the information we need are:

- conduct experiments
- interview people with information
  use the telephone
  speak face-to-face
  send letters of inquiry
- gather information from surveys and polls
- use college and local libraries
  ask the reference librarians
  employ the card catalogue
  find periodicals, including journals, magazines, and newspapers
  consult reference books
  conduct online searches of databases
- use company libraries and reference librarians
- search company records
  check file cabinets and computer files
  examine other internal records in reports, memos, and letters
- view screen images and hear recorded sounds, including visuals, videotapes, movies, radio, taped messages, television, computers, and newer media.

With such a vast range of information sources, it is no wonder that this period is referred to as the Age of Information. The better you are at gaining access to this wealth of information, the more success you will achieve.

This chapter attempts to give you an overview of some, not all, of the resources at your disposal.

4.4 CONDUCTING EXPERIMENTS

Companies and institutions involved in Research and Development (R & D) conduct pure and applied research into new products and processes. (Pure research is research to expand the amount of available knowledge, while applied research concerns how to use products and processes, finding the best applications for what is discovered.) Essentially, R & D involves designing a product or process in theory and implementing the theory by building prototypes and models and seeing what happens. It is important to remember that we gain knowledge when the prototype or model does **not** work, as we do when it does work. The technologies we enjoy today are the result of research and development begun many years ago.

Your career choice may involve such essential first-hand research work in a laboratory. Such research may provide you with the content of the documents you write, which may include laboratory notebook entries, memos, reports, technical directions, and user manuals. If you work in Quality Assurance, for example, you do research on each of the products you test. This research is considered primary because it is first-hand.

4.5 INTERVIEWING

When we are very young, we interview our parents by asking "why?" questions. We learn about the world from the responses that our parents give us. (Obviously, we also learn by observing the world around us and drawing conclusions from our observations.) In the same way, as adults, the ability to ask questions is a key to learning. You can find things out just by asking the right people. These include the people who work with you and outside experts in particular fields. If you are unable to interview the person directly, you can telephone or you can send letters of inquiry. Most people enjoy being asked, and you will probably receive a high rate of response.

Make sure you prepare written questions that ask for specific, precise information to fill in gaps in your knowledge. Do this after you have done your preliminary research. Phrase your questions to elicit a thoughtful, focused explanation rather than a simple yes or no.

EXAMPLE

Rather than asking "Are many of our nation's bridges in danger of collapsing?"
ask "What should we be doing about the threat of collapse of some of our nation's bridges?" or "How will the gel electrode technique allow us to diagnose structural problems with bridges?"

EXAMPLE

Rather than asking "What happens when a train carrying toxic chemicals derails?"
ask "What is the step-by-step process followed by fire fighters when they arrive at the scene of a toxic fire?"

Make sure the person understands how the information will be used. If you plan to tape-record an interview, make sure you inform the person beforehand. If you are interviewing in person or on the phone, take notes and review the notes immediately after the interview when the material is fresh in your mind. Write down clear explanations of the terms and phrases in your notes so that you will be able to understand your notes when you begin to draft your document.

4.6 SURVEYING A TARGET AUDIENCE

Sometimes, you need to conduct a survey or a poll to gather and compile information from a specific population base, which may range from specific clients for a product you manufacture to the general population to learn how to market a product. You should carefully design the questions for the survey based on what you want to learn. Ask the questions in the shaded box as you draft your questions.

> Why do we need this information?
> How can we best phrase our questions to elicit responses from our target audience?
> Where can we find this audience?
> Do our questions have a built-in bias that guarantees that we will get the response we are looking for? If yes, then you need to rephrase the questions. Remember, the reason you are conducting this survey is you need to know what your target audience thinks.

Although conducting a survey may seem to be an easy task, you will find developing an effective and accurate survey very difficult. You need to carefully plan your questions so that the responses will tell you what you want to know. You need to carefully select your audience. You need a broad enough sample so that you can have a truly representative response, but one small enough so that you can handle the tallying and statistical analysis of the data. If the survey is large in scope or vital to the welfare of the organization, you should consider letting specially trained statisticians design and conduct the survey. Whatever the scope of your survey, you need to get it right the first time.

Fig. 4-1 presents examples of five different kinds of survey questions: multiple choice, dual alternative, rank order, completion, and continuum. Each one is appropriate for a different situation, depending upon the kind of response you are seeking.

Multiple Choice

Check the box that best describes you:

- I live on campus. ☐
- I walk to campus. ☐
- I commute by car. ☐
- I commute by train. ☐
- I commute by bus. ☐
- Other ________ ☐

Dual Alternative

If you commute by car, do you park on campus?

Yes ☐ No ☐

Rank Order

Rank order the four parking facilities on campus. Use 1 to describe the safest parking lot and 4 to describe the least safe.

East Lot ______ South Lot ______
West Lot ______ North Lot ______

Completion

Complete the following sentence.
The parking facilities on campus are ____________________.

Continuum

Rate the overall parking facilities on campus based on the continuum presented below. Place an x on the continuum in the appropriate location.

very good — good — average — poor — very poor

Figure 4-1 Five Kinds of Survey Questions.

You can also request an essay response, but you will probably receive fewer responses and it will be more difficult to draw conclusions from the responses. If you ask for essay responses, make sure you ask a specific, focused question.

4.7 UTILIZING THE LIBRARY

Technical professionals need the ability to enter a library and quickly locate specific material. It can, however, be somewhat intimidating to walk into a library for the first time. You may feel out of place in unfamiliar surroundings. In time, however, as you begin to know the physical layout of the library and how to access information, you will become more comfortable, and you will make discoveries about what material can be found there. For this reason, you may want to visit your reference library on a regular basis.

What can you find there? First, you can find reference librarians who know the library facilities and are very willing to help you. They will take the time to work with you in your investigation. You will find books, magazines, journals, newspapers, and general and specialized reference material. You may also find videotapes and audiotapes. And you may be able to access information that exists in other libraries through computerized databases.

4.7.1 Books

A collection of books on a vast array of subjects is the foundation of a library. Information on the book collection is stored in a card catalogue, which can be organized in one of two ways: the Library of Congress System and the Dewey Decimal System. College and university libraries use the Library of Congress system, while public libraries use the Dewey Decimal system. The Library of Congress system divides books into 20 categories identified by a letter followed by specific numbers; the Dewey Decimal System divides books into 10 categories identified by numbers. Under both systems, each book has a individual *call number* to help you locate it on the shelves.

The Dewey Decimal System

000-099	General works
100-199	Philosophy
200-299	Religion
300-399	Social sciences
400-499	Philology
500-599	Pure science
600-699	Useful arts
700-799	Fine arts
800-899	Literature
900-999	History

Each of these divisions is further divided into ten parts. Each division is further divided, and so on to form the specific call number of a book.

The Library of Congress System

A	General Works
B	Philosophy, Religion
C	History, auxiliary sciences
D	Foreign history, Topography
E-F	American history
G	Geography
H	Social Sciences
J	Political Science
K	Law
M	Music
P	Language and literature
Q	Science
R	Medicine
S	Agriculture
T	Technology
U	Military science
V	Naval science
Z	Bibliography

Each of these sections is divided by letters and numbers to form the specific call number of a book.

While most libraries still have actual cards with the data on the book collection, many libraries have gone to computerized card catalogues. You conduct a search by entering certain key words, and the computer will search its memory and locate the material for you. With either a computerized or a card system, you can locate books by title, subject, and author. Many libraries now utilize inter-library loans so that you can have books located at other libraries delivered to your library.

3 Examples of Searches Conducted with a Computerized Card Catalogue

T = The soul of a new machine	(T = Title)
S = computer engineering	(S = Subject)
A = Kidder, Tracy	(A = Author)

Note: The card catalogue is frequently where audio-visual material is filed.

Two helpful guides to books in technical fields can help you limit your search. Each is published annually.

Scientific and Technical Books in Print
Technical Book Review Index

4.7.2 *Periodicals*

The term "periodicals" refers to any publication that is published in intervals, or in periods, such as a monthly or a quarterly. (The intervals need not be evenly spaced.) Therefore, magazines, journals, bulletins, newsletters, fact sheets, and newspapers qualify as periodicals.

The main advantage of periodicals is that they are current. While a book takes time to get published, a recent periodical will likely contain up-to-date information. In many technical fields, staying current is a priority. If you fall behind, you lose your competitive edge. If, for example, your company depends on government funding for which you submit bids and proposals, you need to be informed as soon as possible after a project is announced.

Most libraries have a current periodicals section, where you can locate very recent issues of the periodicals to which the library subscribes. Back issues (usually 12 or more months old) are stored on shelves or transferred to film where they are substantially reduced in size. Most college libraries have specialized periodicals in the fields where students have degree programs.

If you are researching a specific subject and you would like current material, a productive approach is to check general and specialized *periodical indexes*. The indexes organize articles from a variety of periodicals according to subject. A list of the periodicals compiled by an index and the abbreviations for each periodical can usually be found in the front of the index. The box supplies the names of some of the many indexes you may find helpful.

General Indexes
- Reader's Guide to Periodical Literature
- Humanities Index
- General Science Index

More Specialized Indexes
- Applied Science and Technology Index
- Engineering Index
- Social Sciences Index
- Biological and Agricultural Index
- Bibliography and Index of Geology
- Index to Scientific Reviews
- Business Periodicals Index

Abstract indexes can help you save time because they provide you with a summary of each article along with the title, author and details about the article.

Some indexes are accessed through a computer terminal. For example, the WILSONDISC is a computerized service that contains all of the information in the Applied Science and Technology Index stored on a CD-ROM disc. You can conduct a computer search for material organized by subject by entering key words and phrases.

4.7.3 Reference Works

Most libraries have reference sections where you can find a wide range of general and specialized sources: everything from telephone books to business guides to general encyclopedias to atlases to specialized handbooks, encyclopedias and dictionaries. The material in this section cannot be taken from the library. Since the material in your library will be varied, you should walk the aisles to get a sense of what is available. You will find most reference sections organized from general to specific.

Here are some specialized technical sources that you may find valuable.

Thomas Register of American Manufacturers (published annually) contains alphabetical listings of products and services, and company profiles with addresses and telephone numbers. A Product Index and a Brand Names Index are included. (Thomas Publishing Company, New York)

U. S. Industrial Directory compiles product directories and company directories for industrial companies. (A Reed International Publication, Stamford, Connecticut)

Sweet's Catalog File (published annually) is a collection of manufacturer's catalogs and products for general building construction and renovation. (McGraw-Hill, New York)

Annual Book of ASTM Standards contains standard classifications, guides, practices, specifications, test methods and terminology for a wide variety of industries and materials, from metals to geothermal energy. (American Society for Testing and Materials, Philadelphia, Pennsylvania)

The Directory of Federal Laboratory & Technology Resources: Guide to Expertise, Facilities and Services lists federal government laboratories that will share expertise and equipment. The directory includes subject, state, resource name and agency indexes. (National Technical Information Services, Springfield, Virginia)

The Monthly Catalog of U. S. Government Publications lists federal government publications, which are indexed by author, title, subject, series, and classification number. (The United States Superintendent of Documents)

The **SAE Handbook** (published annually) contains standards and recommended practices, and information reports on a wide range of surface vehicles. (Society of Automotive Engineers, Warrendale, Pennsylvania)

The **Visual Dictionary** uses graphic representations instead of written definitions of words and terms. (Facts On File Publications, New York)

As you become more experienced in a particular field, you will want to progress from general reference works to more specialized texts. Most technical fields have specialized handbooks, encyclopedias, and dictionaries, such as the *Handbook of Industrial Robotics* (John Wiley & Sons) and the *Encyclopedic Dictionary of Industrial Technology: Materials, Processes and Equipment* (Chapman and Hall).

4.7.4 Online Databases

Many libraries have increased their *online database* facilities substantially within the last five years. The databases allow users to tap into computerized networks and retrieve information from sources that would far exceed the physical capabilities of most libraries. While database searches can be very helpful, they are also very costly, so consider them only for large and significant research projects.

You will need to work closely with a librarian to prepare for your search. You will use key words and terms to begin your search, so spend time carefully deciding which key words to choose. You will receive a printout of the findings of the search.

A sampling of databases that contain information in technical fields, including the biological sciences, chemistry and chemical engineering, civil engineering, computer sciences, environmental studies, electronics and electrical engineering, geology, health and medical technology, and physics would include SCISEARCH, NTIS, COMPENDEX, and SUPERTECH. Some databases can be highly specialized, such as POLLUTION ABSTRACTS. The Elsevier Science Publishing Co., Inc. (New York) publishes the Cuadra/Elsevier *Directory of Online Databases* annually. They recorded 4465 databases in 1990.

ETHICAL CONSIDERATIONS

It is ethically, and often legally, wrong to include someone else's words, ideas, or data as your own work without credit to your source. You need to acknowledge your sources through appropriate references, including parentheses, endnotes, and footnotes. You must let your audience know when you are directly quoting material by using quotation marks for shorter quotes and indentation for longer (more than 50 words) quotes. Even when you paraphrase a person's idea, opinion, or theory in your words, you should indicate the original source of the material.

A Bibliography is a separate alphabetized list of all of the sources you consulted in assembling the document. A Works Cited list details the sources actually referenced in the document.

4.8 SEARCHING COMPANY RECORDS

Virtually all companies keep files. These files form a history of the business matters of the company: records of business transactions and decisions, the minutes of meetings, company policies and rules, annual and quarterly reports, descriptions of projects, and so forth. Important documents may be centrally located in a company library staffed by professional librarians, or they may be spread throughout the organization in file cabinets and on computer disks.

Wherever you work, you should have an understanding of the filing system, and you should know how to access the information you need. Knowing what was done and how it was done can help save you time and help you avoid errors.

4.9 USING NONPRINT MEDIA

Many individuals limit themselves to the printed word when they are gathering information, perhaps because of a bias against nonprint resources. This bias is gradually being erased as more options become available to us, and as we realize that nonprint media can inform us *and* entertain us. We can gain important information from a television program, for example, as long as we proceed carefully and do not trust our memories to capture each fact. (Transcripts of many television shows are available for purchase.)

We recommend that you take accurate notes as you watch or listen. You may want to contact the station or company responsible for the program to verify key information. Playback features on videocassette recorders allow you to view important or complex material on a videotape many times. Accuracy and precision are crucial when we use nonprint media.

4.10 A FINAL WORD

The world is much too complex for any one individual to know everything; in fact, no one can know all there is to know about any one field. Since we are unable to know all, we should know how to access information when we need it. We learn about the world by doing research, whether it involves asking a friend for advice on how to repair a part in your car's engine or traveling to a library to consult a specialized reference work.

RESEARCH EXERCISES

1. **(a)** Look in an almanac for a listing of Nobel Prize winners. For the years 1960-1980, how many Nobel Prize winners in Physics came from France? Who were they?
   (b) What percentage of Nobel Prize winners in Chemistry for the years 1970 to the present have come from the United States?
2. Find statistics about the amount of oil spilled in the United States in the most recent year. Include coastal waters; spills, however, also include land spills. Write a short report about your findings.
3. Find out how electrical units are measured in Europe. Write a memo explaining how conversions are done.

4. Write a one-page summary of the life and accomplishments of Thomas J. Watson, Sr.
5. F. Robert Gates has asked you to do a preliminary investigation into the area of virtual reality. He has asked you to write a two-page synopsis of virtual reality, including an explanation of what it is, who is developing it, and what are its possibilities. (More on F. Robert Gates, the founder of Gates and Associates, can be found in Chapter 26, Communicating In An Organizational Environment.)
6. Conduct an investigation into the causes of metal fatigue and corrosion. Write a two-page report of your findings. Some information on metal fatigue can be found in Chapter 26, and in particular Section 26.7.
7. Find out about radiation dosimetry and air-capacitors. Write a short report for an informed audience in which you present some background information and recent developments in the field. Include diagrams and illustrations to help your readers.
8. Conduct a survey to find out the opinions of your classmates about a topic of your choice. Design the survey to elicit specific answers. Give the survey, tabulate the results, and present a summary of your findings in a brief report.
9. Make a list of reference works in your area of study contained by the library most available to you. Group the list into a general category and a specialized category.
10. The following exercise is designed as a group activity. It needs to be done in conjunction with a library, and the library should be forewarned about the possible disruptions that may result from the flurry of activity. The questions may, however, be assigned to different individuals to do as homework.

THE RESEARCH SCAVENGER HUNT

Try to answer as many questions as you can. You may not be able to answer every question given the material available in the specific library where you do your research.

(1) What can be found in Sweet's Catalog?
(2) Who is Antoine Lavoisier?
(3) How often is the **Readers' Guide to Periodical Literature** published?
(4) What is your major? What are the call numbers for books in your major?
(5) List three books that discuss the evolution of computers?
(6) What is the address of the United States Patent Office? What does the office do?
(7) Who are the two United States senators from New Mexico?
(8) How is the information organized in the **McGraw-Hill Encyclopedia of Science and Technology?**
(9) What is the definition of **parallax view?**
(10) Where can you find a topographical map of California? Be specific. Include the page number.
(11) What was on the cover of **TIME** magazine in the issue dated in the first week of June in the most recent year?
(12) List three periodicals that your library receives that apply to your major.

(13) What is the phone number of the White House?

(14) What was the temperature range in Paris, France last Thursday?
high ____________ low ____________

(15) Where was Buckminster Fuller born? Who, briefly, was he?

(16) Who was Harry Truman's Vice-President?

(17) Define the following term: **Enola Gay.**

(18) What was Roy Campanella's batting average during the 1953 baseball season?

(19) What is haggis?

(20) What is the official language of Kenya?

(21) What is the French term for a **stop lamp?** (Hint: Check the IEC Multilingual Dictionary of Electricity.) Your library may not have this dictionary. If not, how would you discover this information?

(22) What is the population of Luxembourg?

(23) Who publishes the **Metals Handbook?** What is it?

(24) What is the **Environment Reporter?**

(25) Define **lumbang oil.** (Hint: Consult a dictionary of Scientific and Technical terms.) If your library does not have this source, how would you find out?

(26) Define **degauss.** (Hint: Check the IEEE Standard Dictionary of Electrical and Electronic Terms or another specialized dictionary.)

(27) What nation currently has the highest per capita income? What is the income? (Make sure you record the year and the source of the statistics you are using.)

(28) Who invented air conditioning?

(29) What is the world's busiest airport?

(30) How many silver stars are on the insignia for a Vice Admiral in the United States Navy?

5

Scheduling Your Writing

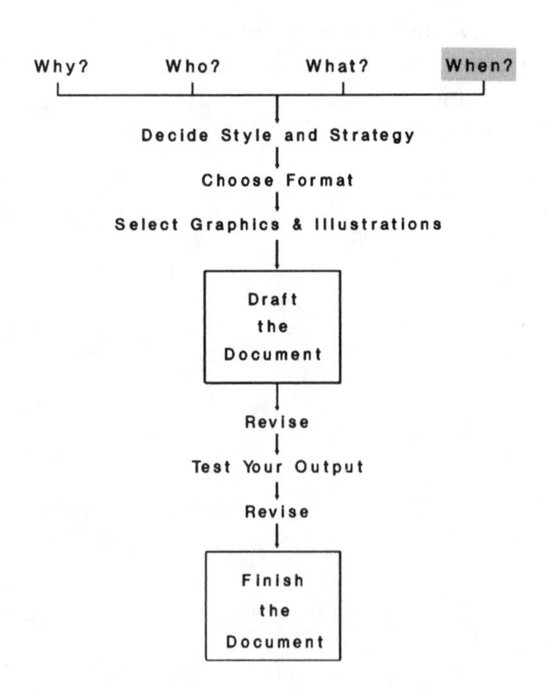

5.1 *INTRODUCTION*

For many of us, a blank sheet of paper is intimidating. We wonder how we will ever fill it up with words, much less the best words to serve our purpose. We worry that we will never be able to complete the writing project and, worse, that the finished product will make us look bad. Unable to put those first key words down on paper, we procrastinate until we are pushed right up against the deadline. Now we have to get it done! So we rush through the writing, hating every minute, and when we are done we don't want to see the paper ever again. Proofreading? Editing? We just want the paper out of our sight.

Writing does not have to be this way. Ideally, we would know our purpose and define our audience, we would gather all the material we needed, we would find a comfortable yet stimulating place to write, and we would begin. The ideas and the phrasing of these ideas would flow freely. Soon we would have filled the page with the exact words to convey our meaning. There would be enough time to proofread our work carefully, to use a spelling checker, and to get input from others to see how they would react to what we have said and the way we have said it. We would make the necessary changes and print our work using the best available technology. The finished product would shine!

The truth is that our writing experience should be more enjoyable than what was described in the first paragraph, and it will never live up to what was described in the second paragraph. All of us must learn to accept our limitations even as we strive to improve our writing. This chapter attempts to provide you with some help to make the process of writing easier, perhaps even to change the way you look at the task of writing. By changing the process, we believe that you can improve the quality of your writing.

5.2 *DEADLINES*

Many writers have difficulty adjusting to the pressure of a limited time-frame, and deadlines put pressure on us because they limit us. To produce a perfect essay, or memo, or set of directions, or formal report, we need more time. If we just had another day, we could make this thing look really good, or so we tell ourselves. The truth is, deadlines help us. No piece of writing is ever perfect; we could always add improvements, no matter how long we have spent working on it. By giving us a time limit, deadlines allow us to finish what we are writing. The same is true for any other form of communicating: we need limits.

The trick to writing with a deadline is to make the deadline work for you. Admit that you have a limited amount of time to accomplish your task, and use the available time as well as you can.

Be aware of how much time you need to complete specific tasks. Allow sufficient time

- for art work and photographs
- for illustrations (tables and graphs)
- for responses, testing your output, etc.
- for proofreading and editing

For any large or particularly important project, you need to have a timetable to inform you when different aspects of the project should be completed. If these deadlines are not imposed on you by someone else, then impose them on yourself. Self-imposed deadlines may be harder to meet, but they are just as important.

Time charts are particularly valuable for organizing large and complex writing projects. As you can see in the following example, a chart allows you to divide a large project into smaller projects.

Preparing a Gantt chart (see Section 8.4) as in Fig. 5-1 requires that you break up the amount of time before the deadline into segments; the time becomes the vertical segments in the chart. The various tasks that you will need to do to complete the document are written horizontally. The bars are an estimate of the amount of time required for each task. By filling in the chart, you gain a perspective on what it will take to finish the writing project. As you complete the individual tasks, you are able to monitor your progress against the final deadline.

Above all, having a timetable allows you to avoid the worst way of writing: the negative cycle we described earlier. If we stall and procrastinate until it is too late to produce a good document, then it does not matter if we **could** have done a good job **if only we had time.**

So start making your deadlines work for you. Realize that a deadline determines what you can do and what you can not do. Let your deadlines free your writing.

5.3 *THE FIRST STEP: PREPARING TO WRITE*

To write well, you need to find a comfortable environment in which to write. The key here is knowing yourself and what conditions help you concentrate. Since this may differ with each person, we will not attempt to advise you. You may write best with a heavy metal CD blaring out of your Kenwood speakers, or you may need the absolute silence of the library stacks to write well. Many people write best when they place themselves in a quiet calm setting where they can focus on a single task, but some people write best when they are working on two or three projects at the same time. By reflecting on your work habits and preferences, you can decide what works best for you, but it is something you should consider.

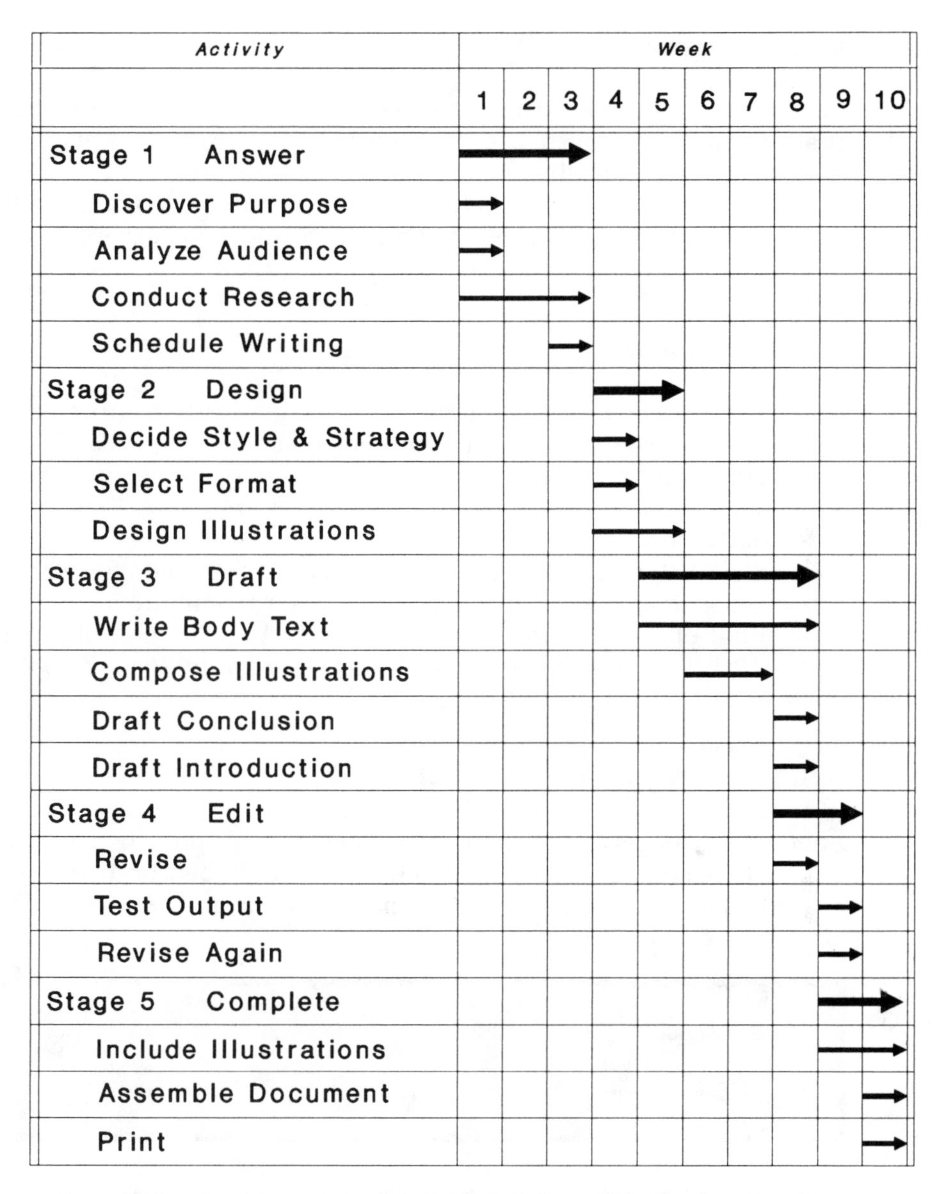

Figure 5-1 An example of a Gantt chart for a writing project. Note how the chart follows the Document Design Process.

One of the problems we have when we are trying to get started is the "censor" we have in our brain that is constantly interrupting, saying "No, not that, don't say it that way." The process of creating words on paper is difficult enough without contending with a voice that is saying "Leave that out!" all the time. The following are some ways of handling our censors and releasing our creative juices.

5.3.1 *Brainstorming*

The process of generating many ideas about a topic quickly is referred to as brainstorming. You need to let yourself go and write down as many **possible** ideas as you can. For the moment, close your critical eye, the one that says that this idea is no good, and just write everything down. Later, when you look at your list, you can open the critical eye and let it go to work. Use these notes as the basis for your rough draft.

5.3.2 *Freewriting*

Freewriting is a technique that may help you get started, particularly if you are blocked or frustrated at the beginning of a document. Instead of thinking about what you will end up with, your final product, concentrate on what you are doing. Become part of what you are doing. Be intent on the way, not the goal.

Exercise 1 at the end of the chapter is an exercise in freewriting. Like any exercise, it only works if you do it. Write for ten minutes straight. If you can't think of anything to write, then write "I can't think of anything to write." No human can keep writing that over and over. Soon you will be producing some other text.

5.3.3 *The Tagmemic Approach*

Another way of getting started is the Tagmemic Approach, previously discussed in Section 2.3. This approach closely resembles brainstorming with a particular focus. By answering the questions in the shaded box, you attempt to define your subject.

> What makes this item unique?
> What possible forms can it take?
> How does it fit into the larger world?

These questions should help you make your subject matter distinct, isolating it from everything that surrounds it. The questions can be broken down further to allow you to probe deeper and make your subject precise:

What makes this item special?
How is this item important?
What are the major features of this item?
What are some of the forms it can take?

What is the most unique form of this item?
Is it different from what it used to be?
Will this item change in the future? How?
What makes this item different from other items?
What else relies on this item?
What causes changes in this item?

When you have finished answering the questions, you should have a much clearer idea of your subject matter, and about its exact identity, its range of forms, and its relationships with the larger world. We recommend that you write down your responses. It will give you some text to begin refining.

5.4 THE NEXT STEP: ORGANIZING

Once your ideas are flowing freely, you will have the basic material with which to build your document. Now you can proceed to the next step: organizing your material.

To organize your material, you need to know your purpose. If you have not yet read *Chapter 2, Discovering Your Purpose,* now is a good time. Once you know your purpose, you can begin outlining your thoughts.

5.4.1 The Outline

If you have followed the exercises described in 5.3, namely brainstorming, freewriting, and writing a tagmemic matrix, you should have a list of ideas about your topic. Now you need to prioritize your list: you need to find out which items are vitally important, which items support other items, and which ones are relatively unimportant.

As you prioritize, look for connections. Is it possible to group certain items under the same heading? Do other items form a sequence or pattern? Which items need to be placed first? Which items would be appropriate at the end of your document? Use notations or symbols to mark your organizational thoughts. Make preliminary decisions about which topics you consider to be primary, which you consider to be secondary, and which you consider to be tertiary. Try to think of headings (primary topics) and subheadings (secondary topics) that would capture the nature of these topics for your readers.

Now turn your list into an outline.

There are many ways to outline. You can employ a Roman numeral outline, a numerical outline, or an outline where distinctions are made by different headings and typefaces. Observe the three possibilities here:

I. Introduction	**INTRODUCTION**	1.0 Introduction
A.	First major topic heading	1.1
B.	Second major topic heading	1.2
II. Development	**DEVELOPMENT**	2.0 Development
A.	First major topic heading	2.1
B.	Second major topic heading	2.2
1.	*First sub-topic heading*	2.2.1
a.	First sub sub-topic heading	2.2.1.1
b.	Second sub sub-topic heading	2.2.1.2
2.	*Second sub-topic heading*	2.2.2
C.	Third major topic heading	2.3
III. Conclusion	**CONCLUSION**	3.0 Conclusion
a.	First major topic heading	3.1
b.	Second major topic heading	3.2

Whatever system you choose, make the outline consistent; the kind of information provided at each level should be equivalent. Each lower level should break down the higher level into separate but equivalent pieces. The entire outline should allow a reader to see the logical progression of the ideas to be presented.

Although the three outlines convey the same material, we recommend that you consider the numerical outline for most technical documents. It is easily understood by everyone; its mathematical nature makes it the most obvious for many technical people. We hope you have noticed that this book is written using a numerical outline. Some companies do require that all company documents follow one particular format; if the company you work for has such a policy, we obviously recommend that you follow it.

5.4.2 Notecards

For longer papers you may want to use notecards as a way of keeping large amounts of information accessible. You can shift cards easily when you need to make changes.

Notecards work particularly well when you are writing a document that requires research. If you are working in a library without access to a word processor, you will be writing material with a pen or pencil. You write down items on separate index cards. When you are ready to organize your material, shuffle the cards into the order you desire. As your document evolves, you can easily move your cards around.

Remember to also write down the sources of your statements, facts, statistics, quotations and expert opinions. You may need them for the final version of your document. You need to acknowledge your sources with footnotes, endnotes or some other accepted format within your technical field. This is more easily accomplished with notecards because you have easy access to the information you need.

When you are ready to begin your first draft, check through the order of the cards. The first draft should go smoothly since you have your material in front of you.

5.5 *THE DRAFT*

Working from your outline or from your notecards, you can now begin to write your draft. By viewing the draft as a preliminary stage in the process, you can relieve some of the stress of having to "get it right." Changes can be made later, and you can concentrate on getting all of the material into the document. Polishing and perfecting will also come later.

A significant advantage of this way of writing is that it allows you to review your writing after time has passed. You may come to see that a different approach or format is appropriate for your document, and the draft allows you to "play" with a variety of potential products. Rewriting is much more than copying over your work with a few minor changes. You need to challenge what you have written, evaluate the strong and weak points, and revise accordingly.

Many writers like to put their thoughts down on paper well before a deadline so that their unconscious mind can ponder what they have written. Without being aware that they have been thinking about the writing task, they return to the work some time later with a whole new set of ideas. A draft allows them to do this. (Please note: we cover the process of revising, editing, and proofreading in Chapter 22, and the process of testing what you have written in Chapter 23. You may want to read those chapters at this point.)

5.6 *RESOURCES: THE BOOKS YOU NEED TO HELP YOU WRITE*

No individual can write without occasionally stopping to check on a detail: the spelling of a word, a rule of grammar, a fact, a date, an interpretation of material, a quotation, or any of a hundred other possible items. When we check our work, we are insuring that the things we say and the way we say these things are precise and accurate. Whenever possible, you want to **verify** your statements.

When we fail to check our work, we open ourselves to error. One major cause of writers failing to check their work is distance. When the books that would help us are far away, we are reluctant to use the time away from our writing. Sometimes we convince ourselves that we will check the material later, but we never do.

For this reason, we recommend that you build a small resource library in a handy location. This is, we admit, much easier to do if you have a job where you spend most of your time at a desk. It is harder to do when you are a student or when your job sends you to many different locations. We recommend the following list as a foundation for your resource library:

General dictionary
Dictionary of Scientific and Technical Terms
Grammar Handbook
Thesaurus
Current almanac
Current atlas

We hope you will find this handbook valuable enough to include it on your shelf as a major resource for your technical writing. You may also want to include one or two reference works from your specialized field. The longer you specialize in your field, the more resources you will add.

If you find that your career path takes you to a position where you are preparing and editing technical documents on a regular basis, you may want to consult a technical editing guidebook, such as George and Deborah A. Freedman's *The Technical Editor's and Secretary's Desk Guide* (McGraw-Hill).

5.7 WRITING WITH A WORD PROCESSOR

We may write the same thing when we write with a word processor as when we write with a pen and paper, but writing with a word processor changes the way we approach writing. For example, we may know the concept "you don't need to begin with your first sentence." Until we see how easy it is to revise a document on a word processor, however, we never do it. When we do, we realize that it's something we should have been doing all of our lives.

Since changes are easy to make, and since we can cut-and-paste at will, we lose the fear of making errors, the sense that we have to get it right the first time. We can test and preview different versions, and nothing is final until the last minute when we issue the print command. And since we have the material stored on disk or on a hard drive, we can return to the document later to make revisions. We can make the revisions and print out the new version quickly and easily.

We can also work more easily from an outline. We can enter the outline and then develop it section by section until we have our finished product. Few of us can write a thirty-page document from scratch, but most of us can write half a page on a specific topic. Outlines enable you to partition your work into small, manageable sections.

You may be reluctant to try to use a word processor if you have never used one. You may have fears about your inability to type, or you may have fears about technology. You owe it to yourself, however, to risk the attempt. You can always return to pen and paper writing if it does not work for you.

5.8 *WRITING ON THE JOB*

If you have never written documents in a work setting, you need to understand some unique elements to writing in the context of a business organization. The first is the idea of a quick turnover. In a business environment, you frequently do not have the luxury of taking your time and polishing an assignment. The document is needed quickly, and you may feel rushed in its preparation. Particularly when fellow workers or customers need information, the pressure is on you to respond quickly and accurately. As with so many other aspects of writing, this one too gets easier with experience.

If you work in a large company, you may have access to resources that can make your life easier. These include a secretarial pool, research librarians, and company files. Each of the above can simplify your writing tasks and help you produce a professional document.

5.9 *A FINAL WORD*

In order to produce a professional document, you need time. This is why you need to question anyone who assigns a writing task. Please understand that this includes yourself, for you assign many of your own writing tasks. Here are some of the questions you should ask.

- How long should it be?
- Who will read it?
- How important is the task?
- How much support will I have?
- What resources are available for me?

Once you have the answers, you can make decisions about scheduling your time, particularly when you need to shift other tasks so that you have the time to prepare. By making sure that you know the expectations of others and the limitations imposed by time, by economics, and by access to resources, you increase the odds that your document will be successful.

SCHEDULING EXERCISES

1. Try freewriting for ten minutes. If you run out of ideas, write that down. You may write "I can't think of anything to write." Soon the ideas will begin to flow again. Stop when the ten minutes is up. (A suggestion: if you are writing with a pen or pencil, hold it loosely in your hand.)
2. Generate a list of ten ideas on the following topic: *Ways To Make My Life Better.* Now give a priority number to each item on the list, with one being highest priority and ten being lowest priority. Send your list to your instructor in a memo.

3. For the next writing project you are assigned, develop a Gantt chart or other timetable. Monitor your progress by marking both the estimated time and the real time on your timetable. Submit the chart with the completed assignment.
4. Make a list of all the possible environments where you have written papers during your life; for example, have you ever written a paper stretched out in bed? At a desk in the stacks of the library? Outside during the summer under the shade of a tree? List all of the places, and think about the document you produced. Where were you able to do your best work? Write an analysis of the conditions under which you write best.

   Now consider other variables, such as time and writing tools. Have you ever worked with notecards or an outline? How many drafts do you do? How do you verify the accuracy of your work? Develop a profile of the writing process you believe produces your best work.
5. Think of an idea that would improve the learning environment of your current classroom. Then use the questions from the tagmemic approach described in Section 5.3.3. Write a summary of your recommendation in a memo to your instructor.
6. If you are using a computer to write a document, try turning off the monitor or dimming the screen so that you are unable to see it. Now when you write, you may find yourself free of the pressure caused by the impulse to look back at what you have just written. Write a brief memo to your instructor in which you relate the results of this experiment.
7. Put the following list into order using a numerical outline:

   attaching part c to parts a and b
   parts
   installation
   part c
   assembly
   function of each part
   part a
   tools
   unpacking the box
   attaching part a to part b
   part b
   installing the assembly into the unit
     locating the correct spot for installation

8. Now change the outline produced in Exercise 5 to a Roman numeral outline.
9. Make a list of the resources available to you in your current educational or work situation. This list should include all of the books, people, and computer software programs you are able to access if you need help.
10. If you have ever had to write within a work setting, write a memo informing your instructor about one particular task and what was involved. Describe the task, the length of the document, the audience, the research you performed, and the time restrictions placed on you by this writing situation. What resources did you utilize? What problems did you encounter? Is there anything that you would have done differently?

6

Choosing Style and Strategy

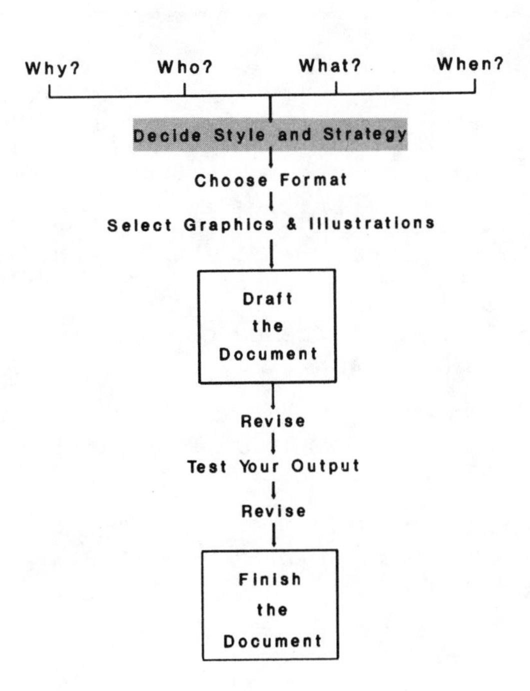

6.1 INTRODUCTION

Style refers to the ways you express your message, the specific words and phrasing you use to convey your ideas. *Strategy* refers to the approach you take to a communication situation, and the way you understand and react to the context of the situation. The first rule of style is to have something to say; now you can shape this something. The first rule of strategy is to know when and where to say it, to use style so that your message will be well and clearly received.

Once you have something to say, you need to find a way to say it clearly and correctly. You also want to think about ways to say or write your message which will help your audience to understand what you want or need. In this chapter, we will discuss some of the ways that style and strategy affect the ways that you communicate important and complex messages.

6.2 WRITING AS DECISION-MAKING

Writing is a process of making decisions. The words you choose, the order of those words, the length and degree of complexity of your sentences, the presence or absence of errors of grammar, and a host of other factors influence the way your audience receives your message. The combined effect of the decisions you make determines your style of writing, the way you convey your message.

Read aloud the following sentences. Try not to think of the meaning of the sentences, but listen to their sound, the combined effect of the decisions the writer made.

The azure waves gently lapped at the golden shore, as frigatebirds lazily pirouetted in the cloudless sky.

Lowest on the governmental totem pole are local governments, which bear the major responsibilities for providing public works services.

Plastics engineers design, develop, and manufacture products that are made out of plastic resins.

The epistemological brilliance of Kant's argument concerning the synthetic a priori leapt from the page and mesmerized the newly awakened student.

Franklin Thomas recommends leaving the HPI pumps on until determination that the hot leg temperature is more than 50°F below Tsat for RCS pressure.

(continued)

> *Customer may transfer the Licensed Software Agreement provided that (i) this Software License Agreement is transferred with the Licensed Software, and (ii) the transferee fully accepts the terms and conditions of this agreement, and (iii) all complete or partial copies of Licensed Software, including copies on data storage devices, are also transferred or destroyed.*

Every communication situation is unique. You need to be flexible to fit your style to the situation, and so you need to be confident in your ability to use a number of styles effectively. In most technical writing situations, you want to be precise and direct. A Shakespearean flourish will not be as effective as a straightforward, accurate statement.

> Style Guidelines
> - Choose shorter words over longer words, when you have a choice, but always opt for the precise words.
> - Place only one major idea in each sentence. Develop this idea fully, then proceed to the next idea.
> - Phrase each sentence so that it conveys your meaning precisely.

Whenever you write or speak with other persons, you make several important choices. Often you will not even be aware of your alternatives, but everything you write or say afterwards will be influenced by these early decisions.

The two key choices you make are deciding on what style to use and what strategy to take in transferring your message. The way you make these choices is similar to the ways in which an experienced builder surveys a work site and makes a quick and reliable estimate of the tools, materials, and costs of the job. The more experience you have, the better you will get at this sort of expert task. So too, when we begin to communicate with other persons we make a series of quick decisions about whom we are talking to, what is an appropriate manner to use, and what is an acceptable approach to take.

The same question you asked in *Chapter 3, Addressing the Audience* will determine how you make these decisions.

- **What is my relationship to this person or these people?**

We talk about relationships with other people as close or distant. Style is also determined by distance. The closer you are to people, the easier it is to communicate with them. Distance, both physical and social, is a factor in how we communicate with each other.

Expressing information clearly to a stranger is more difficult than talking to someone you know very well. An even greater distance can be created by cultural confusion when this stranger does not share your beliefs about what words and pictures mean. Distance explains why it is more difficult to say something in writing. Writing is a way of talking to someone who isn't there, whose reactions you cannot observe.

The very fact that your audience is distant from you while you are writing creates difficulties. You cannot point things out physically. Your audience cannot ask questions. They can not see your gestures nor hear the inflections in your voice. In sum, you have to elaborate and explain what you intend to say.

This chapter attempts to give you a sense of the dimensions which influence your choices. Whether you are older or younger than your audience, more or less important, and even whether you are seated or standing, can affect the ways in which you choose your words, your attitude, and your approach to delivering your message. This chapter is intended to help you get a better idea of what you are already doing.

6.3 TYPES OF STYLE

As human beings engaged in the complex activity of communicating with each other, with all of its rhythms, rules, and confusions, we are constantly making style changes, shifts in language and expression that signal subtle differences and perspectives.

All of us change styles. We use different words and ways of expressing our message, depending on our audience. Certainly it is easier to write a note to a good friend than it is to write an article which will appear in the newspaper where everyone can read it.

What is the difference between these two messages? Quite simply, it is more difficult to talk to people we don't know, people we don't share space and information with. We are constantly adjusting our speech and writing styles to our listeners and our imagined audiences. For example, the city cop who "collars the punk" writes it down as "apprehending the suspect." The ways we use language change with the situation, the subject, and our relation to the audience.

Imagine yourself at the center of a series of circles each of which shows the increasing distance between you and the people you can speak or write with. These styles represent five styles which we all employ.

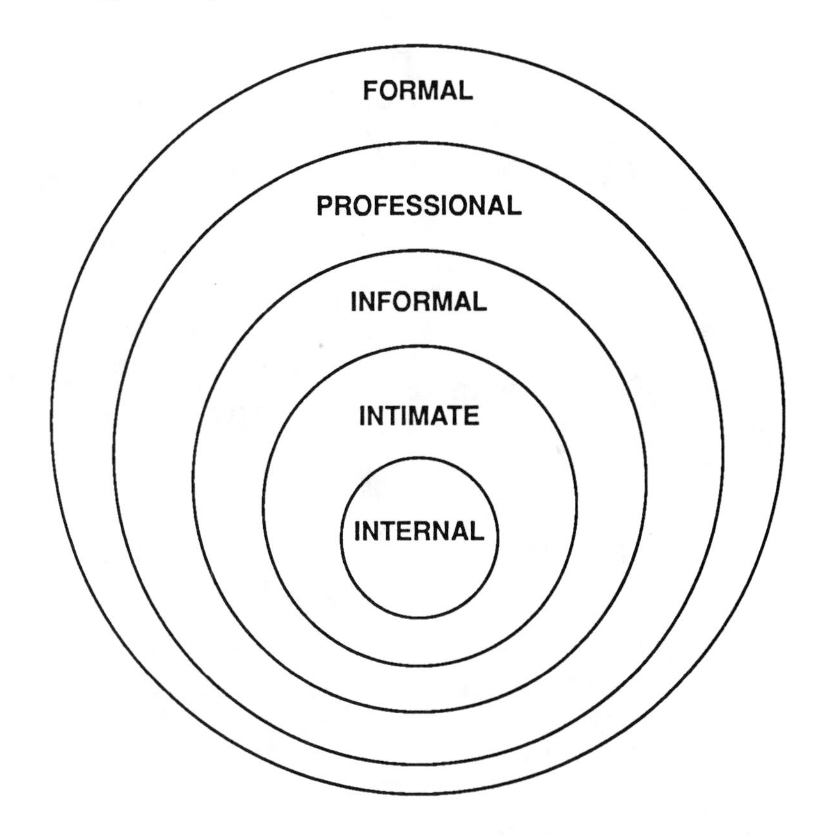

Figure 6-1 Five Types of Styles.

6.3.1 *Internal Style*

The first style we need is internal. This is for inner communication. Whether we talk to ourselves or view things in our mind's eye, each of us develops unique and powerful ways of communicating within ourselves.

This internal style is solid state communication. It takes place inside us. We can jump from one subject to another as quickly as we want. Any type of shorthand or mental indexing will do. This style is immediate and impossible to censor. As soon as we think something, we know what we are thinking.

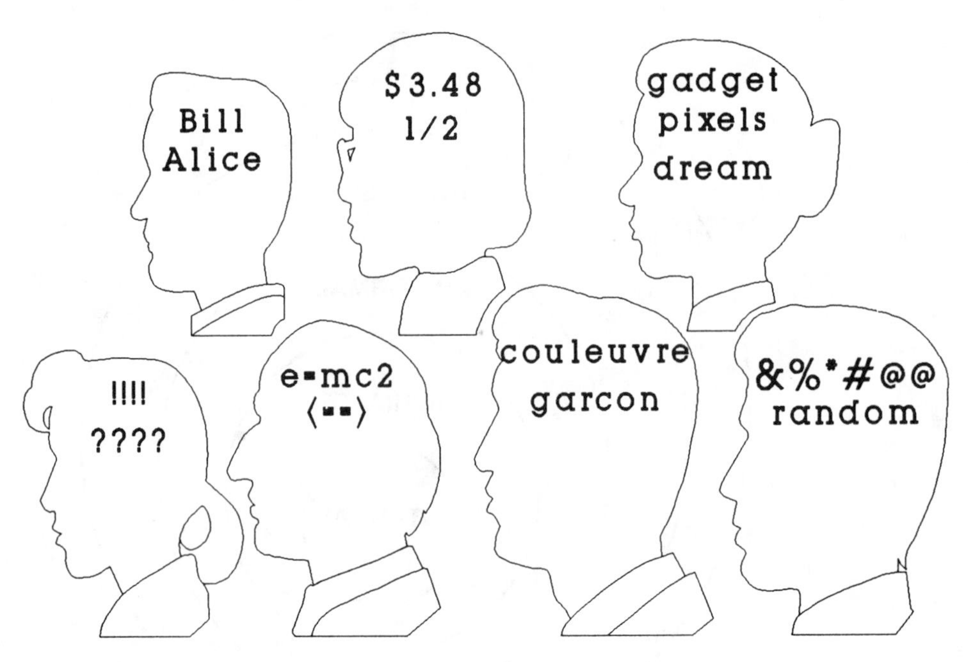

Figure 6-2 An example of internal style.

6.3.2 Intimate Style

We use an intimate style when communicating with the people closest to us. These people, family, lovers, our closest friends, are the people who share our space, our personal history and our experience.

The intimate style is close and quick, marked by shared signals, thoughts, and experience with language. It is free and open, spontaneous, unconstrained, and casual.

Even a few words will communicate a great deal of information when persons share the same close and intimate background. A single word can contain a great deal of meaning because of shared personal associations. This type of language is a restricted code, however, known only to those who have participated in the same experience.

Figure 6-3 An example of intimate style.

6.3.3 *Informal Style*

Informal style is what we choose when we are communicating with friends and acquaintances. Since we are comfortable and familiar with these persons we can use language without being cautious, careful, or rehearsed. We can be relaxed and unceremonious and say what we intend without worrying about how we say it.

You could write a memo in an informal style to one individual within your company with whom you have developed a close working relationship. You need not be extremely close friends outside of work. Occasionally, a business letter to a close acquaintance is written in the informal style.

Again, like the intimate style, the informal is marked by shared personal associations where a word or signal may carry a great deal of meaning. If you know someone quite well you can communicate a lot of information by referring to shared experience. When you say, **"You know."** to your friends, they probably do understand what you mean.

Sheila

Barbara and I want to thank you and Tony for a wonderful evening last Thursday. I will never forget the look on your face when you learned that Tony was a Trollope fan. Barbara will get in touch with you about the golf match.

Bob

Figure 6-4 An example of informal style.

6.3.4 Professional Style

A professional style is more elaborate than either the informal or intimate styles. Great care is taken to explain things fully. This style is cautious and deliberate and marked by strong concern for the audience. In business and industry it is essential that we are understood when we speak and write to each other.

Often this means preparation and rehearsal. A professional style is not spontaneous and immediate. Rather it is expected to be careful and deliberate. Technical descriptions, technical directions, progress reports, and oral presentations are just some of the technical tasks that require a professional style. Such documents and presentations are designed ahead of time, reviewed, and then revised to be precise and exact.

This does not mean that your professional writing and speaking styles cannot be conversational and easy to understand. However, even if your writing resembles offhand conversation, you need to take *extra care* to explain things completely to avoid the chances of misunderstanding.

Using a professional style means taking responsibility for the successful transfer of the message. This means clarifying, explaining, defining, demonstrating, illustrating, making things clear, and spelling them out.

Here is a letter written by a world-noted science fiction author to *Sky and Telescope Magazine:*

False Advertising

I was shocked by your advertisement for "Exciting Full-Size Posters of Mercury, Venus, Earth, and Mars!" (October issue, page 383) and wish to register my strong protest against such an unparalleled assault on our planet's ecology.

Have you stopped to consider the appalling destruction of forests that would be caused by your ill-considered proposal to manufacture sheets of paper 4,900, 12,100, 12,800, and 6,800 kilometers across?

I am sure my concerns are shared by many other lovers of the environment -- and of accurate English.

ARTHUR C. CLARKE

P.S. How did you propose to ship them?

Figure 6-5 An example of professional style. Reprinted with the permission of Arthur C. Clarke.

6.3.5 *Formal Style*

Documents that you would want your attorney to review will almost certainly be in a formal style. Leases, annual reports, contracts, job bids, estimates, technical proposals: these types of documents frequently require very exact formats which are rigid and inflexible. Legal and contractual specifications call for precise and particular forms of language and design.

The best advice we have here is that you follow the specifications exactly. If your document needs to be single-spaced, then make sure that it is. One of the things you can do is to make up a list of the specifications, then check them off as each is completed. When your audience supplies you with an exact format, you are saved the trouble of doing this yourself. So, the same as with a coloring book: stay within the lines.

In the formal style, a single word or decimal point out of place can cause serious complications. You need to proofread very carefully or have someone do this for you. Careful review is essential when your document functions as a contract and obligation.

LICENSE AGREEMENT

DISK MANAGER/DISK MANAGER DIAGNOSTICS (the "Software") is copyright 1985-1990 by ONTRACK COMPUTER SYSTEMS, Inc., and is protected by United States Copyright Law and International Treaty provisions. All rights are reserved. The ORIGINAL PURCHASER ("You") is granted a LICENSE to use the Software only, subject to the following restrictions and limitations.

1. The license is to the original purchaser only, and is not transferrable without the written permission of Ontrack Computer Systems, Inc.
2. You may use the Software on a single computer owned or leased by you. You may not use the Software on more than a single machine without the written consent of Ontrack Computer Systems, Inc. even if you own or lease all of them.
3. You may make back-up copies of the Software for your own use only, subject to the limitations of this license.
4. You may not engage in, nor permit third parties to engage in, any of the following:
   A. Providing or permitting use of or disclosing the Software to third parties.
   B. Providing use of the Software in a computer service business, network, timesharing, multiple CPU or multiple user arrangement to users who are not individually licensed by Ontrack Computer Systems.
   C. Making alterations or copies of any kind in the Software (except as specifically permitted above).
   D. Attempting to disassemble, decompile, or reverse engineer the Software in any way.
   E. Granting sub-licenses, lease or other rights in the Software to others.
   F. Making telecommunication data transmission of the Software.

Ontrack Computer Systems reserves the right to terminate this license if there is a violation of its term or default by the Original Purchaser. Upon termination for any reason, all copies of the Software must be immediately returned to Ontrack Computer Systems, and the Original Purchaser shall be liable for any and all damages suffered as a result of the violation or default.

This agreement shall be construed, interpreted and governed by the laws of the State of Minnesota.

62-1100-430-005

Figure 6-6 An example of formal style.

6.4 *TYPES OF STRATEGIES*

Many people are uncomfortable with the idea that strategy is something to consider when designing a document or presentation. Some of the synonyms for strategy include cunning, tactics, and skillful management. Obviously these words reflect a range of viewpoints toward the idea that messages should be carefully shaped for particular audiences.

The words which each of us substitute for other words often give clear indications of our attitudes, how we feel about a word and what it represents. Some people do not like planning their communications for a specific audience. Cunning and tactics do not belong in clear communications, they insist.

The point of view that we are taking in this book is that the strategy you ought to be using most closely resembles the last synonym mentioned above: **skillful management.** With complex information, often it is not enough to send a direct message, clearly and accurately. Additional efforts are required to make sure that the message is presented in the best and most effective way. You need to contrive your technical communications as carefully as you would engineer any other type of project. You need to design your message to accomplish your purpose. You are writing to get things done.

6.5 *FOUR STEPS TO STRATEGY*

To design an effective strategy, begin by defining your problem and considering your audience. Then you choose or design a format and try to imagine how your solution will affect the persons who hear or read it. Essentially there are four steps in this process:

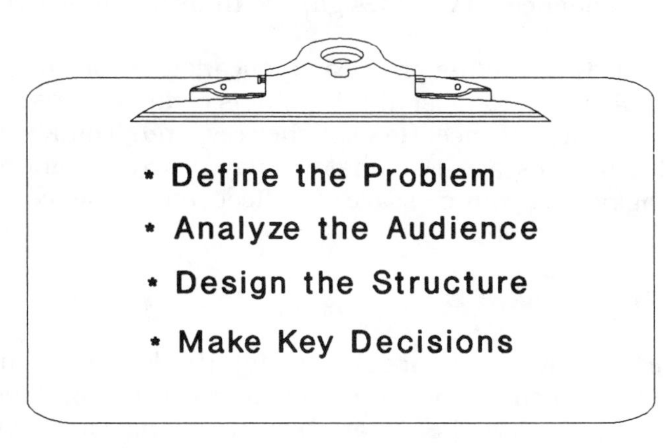

Four Steps to Strategy
• Define the Problem
• Analyze the Audience
• Design the Structure
• Make Key Decisions

Defining Your Problem The first step is formulating the problem. Until you have carefully defined the writing or speaking problems you face in a particular situation, you cannot begin to design your solution. Often, when you are confronted with a writing challenge, you will see just one big mess. As you begin to define the difficulties, design solutions become more clear.

"What is my purpose?" and "What are my basic objectives?" are the key questions in reaching an adequate definition of a writing task. (*See Chapter 2, Discovering Your Purpose.*) You also need to consider the difficulties and opportunities the task presents. What information needs to be supplied, what opinions need extra support and who already agrees with you. All of these elements are factored during the document design process.

Audience Analysis Frequently your audience is the problem you must skillfully manage. This can only be done with good audience analysis. (*See Chapter 3, Addressing Your Audience.*) You need to know whom you are talking or writing to, and what their feelings and attitudes are toward your subject and yourself. You need to know what they find persuasive and the types of detail and information that they find convincing.

Structural Design This is the step in the process where you decide how to arrange what you have to say. You begin to shape your material, organize it and form it into some sort of pattern.

Most of the time you already will be familiar with some sort of structure which will accommodate your message. Ready-made formats or style templates can be used for much technical writing. (*See Chapter 7, Selecting A Format.*) Sometimes, however, you will need to tailor a message for a specific audience.

Decision-making This is probably the most important part of the process of determining a strategy. This is when you look ahead, anticipate the consequences of your design, and think about how your message will be received.

This is where the practical application of your own experience comes in. What is needed most at this point is common sense. Put yourself in the place of your audience. How will they respond? Think about how you would react to the message. Use what you have learned from your own experience to imagine how your message will affect your audience.

6.6 FINDING A BALANCE

Unfortunately many people do not think about finding an effective approach which will help their message get through. Instead, they rely on strategies which worked before, but are no longer appropriate. Many writers and speakers assume that their only task is to express everything they know and let the audience do the work of sorting it out.

What you need to do is to find some balance among your own purpose, the needs of your audience, and the requirements of your material. The best strategy involves the skillful management of what you want, what you know, and what your readers or listeners need to know. Strategies are conscious attempts to get through to your audience, so plan each document carefully.

ETHICAL CONSIDERATIONS

While it is very important that you carefully consider how you present yourself and your messages, it is equally important that you do so honestly. Technical communications need to be accurate. People rely on their accuracy. It is wrong to misrepresent yourself, your company, your product, or your services. Misrepresentation is often illegal and it is always wrong.

The ethical path for choosing both styles and strategies is simple. Stick with the truth. Your reputation for honesty will add credibility to all of your messages.

Your writing cannot respond directly to your readers' indecision, distrust and doubt. A string of unsupported opinions will not influence people to change their own ideas, but will provoke them to argument. Instead you need to describe the facts and estimations that led to your judgments. Even if your audience does not agree with your opinions, at least they will know how you reached your conclusion. Your audience will determine your credibility once they know how you came to your point of view.

We cannot have an opinion about something until we experience it in some way. We respond emotionally to almost everything we experience and this response affects the opinions which we form. If you want someone to share your opinion you need to show them how you arrived at that judgment.

6.7 A FINAL WORD: CHOOSING THE RIGHT TONE

Tone is a manner of speaking or writing that reveals a certain attitude on the part of the speaker or writer. Tone is disclosed in the ways you choose to express yourself. Your style and strategy combine to let your readers and listeners know how you feel about them and your subject.

Enthusiasm, for example, is communicated in the ways we say or write things. So are reluctance and dislike. Your attitude toward your audience and your topic is easily recognized by most people. So it makes sense to choose an attitude while you are choosing your words.

This is style and strategy, doing the best you can to communicate information to a person, not a business or an address but a person who, like yourself, will appreciate the effort to communicate clearly and accurately.

When you are communicating face-to-face with people you can adjust your message as you view their response. When you are writing to persons outside your organization, you need to adopt an attitude that is positive and cheerful. Your words, phrasing, and approach to the reader will signal how you are feeling and affect the response of your audience.

STYLE AND STRATEGY EXERCISES

1. Gates & Associates receives a letter from a client who has been wavering and is indecisive about signing a contract which you have negotiated for her company. This contract will commit the client's manufacturing site to an expensive Superfund Cleanup Schedule. You are sure that you have worked out a good deal for your client.

   Write a letter of reassurance that responds to the tone in the close of her most recent letter to you:

   > "I would be pleased to discuss this matter with you further, with the goal of reaching agreement on terms under which I could give you the approval which you have requested."

2. Bertrand Russell, the British philosopher, devised a parlor game with language based on physical proximity and social distance. Imagine yourself talking to a friend and referring a third person who is not present. Your task is to choose a series of adjectives, increasingly less flattering. These adjectives will describe yourself, your friend who is present, and the person who is not there.

   For example,

   I am heavy.
   You are fat.
   She (not present) is obese.

   I am thin.
   You are skinny.
   He (not present) is emaciated.

As distance increases we are less socially sensitive to the words and phrasings we choose. Find three examples of this to discuss in class. How does this affect the ways we deal with people we perceive to be very far away from us?

3. Write a statement in the five different styles presented in Section 6.3. You need to meet with the company lawyer to discuss a legal issue surrounding a new technical product your company plans to bring to market next Friday. You can assume greater and lesser distances between you and the lawyer as you move through the Intimate, Informal, Professional, and Formal Styles. In the Internal Style, you will be making a note for yourself.

4. Many different dimensions affect the ways we choose our words, our attitudes, and our approach to delivering a message. These dimensions include age, gender, and race, all sensitive subjects in the workplace. Geography, education, and experience also affect the way messages are sent and received. A professional style requires you to avoid expressions and manners of address which offend people. This means observing and anticipating the reactions to communications you design. This means being aware of your audience and the responsibilities you have to them.

Discuss the ethical responsibilities involved in designing technical communications for the general public. Break up into small groups of four or five individuals. Can everyone agree to three responsibilities you have to your audience which apply to every technical communication?

5. Here is a letter written by Samuel Clemens (Mark Twain) to his local gas company. Can you suggest some ways to Mr. Twain that he could make the same points with more tact?

> To the gas company
>
> Hartford, February 1, 1891
> Dear Sirs:
>
> Some day you will move me almost to the verge of irritation by your chuckle-headed Goddamned fashion of shutting your Goddamned gas off without giving any notice to your Goddamned parishioners. Several times you have come within an ace of smothering half of this household in their beds and blowing up the other half by this idiotic, not to say criminal, custom of yours. And it has happened again to-day. Haven't you a telephone?
>
> Ys
> S L Clemens

6. The most common strategy for delivering bad news in a letter is this simple formula: **Thanks, Sorry, Because, Thanks.** (See Section 10.10.) Suggest strategies for communicating in these situations:

(a) You are designing a FAX message for a Mr. Lin Eng who works in Kuala Lumpur, Malaysia. From his message to you, it is apparent that his English skills are fairly limited. Make a list for strategies to help get your technical message across to him.

(b) You are designing a weekly progress report for Mr. Charles Auburn, President of a financial management firm. Mr. Auburn is 60 years old, a self-made man who has a temper and low tolerance with frustration. You know from past experience that Mr. Auburn has no patience with technical details. Unfortunately, your original estimates for the installation of an electronic mail network have been thrown off because of compatibility problems among his varied microcomputers. Describe your style and strategy for this important client.

(c) You have been assigned to lead a tour of visiting electronics researchers who are attending a convention in Washington, D. C. The fiber optics lab is off-limits to foreign nationals because of the classified research conducted in the lab. You want the American researchers to visit this lab because they might participate by duplicating some of our results in their own facilities.

You have to divide the tour group; citizens of the United States will be separated from the foreign nationals. What styles and strategies would you rule out in this situation and which would you choose?

7. Choosing your words carefully is very important to both style and strategy. One way to learn about correct language is by playing with your word processor's thesaurus or synonym finder. Choose a short article from *Time* or *Newsweek* and type it into your word processor. Now use the thesaurus to replace every adjective with a synonym that is not quite correct.

   We recently replaced *word* with *tidings* and *processor* with *slaughterer*. We found tidings-slaughterer a good description for some types of writers, those who think any word will do. (You can also try this technique with song lyrics or a poem.)

8. A number of companies offer software which contains pre-written business letters for all sorts of occasions. You just bring the letter into your word processor and add the name and address. In groups of four or five, discuss whether these letters would prove useful. If you can, arrange to try one of these programs and report the results to the others in the group.

9. Read the letter written by Bob Curtain in Section 3.6. What impression do you have of Mr. Curtain? Would you want him to be the reporter for a project in which you were involved?

10. Changing a document from one style to another can be like transposing musical chords. The results can be surprising and illuminating. Try changing a document from one style to another. Write a lease in an intimate style. Shift a presidential speech into informal. Design a marriage proposal in professional. What happens to the message if the style contradicts the message?

11. Read the following passage. How effective is the strategy employed by the writer of this message?

> On July 11, you wrote to us about the problems you are experiencing with the computer you purchased from us.
>
> We failed to live up to our own standards of quality, and we failed to live up to your standards.

7

Selecting a Format

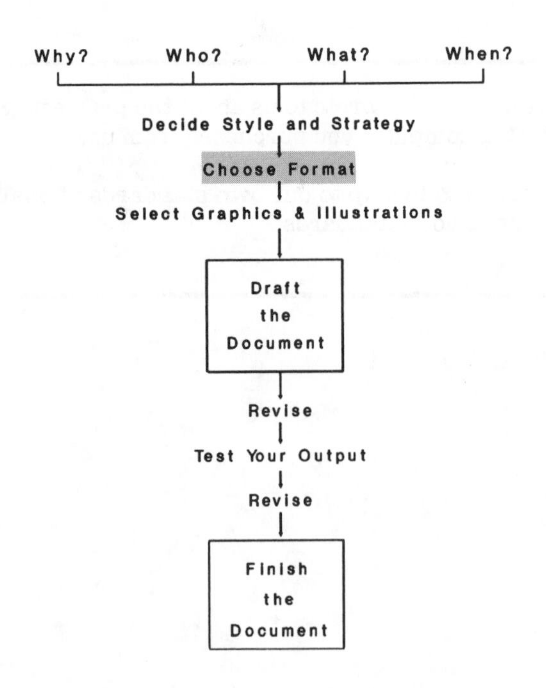

The Document Design Process
Why?
Who?
What?
When?
Decide Style and Strategy
Choose Format
Select Graphics & Illustrations
Draft the Document
Revise
Test Your Output
Revise
Finish the Document

7.1 INTRODUCTION

First impressions are very important. Before your audience reads your words, they will look at your document and form an opinion about whether to continue. If your document does not imply quality and importance, it may not be read or remembered. The audience's first question is the one you need to answer: "Is this worth my time?"

Let's say that you are in charge of hiring a technician or a manager for your company. Would you ever hire someone who sent you a hand-written resume? No matter what the individual's credentials were, you would assume that the sender of the resume did not really want the job. If she or he had wanted the job, they would have taken the time and effort to make a positive presentation.

Any document that you submit to others with your name on it represents you. Particularly in a work setting, the people who receive your documents will draw conclusions about your performance from these documents. And the better the documents look, the more favorably they will be received. A good-looking letter or report says that the author took care in designing it.

In this chapter we encourage you to examine how your writing looks to your readers, and we will address the following topics:

- How to visualize blocks of text.
- What format options you should consider.
- How to enhance the design of your pages.

While there are no absolute guidelines that will make the selection of format an automatic process, there are some concepts that should help you determine the correct format for your work.

7.2 DIFFERENT FORMATS CONVEY DIFFERENT MESSAGES

We use different formats for different kinds of writing tasks. Memos, business letters, resumes, technical directions, long reports, and so forth appear differently because our culture has developed a variety of formats to make the different tasks distinct. When you pick up a memo, for example, you know that it is a memo, a communication within a company, and not a business letter, a communication between companies, before you read the text. The way the document appears tells you what it is, and this first impression sets the stage for your audience's response. Notice how quickly you can tell the differences between the four documents in Fig. 7-1.

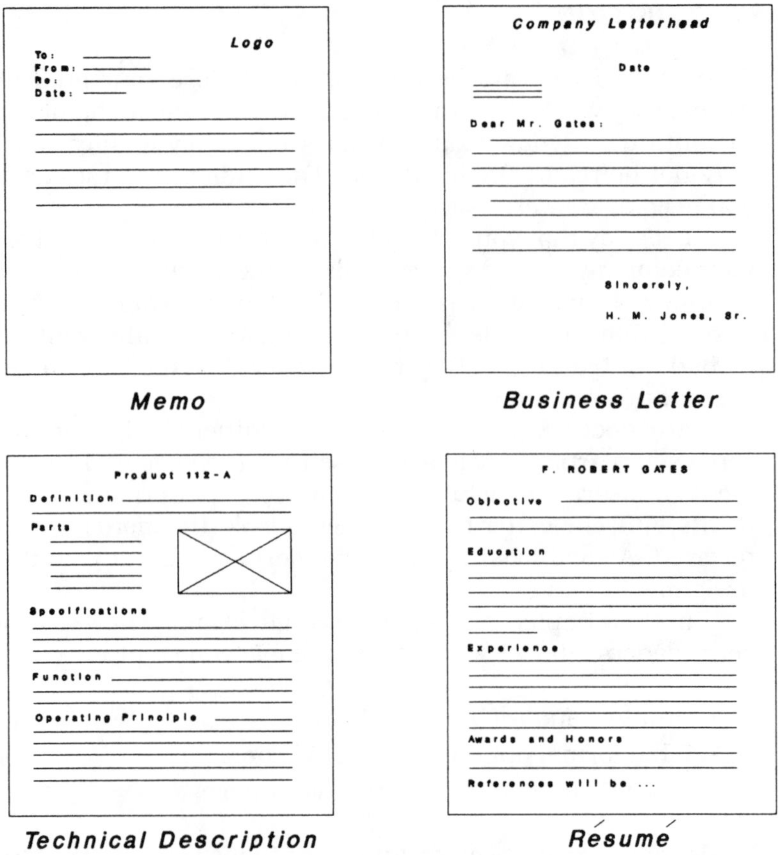

Memo

Business Letter

Technical Description

Résumé

Figure 7-1 A comparison of formats.

Individual companies and industries have developed their own formats for certain documents, and you should be sensitive to these standards as you pursue your career goals. You should conform to the accepted practices; within these restrictions, there is still plenty of room for variation and creativity. Most resumes, for example, follow a similar format, but you can make your resume distinctive even while you adhere to the accepted standards.

Chapters 9 through 21, the task chapters in this book, address many of the communication tasks you may do in your career. We have included examples of useful formats in the task chapters, and we urge you to be aware of the different formats as you read these chapters. We want you, however, to think of formatting as more than simply selecting a style sheet or a template. Most of the time you can use the formats presented in these chapters, but you should be aware of how your choices affect your message, and what your choice of format will convey to your audience.

Selecting the Proper Format

Task	*Section(s)*
Memo	9.3
Business Letter	10.3
Resume	11.7
Meetings Agenda	13.8
Presentation Outline	15.7
Technical Definition	16.4
Technical Description	16.3, 16.12
Technical Directions	17.8, 17.9
Short Report	18.5
Long Report	19.5
User Manual	20.3
Technical Proposal	21.5

The particular format you select should reflect the particular task and, in some cases, your personality. There may be times, of course, when you need to invent a format for yourself, but these will be rare occurrences. Most of the time, you will be modifying pre-existing formats to fit your particular needs.

7.3 BASIC FORMAT ELEMENTS TO CONSIDER

You do not need to be a design professional to improve the way your documents look. Simply being aware of the importance of appearance may be enough to increase the appeal of your documents dramatically. And with so many word- processing options on the market, you need not know a sophisticated desktop publishing program to be able to produce quality documents. Here are some basic format elements to consider, none of which involve expensive or hard-to-learn technology:

- ***Paper quality*** Use high quality paper with at least 25% rag content for business letters, formal short and long reports, proposal submissions, and any other document where you want to make a good impression. Internal memos and informal reports can be produced on lesser-quality paper.
- ***Cover*** An important business document, such as a long report, a bid on a proposal, or a user manual, should have a cover. This may be a basic title page or a cover sheet, but some care should be paid to producing a quality cover for your document. A word of caution: A cover can be too expensive-looking or too fancy. If you want to emphasize the economically conservative nature of your company, avoid flashy covers. The cover should reflect the document it introduces.
- ***Paragraph structure*** Vary your paragraph length for greater effect, but remember that long paragraphs (more than 8 sentences) are difficult for readers to follow. Structure your paragraphs for the reader.
- ***Justification*** This refers to the way right-hand margins align. With justified text, the margins are uniform. Unless you are using a desktop publishing program that allows kerning (proportional spacing of text), use ragged right, or margins that are not justified. Most readers feel comfortable with ragged right margins.
- ***Margins*** The white space that surrounds the text in a document allows you to frame the document. For a standard 8 1/2″ × 11″ sheet of paper, allow margins of at least one inch at the top and bottom, and 3/4 of an inch to an inch on the right and left sides.

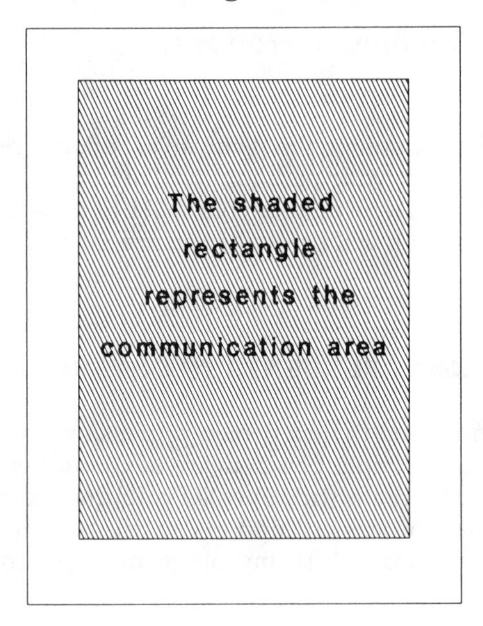

- ***Line spacing*** This is connected to the concept of white space. Include lines without text for your reader. Most technical communication is single-spaced, so separate the different sections of your document with line spaces. You may make the document longer, but what good is a short document that no one wants to read?
- ***Pagination*** Number your pages for your reader, and give some thought to where the page numbers will appear. Place page numbers outside the margins. Don't number the title page or the cover letter in a formal report. If you attach an appendix, use a separate numbering system for the appendix.
- ***Headings*** If you have learned to write standard essays in school, you may have learned to avoid headings in your documents; headings, however, are an important part of technical writing. They allow readers to skim through a report and comprehend its structure, and they enable readers to locate quickly key information.

While attention to these eight items will not guarantee that your documents will be well designed, you will be surprised at how much better they look. And this is only the beginning of the options available to you.

For example, let's say that you had one piece of very important information in your document. You want your readers to grasp this information above all else. What could you do to highlight this information? Let's explore some options. You could:

boldface isolate the information

underscore

place within *italics*

place the information in ALL CAPS

place a box around it

shadow

You can, we are certain, think of other ways.

Save emphasis for key information. A long underscored passage or a page in capital letters is difficult to read, and the effectiveness of any technique is lost when you use it too frequently.

7.4 PAGE DESIGN

For most people, reading at work is just that, more work. They are busy, hurried, and distracted. The documents you write will receive only part of their attention, so use effective formats to draw the readers in and hold their focus. To create visual interest, follow a basic approach to page design. In order to do so, you will need access to a desktop publishing program or a sophisticated word processing program and you will need to learn a few new terms.

Glossary

Dingbats: Symbols such as bullets, arrows, stop signs and so forth used for emphasizing information.

☞ ☎ ☺ ♫

DPI: (dots per inch) The number of dots that form the resolution of a printer or other output source.
Footer: Material appearing at the bottom of the pages, outside of the margins.
Font: A complete alphabet, including a wide range of characters, of one size of one typeface.
Grid: A page template (layout) containing pre-defined margins and columns.
Header: Material that will appear at the top of a page or pages, usually as part of a general design throughout a document.
Kerning: Adjusting of the white space between two adjacent letters or characters.
Leading: Adjusting of the spacing between lines of type. With some software programs you can adjust the spacing as desired. Leading is a vertical adjustment; kerning is horizontal.
Pica: A standard unit of measurement in typography, equal to 12 points. There are six picas to an inch.
Pixel: A single dot on a computer display, short for "picture element."
Point: A standard unit of measurement in typography representing the height of a typeface. There are 72 points to an inch.

Serif: A short horizontal stroke at the end of the main stroke of a letter.

> This is an example of New Century Schoolbook at 12 points. Notice the serifs extending from the letters.

Sans serif means without these strokes.

> This is an example of Helvetica at 12 points. Look for the serifs.

Typeface: The specific design of an alphabet, complete with other standard characters (numbers, symbols, and so forth)
WYSIWYG: ("what you see is what you get") Used to describe a computer monitor image equivalent to the final output.

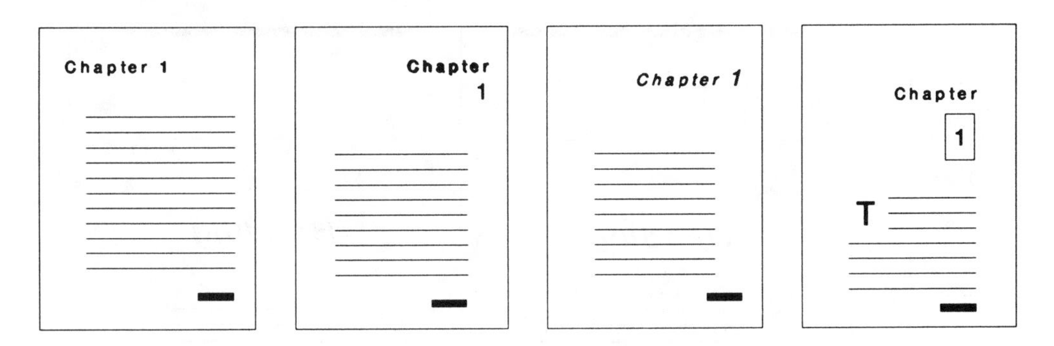

Figure 7-2 A comparison of single-page layouts.

First, begin to experiment with the elements of your page and the location of the elements. Move them around to see how they look. (If you are working with a word processor or a desktop publishing program that does not have WYSIWYG, it is necessary to print the versions of the document for comparison.) You can change the appearance of a page by manipulating a few elements. Notice how differently the page in Fig. 7-2 appears.

A simple, uncluttered format is inviting because it says the document has been designed for use. Readers can quickly understand how the document is organized and which information is most important. When you design pages, you have to help your readers find their way through the document.

One of the most important principles for you to learn is the principle of white space. Think about what you are putting on a piece of paper, but also think about the places where no text appears. These places are important. Your reader needs the spacing to be able to comprehend your message. Imagine a page of text with no margins, no paragraphs and no breaks in the text besides the punctuation: in other words, a solid page of text. Would you want to read it? Notice how Fig. 7-3 uses approximately 50 percent of the entire two-page spread.

The best way to learn about white space is to visualize your information in terms of blocks. These can be blocks of text, or an individual photograph, or some other visual. Let your mind think of the blocks as individual units to be placed on the page. Try to balance these blocks on the page so that the reader is guided along a path. In Fig. 7-4, notice how the text wraps around the illustrations on the left-hand page.

Make a list of the elements to be included in the document. In Fig. 7-4, the elements are text, headings, page numbers, a bar graph with a caption, two photographs with captions, and a bullet list. Your task is to create a visually pleasing arrangement of the elements.

If you are preparing a pamphlet or book-like document such as a formal report, then think in terms of double-page grids. Your readers will open

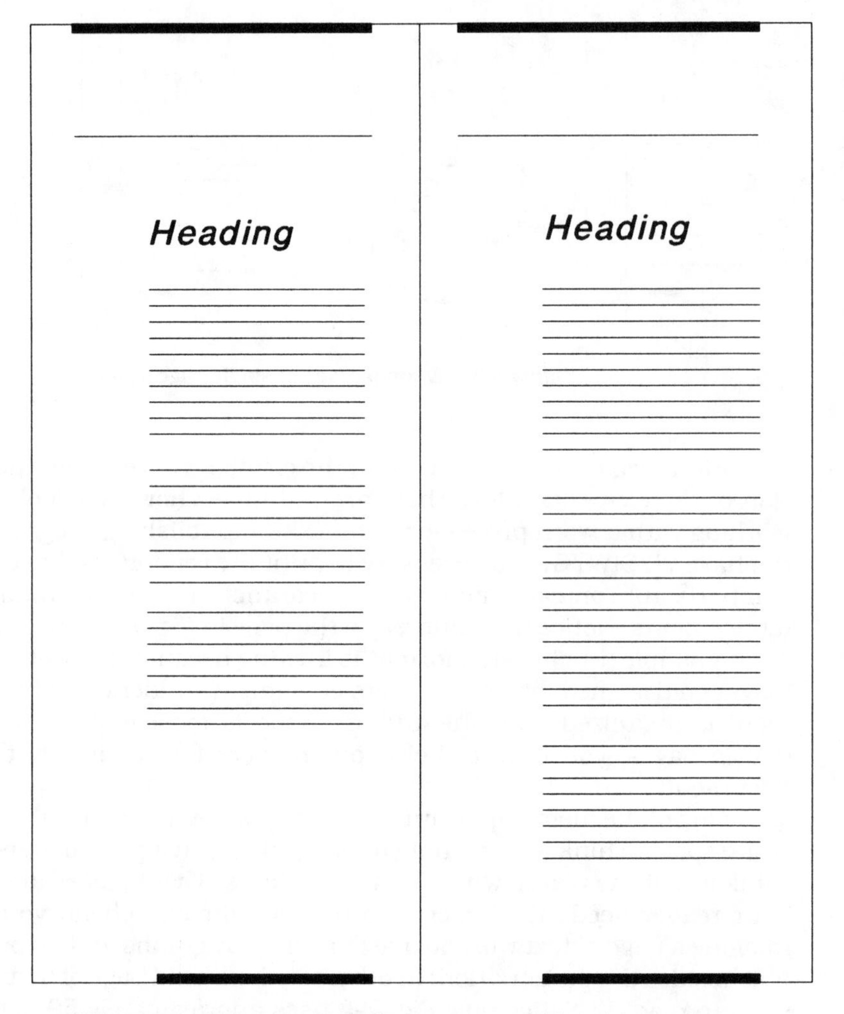

Figure 7-3 **This is a grid design for pages within a thin brochure or booklet. (When folded, the brochure could fit in a business envelope.)**

the text and see two facing pages, and so you should design the pages accordingly.

View the document from the perspective of the readers. What will the readers see when they look at the text for the first time? Remember, you see things before you begin reading and what you see affects how and whether

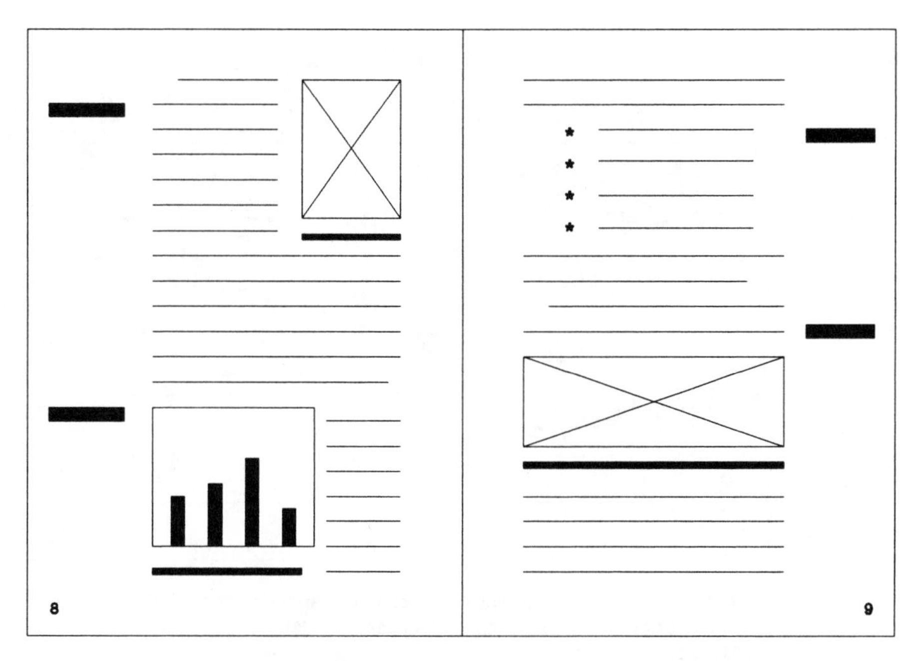

Figure 7-4 An illustration of a double-page spread. The boxes with the crosses inside represent photographs.

you will read. When you read a newspaper, for example, you generally read only about five to fifteen percent of the available text. What you do read often depends on the presence and position of headings, margins, illustrations, and other format elements.

Here is *PC World's* more detailed explanation of the changes done to the document on the right in Fig. 7-5: *To relieve the wall-to-wall text in the page on the left, the designer split the header and subhead into two lines each, adding space above and below, and increased the subhead's point size. The margin width was increased, and an extra 2 points of leading were added to the body text.*

The above description is highly technical. For most of the documents you produce, you do not need to be so technical. As electronic technolgy improves, however, such vocabulary may become commonplace.

You may also want to learn more about the possibilities available through computer technology. Most sophisticated word-processing pro-

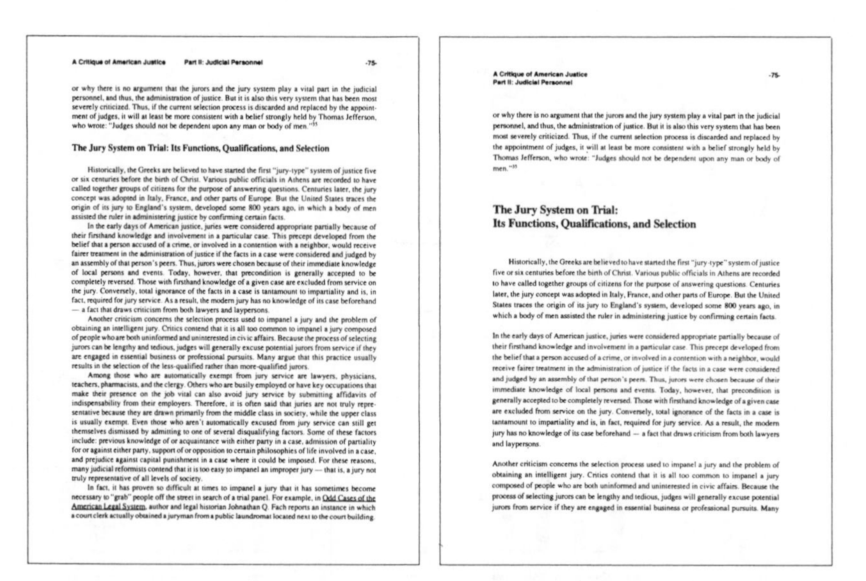

A Critique of American Justice Part II: Judicial Personnel -75-

or why there is no argument that the jurors and the jury system play a vital part in the judicial personnel, and thus, the administration of justice. But it is also this very system that has been most severely criticized. Thus, if the current selection process is discarded and replaced by the appointment of judges, it will at least be more consistent with a belief strongly held by Thomas Jefferson, who wrote: "Judges should not be dependent upon any man or body of men."[35]

The Jury System on Trial: Its Functions, Qualifications, and Selection

Historically, the Greeks are believed to have started the first "jury-type" system of justice five or six centuries before the birth of Christ. Various public officials in Athens are recorded to have called together groups of citizens for the purpose of answering questions. Centuries later, the jury concept was adopted in Italy, France, and other parts of Europe. But the United States traces the origin of its jury to England's system, developed some 800 years ago, in which a body of men assisted the ruler in administering justice by confirming certain facts.

In the early days of American justice, juries were considered appropriate partially because of their firsthand knowledge and involvement in a particular case. This precept developed from the belief that a person accused of a crime, or involved in a contention with a neighbor, would receive fairer treatment in the administration of justice if the facts in a case were considered and judged by an assembly of that person's peers. Thus, jurors were chosen because of their immediate knowledge of local persons and events. Today, however, that precondition is generally accepted to be completely reversed. Those with firsthand knowledge of a given case are excluded from service on the jury. Conversely, total ignorance of the facts in a case is tantamount to impartiality and is, in fact, required for jury service. As a result, the modern jury has no knowledge of its case beforehand — a fact that draws criticism from both lawyers and laypersons.

Another criticism concerns the selection process used to impanel a jury and the problem of obtaining an intelligent jury. Critics contend that it is all too common to impanel a jury composed of people who are both uninformed and uninterested in civic affairs. Because the process of selecting jurors can be lengthy and tedious, judges will generally excuse potential jurors from service if they are engaged in essential business or professional pursuits. Many argue that this practice usually results in the selection of the less-qualified rather than more-qualified jurors.

Among those who are automatically exempt from jury service are lawyers, physicians, teachers, pharmacists, and the clergy. Others who are busily employed or have key occupations that make their presence on the job vital can also avoid jury service by submitting affidavits of indispensability from their employers. Therefore, it is often said that juries are not truly representative because they are drawn primarily from the middle class in society, while the upper class is usually exempt. Even those who aren't automatically excused from jury service can still get themselves dismissed by admitting to one of several disqualifying factors. Some of these factors include: previous knowledge of or acquaintance with either party in a case, admission of partiality for or against either party, support of or opposition to certain philosophies of life involved in a case, and prejudice against capital punishment in a case where it could be imposed. For these reasons, many judicial reformists contend that it is too easy to impanel an improper jury — that is, a jury not truly representative of all levels of society.

In fact, it has proven so difficult at times to impanel a jury that it has sometimes become necessary to "grab" people off the street in search of a trial panel. For example, in Odd Cases of the American Legal System, author and legal historian Johnathan Q. Fach reports an instance in which a court clerk actually obtained a juryman from a public laundromat located next to the court building.

A Critique of American Justice
Part II: Judicial Personnel -75-

or why there is no argument that the jurors and the jury system play a vital part in the judicial personnel, and thus, the administration of justice. But it is also this very system that has been most severely criticized. Thus, if the current selection process is discarded and replaced by the appointment of judges, it will at least be more consistent with a belief strongly held by Thomas Jefferson, who wrote: "Judges should not be dependent upon any man or body of men."[35]

The Jury System on Trial:
Its Functions, Qualifications, and Selection

Historically, the Greeks are believed to have started the first "jury-type" system of justice five or six centuries before the birth of Christ. Various public officials in Athens are recorded to have called together groups of citizens for the purpose of answering questions. Centuries later, the jury concept was adopted in Italy, France, and other parts of Europe. But the United States traces the origin of its jury to England's system, developed some 800 years ago, in which a body of men assisted the ruler in administering justice by confirming certain facts.

In the early days of American justice, juries were considered appropriate partially because of their firsthand knowledge and involvement in a particular case. This precept developed from the belief that a person accused of a crime, or involved in a contention with a neighbor, would receive fairer treatment in the administration of justice if the facts in a case were considered and judged by an assembly of that person's peers. Thus, jurors were chosen because of their immediate knowledge of local persons and events. Today, however, that precondition is generally accepted to be completely reversed. Those with firsthand knowledge of a given case are excluded from service on the jury. Conversely, total ignorance of the facts in a case is tantamount to impartiality and is, in fact, required for jury service. As a result, the modern jury has no knowledge of its case beforehand — a fact that draws criticism from both lawyers and laypersons.

Another criticism concerns the selection process used to impanel a jury and the problem of obtaining an intelligent jury. Critics contend that it is all too common to impanel a jury composed of people who are both uninformed and uninterested in civic affairs. Because the process of selecting jurors can be lengthy and tedious, judges will generally excuse potential jurors from service if they are engaged in essential business or professional pursuits. Many

Figure 7-5 Notice how the added white space improves the appearance of the document on the right. **Reprinted with the permission of** ***PC World.***

grams and virtually all desktop publishing programs allow you to do two- and three-column layouts.

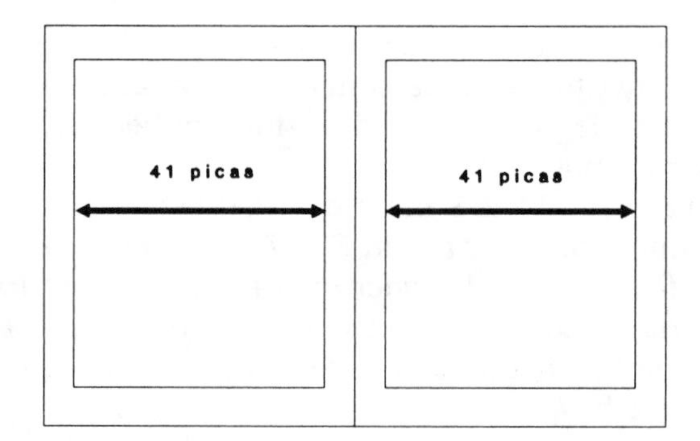

Each of the three grids uses a standard 8 1/2″ by 11″ (210 by 297 mm) page. Each page is 51 picas across, so the margins are 5 picas on each side. Assume that you wil be able to include 60 to 70 characters on a one-column line.

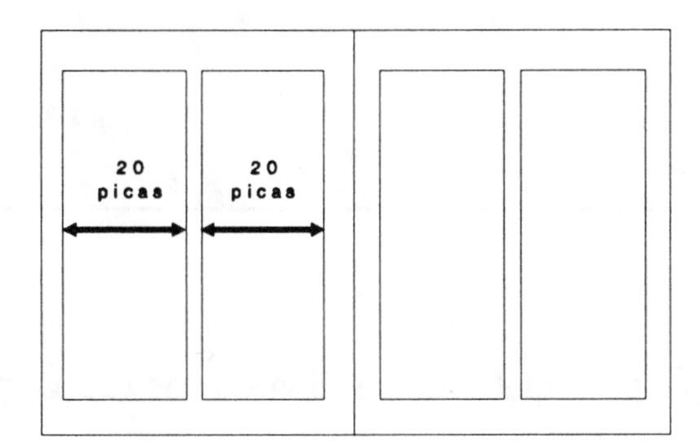

In this example, there is one pica between the columns. For a standard page, there are 66 picas from top to bottom.

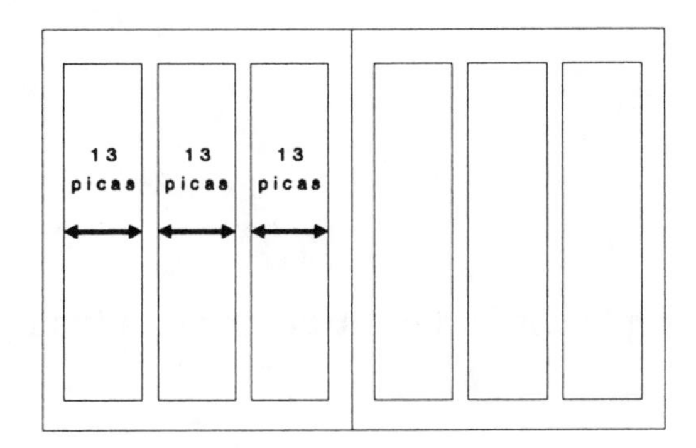

A general rule, when dealing with small columns, is use a point size one-half of the number of picas in the column. Here you would use 6 or 7 point type.

Most sophisticated word-processing packages also now give you the option of changing typefaces and point sizes. Along with a quality printer, a word-processing package such as WordPerfect, Microsoft Works or MacWrite gives you the option to change typefaces and point sizes with relative ease.

Examples of Different Typefaces (at 12 point)

The quick brown fox jumps over the lazy dog. (Palatino)
The quick brown fox jumps over the lazy dog. (Helvetica)
The quick brown fox jumps over the lazy dog. (Courier)
The quick brown fox jumps over the lazy dog. (Times Roman)

Examples of Different Point Sizes (using Palatino)

The quick brown fox jumps over the lazy dog. (6)
The quick brown fox jumps over the lazy dog. (8)
The quick brown fox jumps over the lazy dog. (10)
The quick brown fox jumps over the lazy dog. (12)
The quick brown fox jumps over the lazy dog. (24)

Figure 7-6 Here are some examples of different typefaces and different point sizes. Typeface refers to the style or design of characters. Point size refers to the size of characters. See the Glossary.

Be careful. You do not want to overdo it. Use the table as a guide.

TABLE 1

Mixing Type Effectively

Wondering which typefaces work together best? Start with these suggestions when designing newsletters, reports, brochures, and fliers.

Headline Typefaces	Text Typefaces	Applications
Bitstream Charter Bold	Bitstream Charter	Business correspondence, reports
Futura Bold	ITC New Baskerville	Newsletters, brochures, magazines
Helvetica Black	ITC Garamond	Newsletters, brochures, books, magazines, fliers
Helvetica Bold	Times Roman	Any document; a safe but common choice
ITC Avant Garde Gothic	ITC Lubalin Graph	Business reports, presentations
ITC Franklin Gothic Heavy	Century Old Style	Newsletters, brochures, fliers
ITC Franklin Gothic Heavy	ITC Galliard	Newsletters, brochures, fliers
ITC Lubalin Graph Bold	Helvetica Condensed	Presentations, brochures, fliers
Palatino Bold	Helvetica	Business reports, newsletters
Palatino Bold Italic	Palatino	Business reports, books
Stone Informal Bold	Stone Informal	Business correspondence, reports
Stone Sans Bold	Stone Serif	Newsletters, manuals, business reports
Univers Bold 65	Lucida Roman	Forms, schedules

Now that you are becoming more familiar with some design concepts, we would like to offer five suggestions for improving your documents.

Reach out for your readers. Use informative headings so readers anticipate your message. Tell them what they will be reading if they give your document their attention. Lots of sub-headings help readers search for specific information.

Emphasize important information. The most effective way to focus your readers' attention on one important point is to isolate it from the rest of the page. You can do this by surrounding the information with white space, by using boldface, underlining or italics, or by putting a box around it. If you have the option of using different colors, you can select a vibrant, rarely used color to highlight a key point.

Keep it subtle. You can make a document look ridiculous by using too many different forms of highlighting. Minimize the use of boldface, underlining, italics and shadowprint so that you can grab your readers' attention when you do use highlighting. Don't overuse typefaces simply because you have them.

Keep your pages balanced. Visuals and text should complement, not compete with each other.

Be consistent. Whatever you do in one part of a document, you should do in other parts. If you use particular symbols, labels, margins, styles, figures, and so forth at one place in a document, use them again in similar situations.

7.5 *AUDIENCE ANALYSIS REVISITED*

Analyzing your audience was the subject of Chapter 3. We would like you to keep your focus on your readers as you develop the design of your pages.

First, as you plan the layout of your document, consider what visual aids you can include for your readers. Tables, charts, graphs, drawings, and other illustrations enable readers to see what you are telling them. They may be expensive and time-consuming to create, but good illustrations are well worth the effort. *See Chapter 8* for a more complete discussion of specific graphics and illustrations.

Rules

- Place your illustrations near the first reference in the text.
- The size of the illustration should be roughly equivalent to its importance.
- Enclose your illustrations with a border.
- Keep your illustrations vertical.
- Try to balance your illustration blocks on your grid.
- If possible, position an illustration in the top right-hand column of the first page of a section of your document.

(continued)

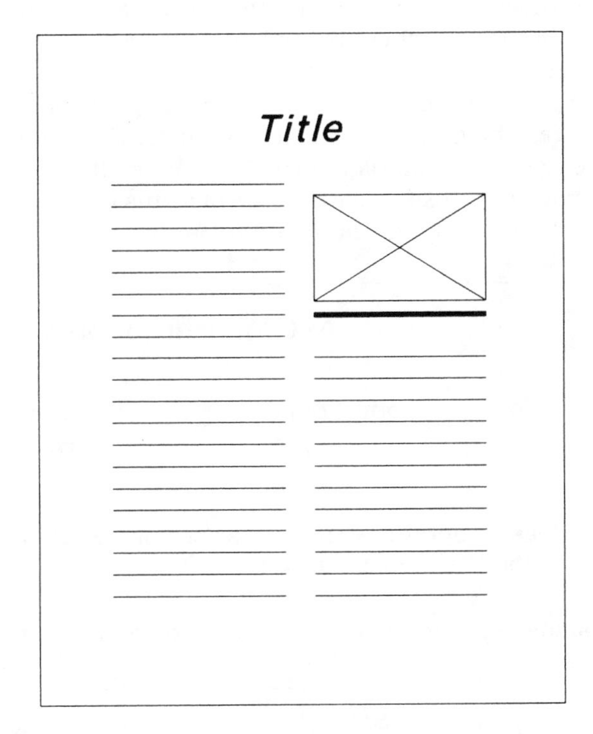

Figure 7-7 An example of an ideal position for an illustration.

Your purpose in most of the illustrations you will do in your technical career will be to inform rather than to attract attention and provoke an atmosphere or feeling. Make sure that the design of the page supports your purpose.

Besides including visual aids, you can help your audience by asking, **Where will your readers be when they read your message?**

The answer to this question can help you create a successful document. For example, there are fish-viewing guides for snorkelers and divers that are sealed in clear plastic so that they can be taken into the water. There are car repair manuals designed with large type and a page surface that grease does not stain.

This question is particularly important if you are designing a form to be completed by your readers on the job. People who work in offices and in laboratories may fill out a variety of standard forms as part of their daily responsibilities, or whenever they need to complete a particular task. Engineers, technicians, architects and other workers who spend most of their time on a job site need to write down their observations and record what procedures they followed in field reports. Such forms should be easy to comprehend and complete. The Test Procedure form and the Photocopy Request form are examples of types of forms that individuals complete hundreds of times a day in a large organization.

Request for Photocopying

To: *Copy Center* Date: ______
From: ______________ Dept. ______

Please reproduce the attached material in the quantity and manner described below.

Number of copies: _____ Date/time needed: ________

Size	*Options*
☐ 100%	☐ Collate
☐ 98%	☐ Staple
☐ 74%	☐ Cut
☐ 65%	☐ Bind
	☐ 3-ring
Stock	☐ GBC
☐ 8 1/2 x 11 (Plain)	☐ Acco
☐ 8 1/2 x 11 (3-Hole)	☐ Other (Specify)
☐ 8 1/2 x 14 (Legal)	________
☐ 8 1/2 x 11 (Color) ________	________
☐ Cover Stock; Color: ________	☐ Single-sided
☐ Letterhead; Type ________	☐ Double-sided

TEST PROCEDURE

Class: TP Number:
Revision: Page ______ of ______

Title: ______
Program: ______
Special Instructions: ______
______

Test Start Date: ______
Test Conductor: ______
QC Monitor: ______
Test Engineer: ______ Date: ______
Test Operations: ______ Date: ______
Quality Assurance: ______ Date: ______
System Safety: ______ Date: ______
Environmental
Engineer: ______ Date: ______

Test Engineering Supervisor: ______
Release Date: ______ No. of copies ______

______

Record of Test Procedure

Attach other pages as necessary.

7.6 A FINAL WORD

We have one final recommendation for you. You may be a student as you read this chapter; you have assignments and projects in all of your courses. We believe that it makes sense for you to submit all of the work for which you will receive a grade in a professional manner: that is, either use a typewriter or a high-quality printer to produce your documents.

In today's modern office, the availability of photocopiers, laser printers, plotters, scanners, and high-quality computer software illustration packages allows everyone the opportunity to produce sharp-looking documents. The technology keeps getting better. The most obvious example is the laser printer. It produces output that only independent printing presses could have achieved as recently as seven years ago. The dramatic reduction in the cost of laser printers and widespread service bureaus make it possible for everyone to produce attractive documents.

Someone in business would never submit a handwritten business letter or report to another company. In the same way, you should think twice before you submit a handwritten assignment. This may require you to locate the particular facilities available to you at your school, and to learn a new way of preparing your documents, but we believe it will be time well spent. You will probably be able to translate many of your new skills to the workplace, and the skills may help you find and keep a job.

FORMAT EXERCISES

1. This exercise calls for scissors, a ruler, a pencil, some scotch tape or paste, and two sheets of 8 and 1/2 by 11 inch paper. Mark out a 3″ × 3″ square, 2″ × 3″ rectangle, and a 1″ × 2″ rectangle. The square will represent a photograph and the two rectangles will represent graphs. Cut them out. Now paste them on the second piece of paper; imagine that text flows around the illustrations. Use straight lines to represent the text. Include sample headings and a page number; add any other design elements that will enhance the appearance of the document.
2. Get a copy of your resume. (If you have never created a resume, *see Chapter 11, Applying for Work* for guidelines on how to design a resume, and then create one.) Use scissors to cut up the resume into sections. Paste the sections of the resume back together. Did you discover anything about the design of your original?
3. See Exercises #2, #3, and #4 at the end of Chapter 1. Follow the instructions. Now take your logo design and design the company letterhead for your company. Place the logo on the company stationery along with a fictitious address, telephone number, and fax number.
4. Take either a famous quotation or one of your favorite lines from a song, and design an overhead transparency that includes only the quotation or line and the name of the author. Attempt to design the overhead to have an impact on your intended audience. (If you are unable to make an overhead transparency, submit an 8″ × 10″ sheet of unlined paper.)
5. Locate a technical document that has an effective design. Bring the document to class and prepare to explain why the design is effective. (If you are having difficulty finding a document, look at the technical periodicals in your library.)
6. Locate a popular magazine, such as *Sports Illustrated, Time,* or *Newsweek.* Find a double-page spread where there is no more than one advertisement. Construct a sample grid of the pages, using a solid bar for titles and headings, thin lines for text, and boxes with crosses for photographs. Identify all of the elements on the two pages. On a separate sheet of paper, analyze the design. What features appeal to you? What suggestions would you make to improve the design?
7. Design a two-page layout for a technical report for Gates and Associates. As a member of the Special Project Investigation Team, you need to file reports of the findings of your investigations. Some of these reports can run well over 100 pages, and can include many diagrams, photographs, and illustrations.

Design a standard format for the reports, including at least one illustration on each of the two pages. Be sure to specify the kinds of headings and subheadings you would use, as well as the page numbers and any other detail you think is appropriate. Consult Chapter 26 for a description of the kind of work you do as part of Gates and Associates.

8. Sheila O'Brien has asked you to write a memo to all employees of Gates and Associates in order to gather information for a company database. You need information on name of employee, length of service with the company, age, home address, person to call in case of an emergency, what department they work in, home phone number, whether they are covered by life insurance through the company, what their extension is, whether they participate in a car pool, and their job title. (Please note: the above information is not presented in order of importance. In fact, the list is not in any order.) *Design your memo so that people can respond easily to your request for information.* If you have never written a memo, look at Chapter 9, Designing Memos for a correct format.
9. F. Robert Gates is interested in the possibility of utilizing color in the documents produced by Gates and Associates. At a staff meeting, he asks you to conduct a preliminary investigation of the costs of color printers and color photocopiers. Write a memo to Gates in which you detail your findings.
10. Buy a packet of dry transfer lettering. Look in a college bookstore, a stationery store, a photocopy center, or a graphic arts supplier. Now design a title page for a formal report using the dry transfer lettering. Use a rounded edge of a pencil to transfer the letters to the page. (Graphic designers use a tool called a *burnisher.*) The formal report will have these elements: a title, your name, the date of submission, and the name of the receiver of the document.

8

Including Graphics and Illustrations

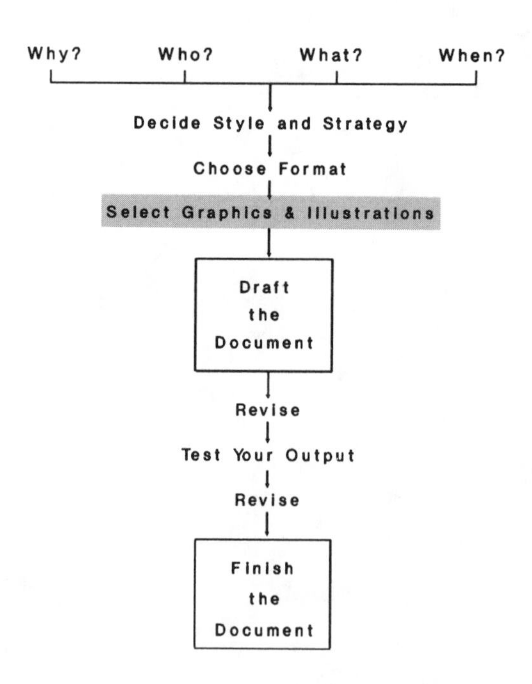

8.1 INTRODUCTION

In the past, there were few colleges that taught individual courses in graphics and illustrations. The colleges that did teach these courses were specialized: engineering, design and art schools. One reason for this was the difficulty of producing quality illustrations without professional-quality drawing ability or equipment too expensive for the average individual or the average small business to afford. It was assumed that most people would not need the ability to create illustrations. Personal computers have changed all of that.

Now you can generate high quality graphics on a personal computer for a remarkably low cost. Relatively inexpensive software programs allow the user to create sophisticated graphs and charts quickly and easily. You will need some time to learn a program, but once you have mastered the basics, you can produce documents that only professionals could have attempted twenty years ago.

Essentially, the advantages of producing your illustrations on a computer are these:

- You do not have to do all of the calculations yourself. You can enter numbers and the software will make the calculations and figure out the correct proportions.
- The computer will produce neat, crisp lines. The quality of these lines will depend on your output source, whether dot-matrix printer, laser printer, plotter, or Linotype printer. The higher the quality you desire, the more expensive the investment.
- The computer will allow you to make revisions easily. For example, you can change a number on a graph and redraw the graph effortlessly.
- With many software programs, you can easily insert a graph or illustration within a body of text. With more sophisticated programs, you can make the text flow around the graphic.
- The computer enables you to include many special features. For example, you can include clip art (ready-made artwork), or you can use a scanner to insert a photograph in your text.

This does not mean that we recommend that you rely solely on computers for your graphics and illustrations. Computers, printers, scanners and so forth are expensive, and not everyone has access to them. Hand-drawn graphs and tables can, when done by talented individuals, look every bit as good and sometimes better than what an individual can do with a computer. If you have talent, you should put your talent to use. There will be times when pens and pencils are the right tools to use. You may find, however, that you can also put this talent to use on a computer, and that the computer will save you valuable time.

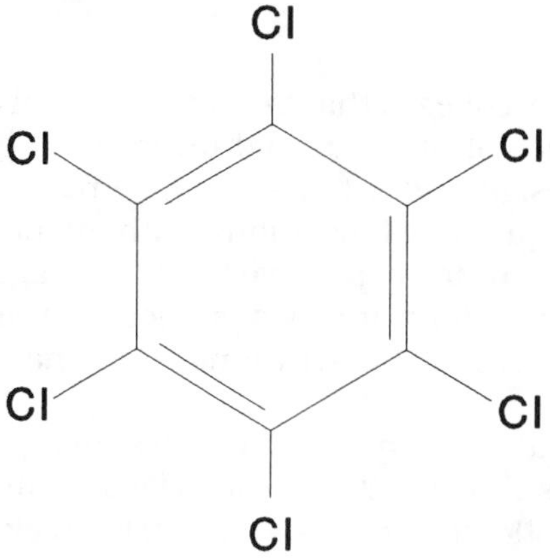

Figure 8-1 Structural illustration of hexachlorobenzene, C_6Cl_6.

However you produce your graphics and illustrations, you should understand the options you have at your disposal as you work to produce good-looking, clear, accurate tables, charts, graphs, drawings, and other illustrations. As an aid to understanding what you write and what you say, or standing on their own, good illustrations are invaluable.

Many people understand material better when they can **see** it. Words do not help these individuals as much as pictures do. See Fig. 8-1, a representation of molecular bonding for the chemical compound hexachlorobenzene. Actual and symbolic representations allow us to grasp concepts. So include illustrations to support your text and clarify concepts for your audience.

8.2 SOME BASIC PRINCIPLES

This chapter covers a variety of different kinds of illustrations, and yet we have not covered all of them. With so many possibilities, building general rules that apply to all of the different kinds of graphics and illustrations is difficult, but there are indeed a few. Fig. 8-2 offers some basic principles that you should be aware of whenever you plan to include an illustration.

When you construct a graph, remember this: tables, charts, graphs, and other illustrations tell a story. You must tell the story in such a way that your audience can read and appreciate its full message. A graph that is unclear, incomplete, or inaccurate, no matter how nice it looks, is virtually useless and sometimes harmful. Thus, you should design your illustration **for your readers.** Follow the Document Design Process step by step with each illustration; that is, consider your purpose, audience, data, deadlines, style and strategy, and format with each illustration. After you have composed the illustration, edit and test your output before you produce the final copy. (If you have entered the text here, you can find an overview of the Document Design Process in Chapter 1.)

Six Rules
for Illustrating Technical Documents

- Consider the audience.
- Keep illustrations simple.
- Illustrate one main point.
- Title each illustration.
- Label each illustration.
- Add a caption.

Figure 8-2 This is also an example of a bullet list.

The vast majority of the time, your illustrations will accompany a longer document: a chart in a formal report, a graph as part of an oral presentation, a schematic or other drawing as part of a technical description, an illustration in a set of directions. Let your graphics enhance the full document. Position the illustrations in such a way that your readers will be led to it from the text, and so that it will lead readers to the text if they read the graphic first.

8.3 TABLES

Tables are particularly effective ways for people to visualize information. A monthly calendar is a table. Try to imagine a month without visualizing it in terms of a table. Tables are also effective ways of presenting comparisons and contrasts. The *lines* or *rows* (horizontal) and *columns* (vertical) allow you to present a great amount of information, particularly numbers, in a condensed form which is easily understood and remembered. (The rectangles formed by the rows and columns are called *cells.*) The essential details (dimensions, specifications, capacities, and so forth) of three or more items can be quickly compared by a reader.

Tables have one clear advantage over graphs: you can put much more information into a table. Notice how much information is presented in Figure 8-3. Imagine putting this much data into a single graph.

Model	Nominal Voltage V	Nominal Capacity @ 20 hr. rate A.H.	Discharge Current @ 20 hr. rate mA	DIMENSIONS								Weight	
				Length		Width		Height		Ht. Over Terminal			
				in.	mm	in.	mm	in.	mm	in.	mm	lbs.	kg.
PS-1207	12	0.7	35	3.78	96	0.98	25	2.42	62	2.42	62	0.8	0.35
PS-1212	12	1.2	60	3.82	97	1.65	42	2.00	51	2.13	54	1.3	0.6
PS-1219	12	1.9	95	7.01	178	1.34	34	2.36	60	2.56	65	1.9	0.9
PS-1226	12	3.0	150	7.68	195	1.85	47	2.76	70	2.95	75	3.1	1.4
PS-1230	12	3.0	150	5.23	134	2.64	67	2.36	60	2.60	66	2.6	1.2
PS-1242	12	4.0	200	3.54	90	2.76	70	3.98	101	4.13	105	3.8	1.7
PS-1250	12	5.0	250	5.94	151	2.56	65	3.70	94	3.89	99	5.0	2.3
PS-1265	12	6.5	325	5.95	151	2.56	65	3.70	94	3.86	98	5.7	2.6
PS-1282S	12	8.0	400	4.40	112	3.86	98	4.65	118	4.65	118	6.7	3.0
PS-1282L	12	8.0	400	7.72	196	2.20	56	4.65	118	4.65	118	6.7	3.0
PS-12100	12	10.0	500	5.95	151	4.00	102	3.70	94	3.86	98	9.2	4.2
PS-12120	12	12.0	600	5.95	151	3.86	98	3.70	94	3.86	98	8.8	4.0
PS-12150	12	15.0	750	7.13	181	2.99	76	6.57	167	6.57	167	12.8	5.8
PS-12200	12	20.0	1000	6.89	175	6.54	166	4.92	125	4.92	125	17.6	8.0
PS-12240	12	24.0	1200	6.52	166	4.92	125	6.89	175	6.89	175	19.8	9.0
PS-12260	12	26.0	1300	6.89	175	6.54	166	4.92	125	4.92	125	18.7	8.5
PS-12300	12	30.0	1500	7.75	197	5.19	132	6.12	155	7.31	186	21.8	9.9
PS-12400	12	40.0	2000	7.75	197	6.50	165	6.69	170	6.69	170	30.9	14.0
PS-12600	12	60.0	3000	10.25	260	6.60	168	8.20	208	9.45	240	39.7	18.0
PS-12800	12	80.0	4000	12.00	305	6.60	168	8.20	208	9.45	240	50.0	22.7

TOLERANCES: Length and Width: ±0.04 in. (1mm) – Height: ±0.08 in. (2mm)

Figure 8-3 An example of a table.
Reprinted with the permission of Power Sonic Corp.

Table Number *and* Title

	Column Description			
Line Description	Column Heading	Column Heading	Column Heading	Column Heading
Line Heading				
Line Heading		(cell)		
Line Heading				

Figure 8-4 You can add sub-headings if necessary. The column and line descriptions are optional.

Fig. 8-4 shows the basic parts of a table.

Sometimes the line (or row) descriptions and headings and the column descriptions are self-explanatory and therefore unnecessary. Here are some general rules to help you create clear, readable tables.

> Use lines for the grid effect.
> Make sure your rows are straight.
> Make sure you include an overall title above the table.
> Provide column headings and line headings.
> Place data to be compared in the horizontal rows, the direction our eyes move when reading.
> Use decimals instead of fractions.
> Convert your data to a consistent unit of measurement.
> Make the information easy to grasp.

A *ratings table* (or matrix) allows the reader to compare different items against certain established criteria, such as speed, performance, quality, price, and so forth.

A particularly effective use of this type of illustration is to create easy-to-read trouble-shooting and problem-solving tables to accompany machinery, tools, and other technological products. Your readers will thank you for a table that tells them how to resolve their problem when something goes wrong.

PROBLEM	ANALYSIS	SOLUTION(S)
If this occurs	then it means	try this
x	y	z

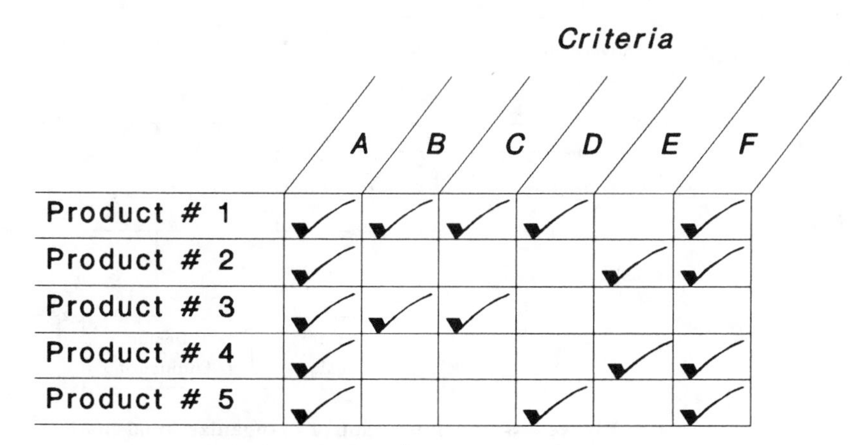

	Criteria					
	A	B	C	D	E	F
Product # 1	✓	✓	✓	✓		✓
Product # 2	✓				✓	✓
Product # 3	✓	✓	✓			
Product # 4	✓				✓	✓
Product # 5	✓			✓		✓

Figure 8-5 A standard format for a ratings table.

Note: It is very difficult and frustrating to construct a table on a typewriter. To make almost any adjustment requires retyping. Tables are easier with a word processor, and easier still with desktop publishing software.

8.4 CHARTS

There are many different types of charts, including organizational charts, flow charts, and time charts. Many people also refer to graphs as charts. As a guide to understanding, we intend to make a distinction between charts and graphs in this handbook; in general, charts deal less with numbers and more with abstract processes and structures. What we refer to as a bar graph in this book may be called a bar chart in another text. Both terms refer to the same thing.

Fig. 8-6 presents an organizational chart of a large Research and Development Company. Notice how this chart focuses your attention on the positions and their relative importance within the hierarchy of the company. Other organizational charts, as in Fig. 8-7, add the names of the people who are currently in charge of the departments.

Flow charts allow us to visualize processes and systems. Flow charts are particularly valuable guides to understanding decision-making processes and technical processes where alternative paths may be taken. The symbols in a flow chart represent specific activities, such as operations,

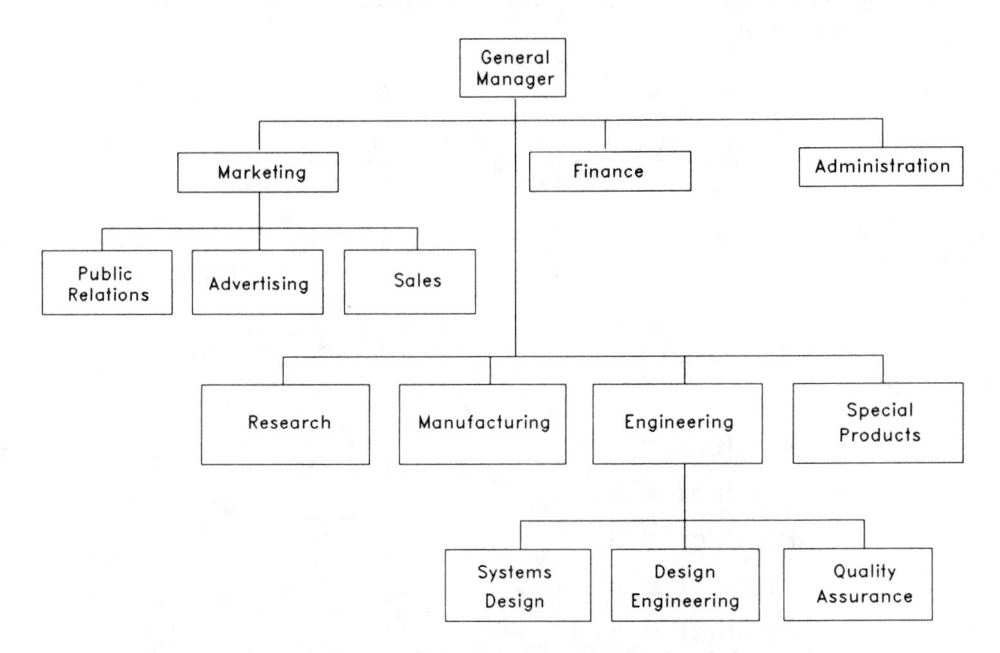

Figure 8-6 An illustration of an organizational chart.

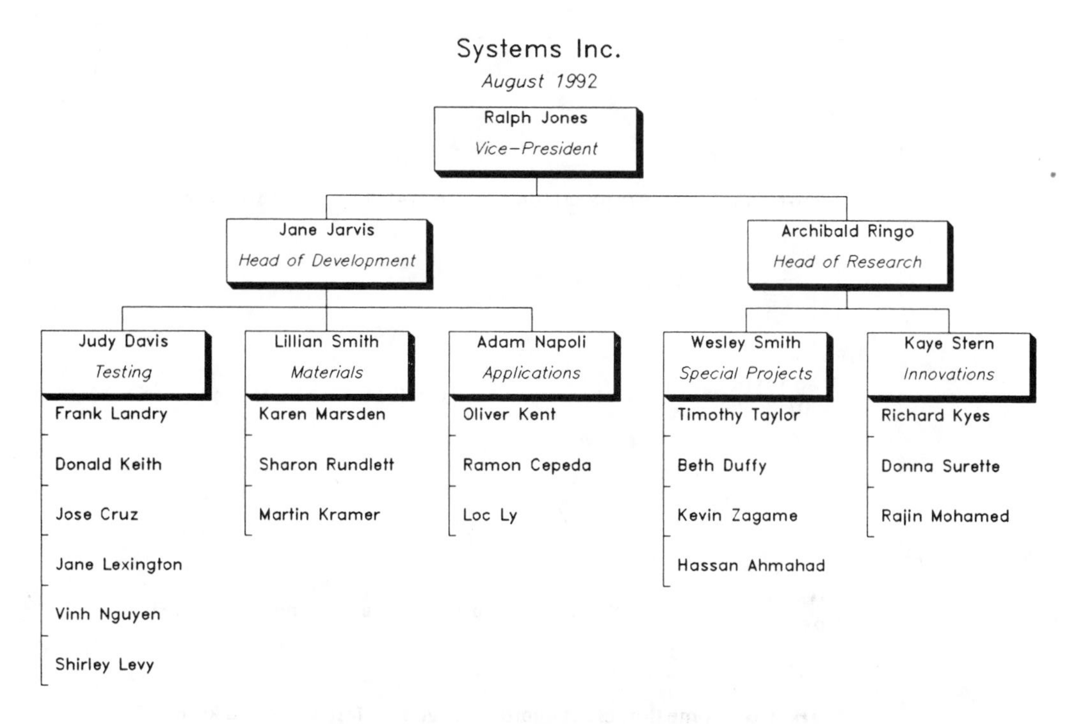

Figure 8-7 An organizational chart where the focus is on individuals.

data, or material flow; for example, a diamond symbolizes a step in a process where a decision needs to be made, while rectangles represent activities in the process. Lines and arrows represent interrelationships among the components. See Figure 8-8 and Figure 8-9.

Time charts allow us to visualize a schedule of work to be done. They are particularly helpful when you have a large project to do which requires simultaneous efforts by many people or organizations. Time charts are an especially valuable planning tool because they make us aware of all of the tasks that need to be accomplished to complete a project.

Gantt charts and PERT (Program Evaluation and Review Technique) charts depict the status of each part of a project, and are particularly valuable for project management. In a Gantt chart, each division of space represents a time interval and bars show the work to be done during the intervals. See Fig. 8-10. Also see Fig. 5-1 for a timetable for scheduling a writing project. Gannt charts can also be used to compare planning estimates and actual work done. PERT charts are much more complex, and require sophisticated computer software to compare current progress vs. planned objectives with an emphasis on time performance and cost. A PERT analysis estimates the probability of a task ending on a particular date at a specific cost. A very simplified version of a PERT chart is presented in Fig. 8-11.

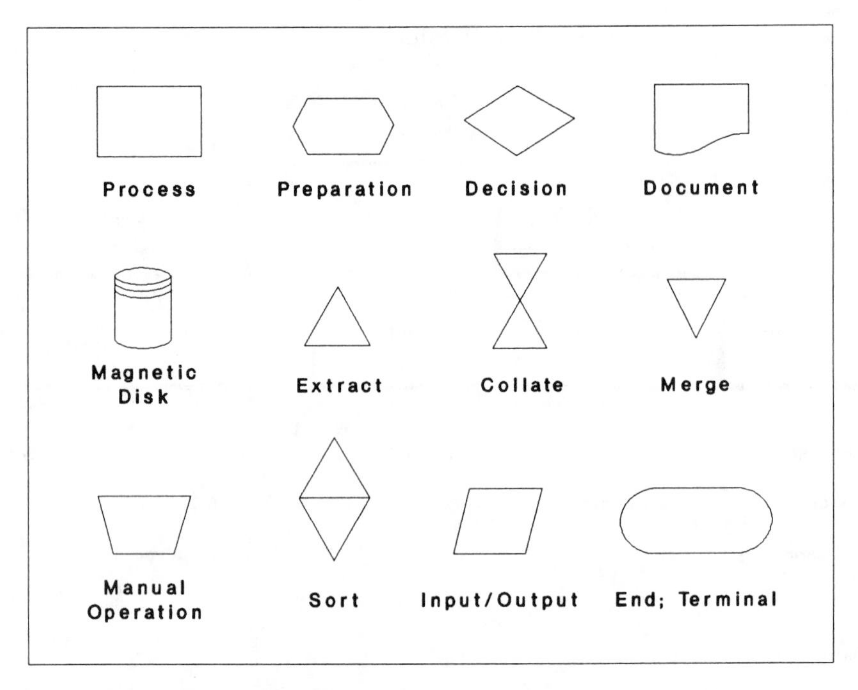

Figure 8-8 Some flow chart symbols. A good reference for all kinds of symbols is *Symbol Sourcebook: An Authorative Guide to International Symbols* by Henry Dreyfuss (McGraw-Hill)

Fig. 8-11 analyzes a six-step process, from the authorization of a new laboratory to the beginning of operations. Some PERT charts analyze thousands of steps.

8.5 GRAPHS

We can visualize simple and complex relationships between numbers with graphs. We can take, for example, 21 different numbers that would be virtually incomprehensible in a list, and turn that sequence into an easily understood illustration that has great impact.

The personal computer has changed our ability to produce graphs. As previously mentioned, we can now take a complex set of statistics and rapidly transform this set into a clear and accurate graph. And since creating graphs has become easier, we now see more of them. Your ability to read and understand graphs and your ability to create clear and accurate graphs will be very important in your technical career.

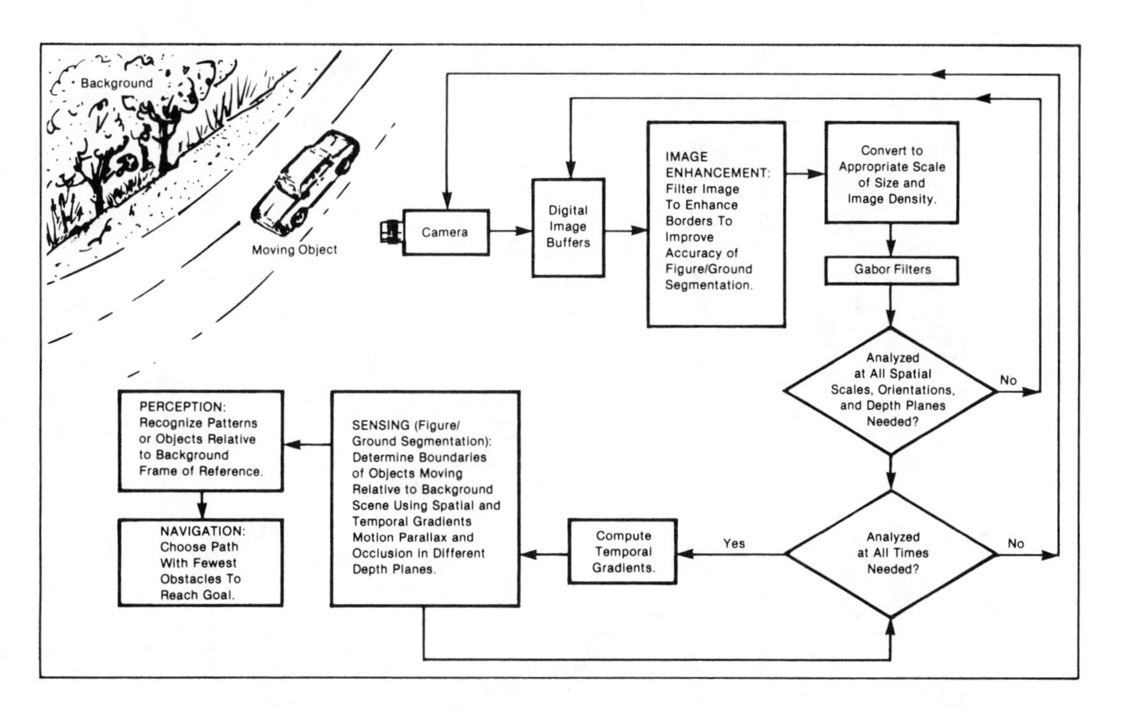

Figure 8-9 An example of a flow chart for a robotic vision system. **Reprinted with the permission of** *NASA Tech Briefs*.

8.5.1 Parts of the Graph

Graphs can differ tremendously. There are, however, some standard features common to almost all graphs. This section examines some of these standard features.

Except for pie and column graphs, graphs have a *horizontal axis* (or *x-axis*) and a *vertical axis* (or *y-axis*). You should divide each axis into equal parts. The two axes will represent different kinds of quantities, such as months (time) vs. dollars (profit), so the axes need only be consistent within themselves. In other words, you could make each division (or *increment*) in the y-axis one inch apart and make each division in the x-axis three inches apart. The x- and y-axes should be clearly and correctly labeled.

Every graph should have a *title*. The title is one more aid for your reader. The title should be straightforward and accurate, conveying in a few words, ideally less than eight, the subject of the graph. A sub-title can be included to add information to the title. If you are producing a series of graphs, you should have a cover page with an overall title.

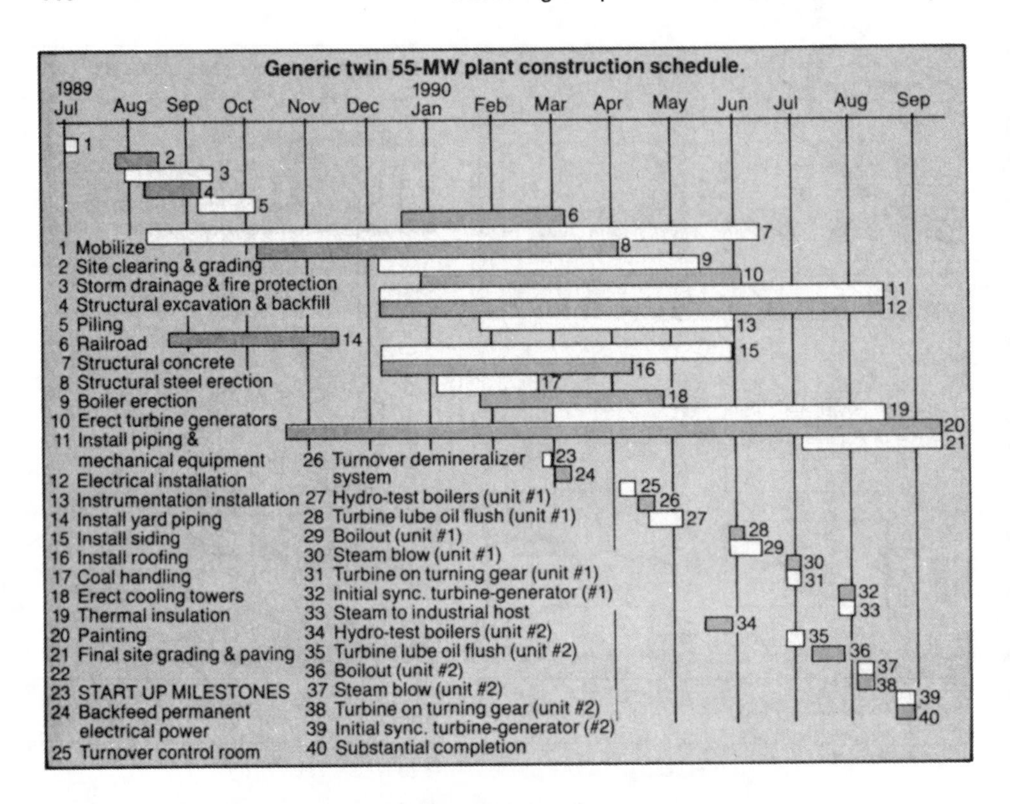

Figure 8-10 An example of a time chart.
Reprinted with the permission of *Power Engineering.*

A *legend* or *key* is needed whenever you are graphing more than one item. Whenever you are comparing two or more items, you will need to represent the items in different ways: different colors, shadings, shapes, or interior designs.

Finally, each graph should be clearly labelled. The *label* should include a *figure number* and some explanation of what the graph represents. This explanation is frequently referred to as the *caption*. Use well this opportunity to direct the way your audience reads and understands your graph. Describe what the graph conveys. Remember, some of your audience may understand the words better than the visual presentation. The figure numbers should be arranged in chronological order; for example, Fig. 1, Fig. 2, Fig. 3, and so forth, or Fig. 1-1, 1-2, 1-3 and so forth.

Note: If you need to abbreviate in order to save space, make sure you use standard abbreviations. In other words, consult a dictionary.

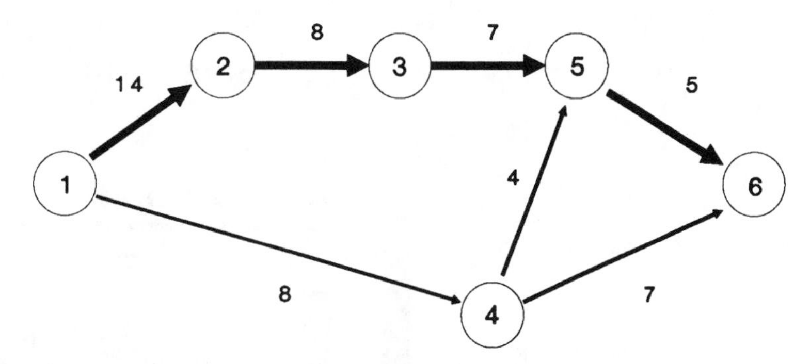

1 ▪ Authorization of Project
2 ▪ Acquisition of Equipment
3 ▪ Installation of Equipment
4 ▪ Transfer of Staff to Lab
5 ▪ Training of Staff
6 ▪ Beginning of Operations

Figure 8-11 In this illustration of a PERT chart, the heavier black arrow represents the path that takes the most days from start to finish. The numerals above the arrows illustrate the number of days projected to complete steps in the process. This chart does not analyze costs.

See Fig. 8-12 for an example of the basic parts of the graph.

At times, in order to save space, you may want to skip some of the numbers or other increments on one of the axes. Usually you will do this with the y-axis. To let the reader know that you have done this, use the symbol ÷ or ∔ .

8.5.2 Graphs Can Lie

Statistics can lie. They can be presented in such a way that readers can be deceived; the readers receive an impression that may be radically different than the actual situation. In the same way, graphs can lie.

ETHICAL CONSIDERATIONS

As both a reader and a creator of graphs, you should know that graphs can lie. As a reader of graphs, you should study the graph closely to make sure that you know exactly what it says, which may be different than what the maker of the graph wanted you to see. As a creator of graphs, you need to be aware of what you should and should not do to influence your audience in a particular way. Obviously, you want to affect the way your readers look at your graphs, but you do not want to cross over the fine line between influence and misrepresentation.

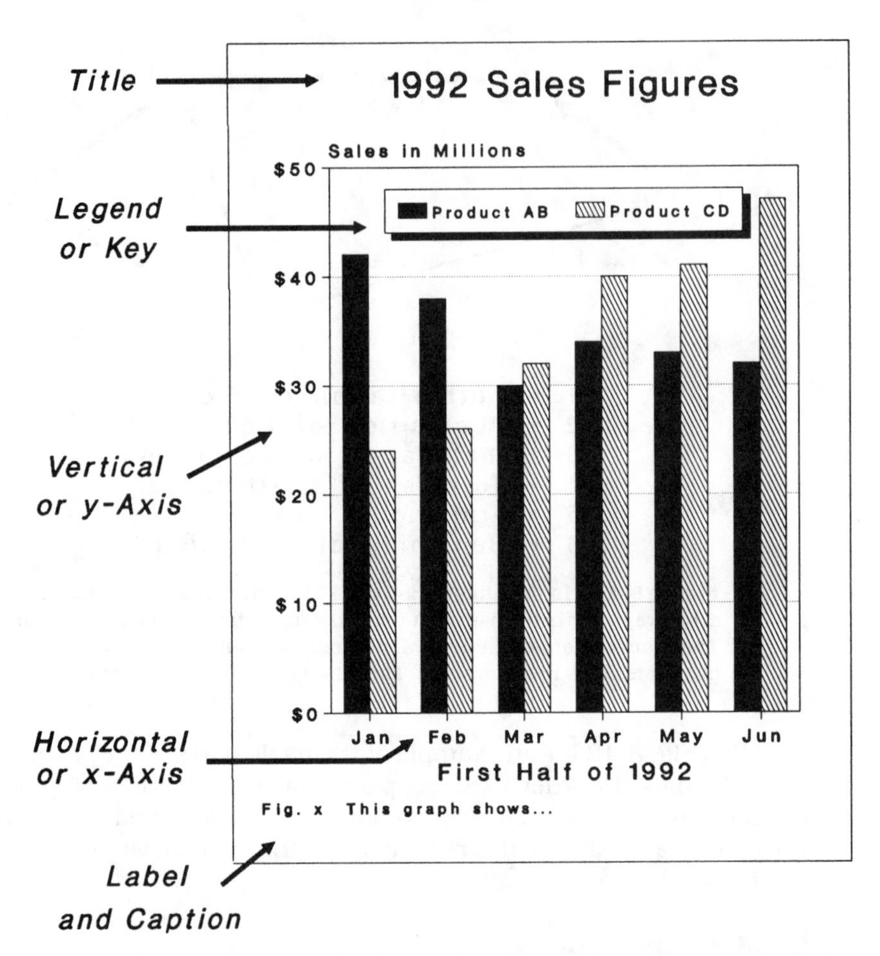

Figure 8-12 This vertical bar graph exemplifies the basic parts of graphs.

Examine Figures 8-13, 8-14 and 8-15. Notice that the **only** difference between the three graphs is that the increments on the y-axis have been changed. In Fig. 8-13 the y-axis goes from 25 to 50 thousand dollars; in Fig. 8-14 the y-axis goes from 0 to 50 thousand dollars; and in Fig. 8-15 the y-axis goes from 0 to 200 thousand. Everything else about the graphs is exactly the same. Yet, if you were to take a quick glance at the three graphs, you would assume that the graphs showed completely different sales figures: sales seem to increase dramatically in Fig. 8-13 while sales seem to hardly change at all in Fig. 8-15.

Which one is the real graph? They all are. Each of the three is an accurate representation of the data. (The graphmaker should highlight the fact that the y-axis in Fig. 8-11 begins at $25,000 and not at zero.) What do the three graphs show? You must study each graph closely so that you can interpret what it says.

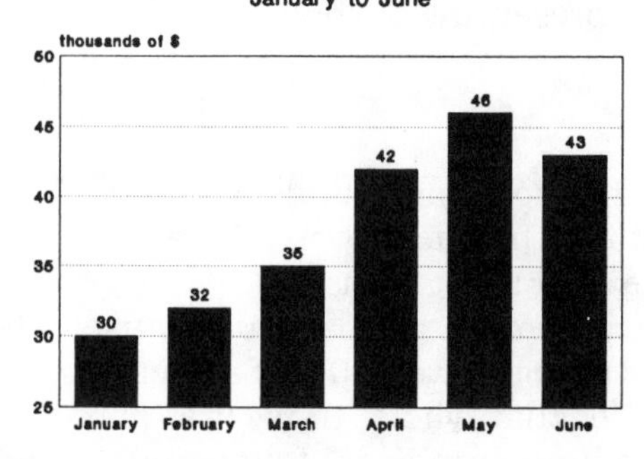

Figure 8-13

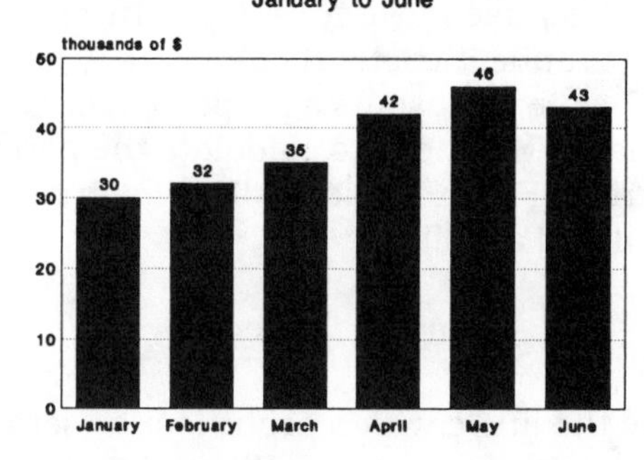

Figure 8-14

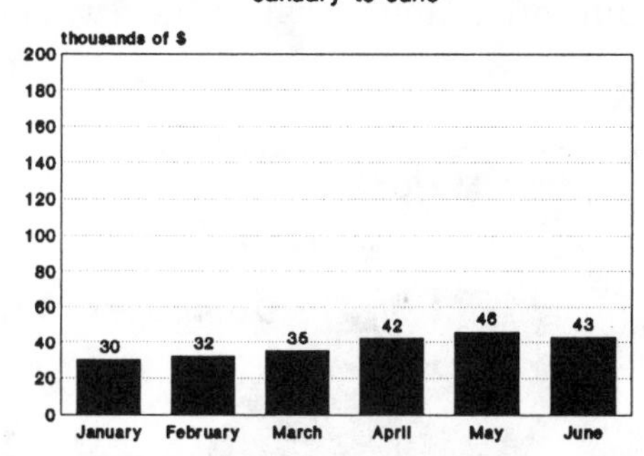

Figure 8-15

This is obviously not the only way to use the structure of the graph to influence your readers. Some of the things to avoid when creating graphs are:

- distorting the bars in a bar graph. For example, making a bar twice as tall and twice as wide creates a visual image four times as large.
- using a non-uniform time scale.
- omitting key information. (Figure 8-13 would give a very different impression if the month of June were omitted.)
- using outdated data.

You owe it to your audience to represent the information accurately and clearly and to avoid misrepresenting the data.

8.5.3 Types of Graphs

This section illustrates the most basic kinds of graphs. Many other kinds exist, but if you understand the ones covered here, you have a good basis for tackling more complex kinds of graphs.

Pie graphs and *column graphs* are particularly useful in depicting the parts of a whole. Whenever you have 100% of something, or one unit, and you want to show parts of that whole, think first about pie and column graphs. (Note: pie and column graphs are frequently referred to as charts. We are covering them in this section because they are so closely related to bar and line graphs, and because most graphing software programs allow you to do pie and column graphs. Strictly speaking, they are graph charts.)

Since each slice of the pie represents a proportion of 100%, pie graphs are difficult to draw by hand. You need to calculate the percentage of each slice in terms of 360 degrees. For example, a slice which represents 20% of the whole would be calculated as follows:

$$\frac{20}{100} = \frac{x}{360} \text{ or } x = 72 \text{ degrees}$$

You will need to measure the angle with a protractor or some other instrument, and draw the circle with a compass. (When drawing a pie graph by hand, you should begin measuring the angles at the noon position on a clock.) A computer program with graphing capabilities will handle all of the

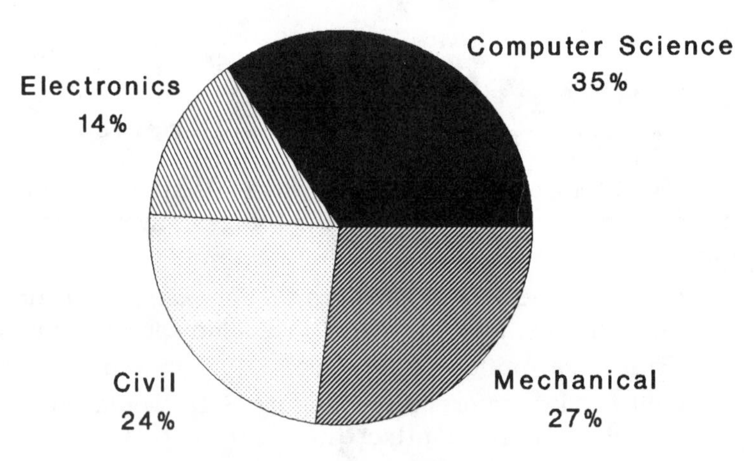

Figure 8-16 An example of a pie graph.

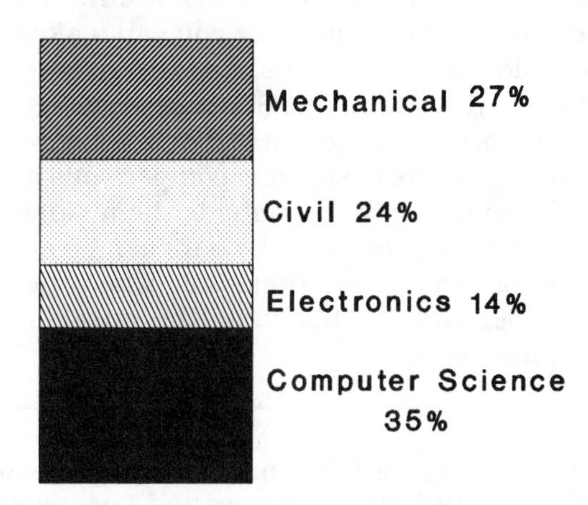

Figure 8-17 A column graph illustrating the same data as Fig. 8-16.

calculations and the drawing of the graph in milliseconds. The same principle is true of column graphs; a computer program will calculate the correct proportions for you.

One advantage of a pie graph over a column graph is that you can more easily emphasize one piece of the whole with a pie graph: you can separate one slice from the rest of the pie.

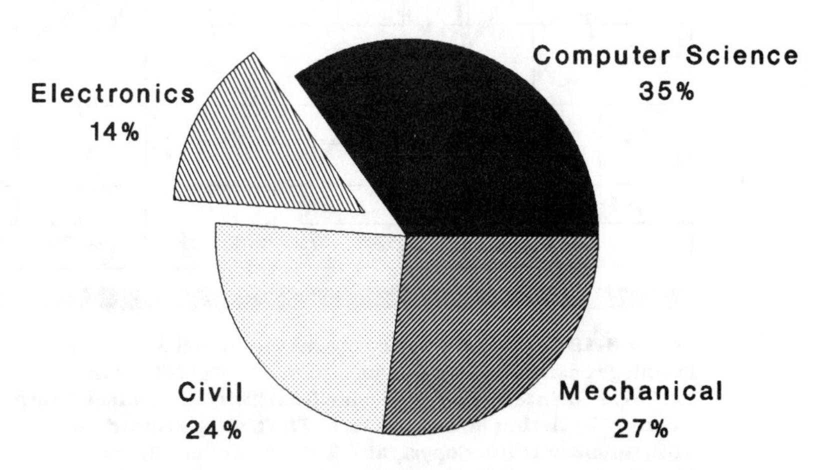

Figure 8-18 Notice how a slice has been cut out for emphasis.

Line graphs are good ways to show trends. Generally easy and quick to plot, they allow the reader to see the direction or progress of a variable against time or some other constraint. You can connect points on a line graph with straight lines, or you can make a single line to show a trend, or, if you do the calculations, you can show a curve of the changes between points on the graph. Fig. 8-19 connects the points.

Bar graphs represent information in bars or columns. They are particularly clear ways of showing comparisons between two or more variables. A helpful feature of bar graphs is the inclusion of the values of the bar just above or inside the bar. The bars can be either vertical or horizontal. Horizontal bar graphs are preferable when the bars do not begin at 0; in Fig. 8-20, the bars are horizontal because this orientation allows for easier comparisons.

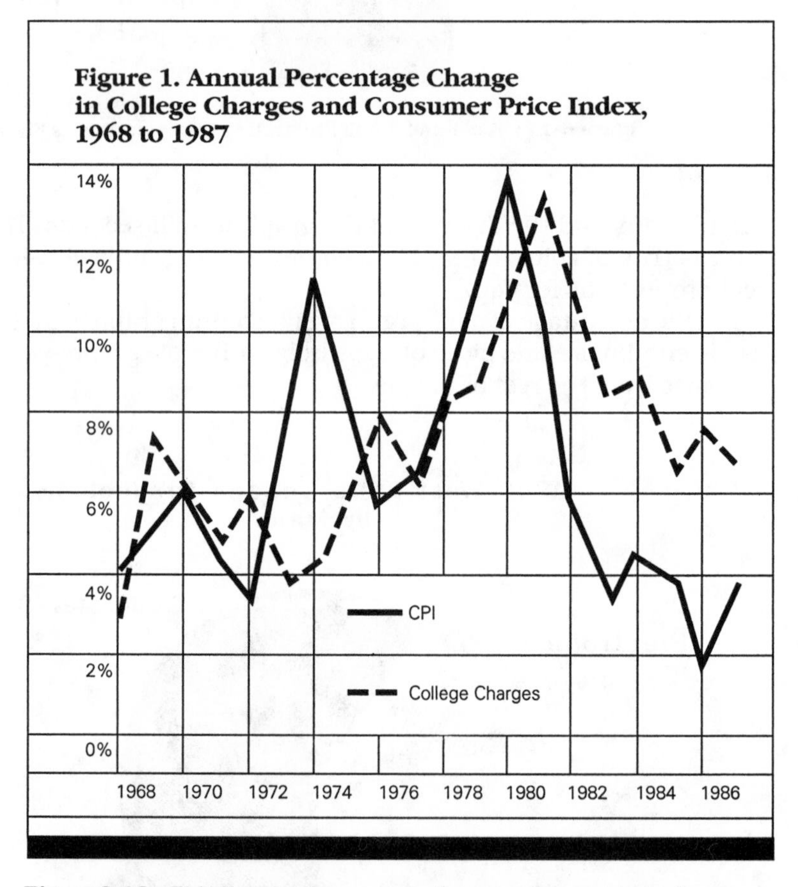

Figure 8-19 This line graph compares the annual percentage changes in college charges and the Consumer Price Index (CPI) for the years 1968 to 1987. **Reprinted with permission from "Why Are College Charges Rising?" by Arthur M. Hauptman in *The College Board Review* No. 152, Summer 1989, Copyright © 1989 by College Entrance Examination Board, New York.**

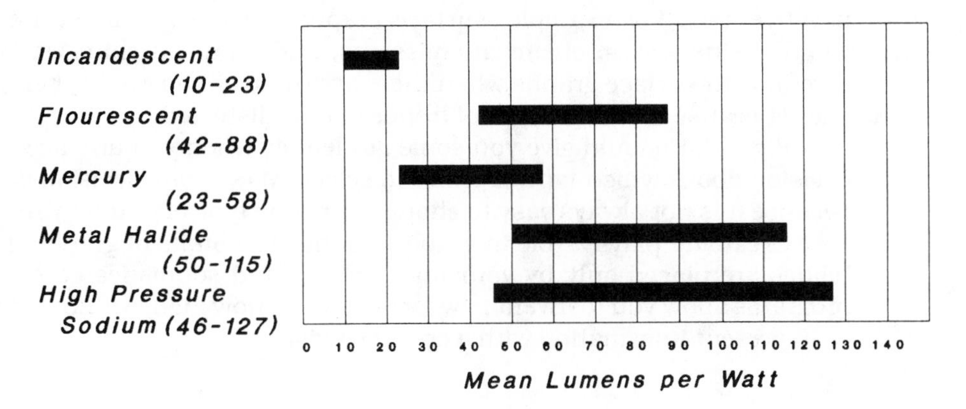

Figure 8-20 A horizontal bar graph. Fig. 8-12 and Fig. 8-26 are illustrations of vertical bar graphs.

Stacked bars are bars that are divided to show how different variables contribute proportionally to the whole. Each stacked bar is similar to a column graph; the stacked bars are plotted on a graph to show comparisons. See Figure 8-21.

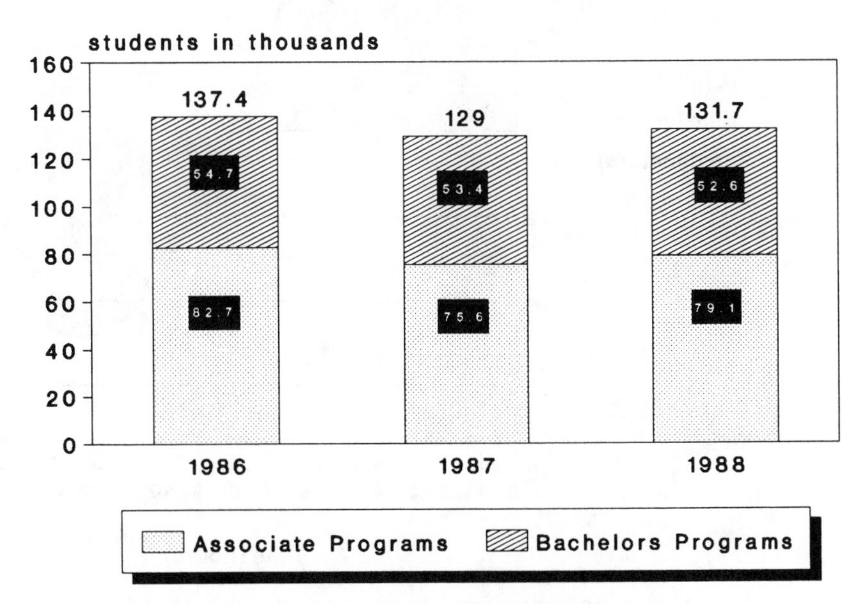

Figure 8-21 A bar graph with stacked bars.

Surface graphs (or area graphs) depict fluctuations over time, and are particularly good for showing changes over many time periods. Use surface graphs for illustrating changes in volume and changes across a positive and negative axis. For example, surface graphs are used to illustrate wavelengths, sine waves, chemical emissions, and financial changes. Be very careful with surface graphs when there are three or more items being compared because the magnitude of trends can be distorted.

Fig. 8-24 should give you some guidelines when you are faced with a decision about which type of graph to select. Make your decision carefully because it is not always easy to change from one type of graph to another.

Creativity plays a role in graph-making. The limits of graphing possibilities are placed only by your imagination, and sophisticated computer programs allow you to invent new forms. Notice how Figure 8-25 and Figure 8-26 present information with a creative twist.

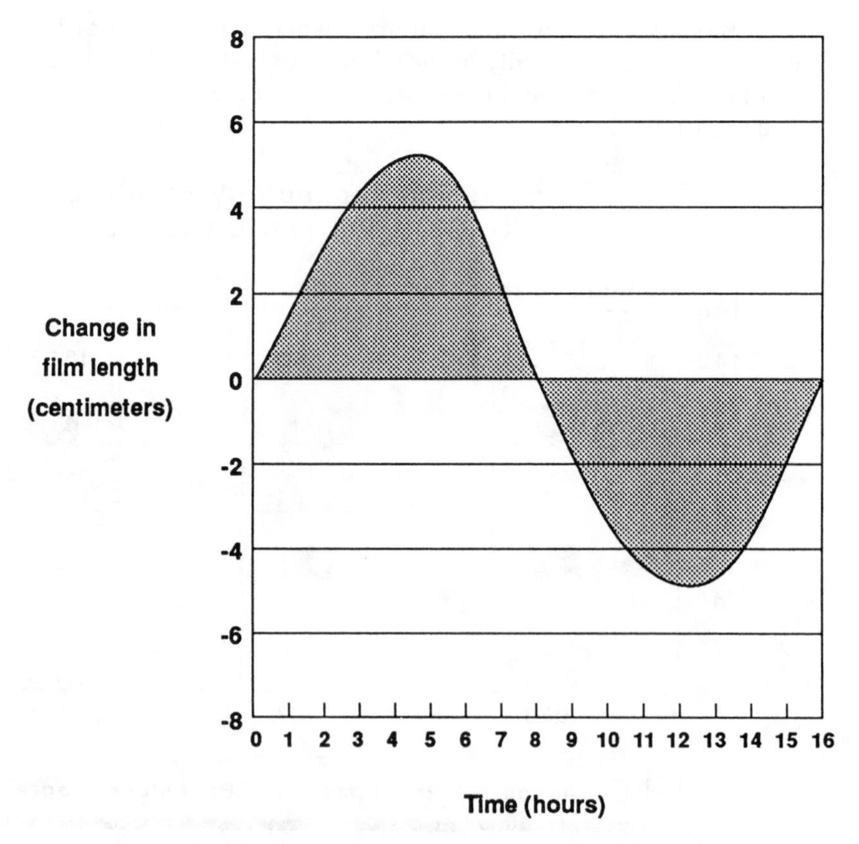

Figure 8-22 This surface graph shows changes above and below zero.

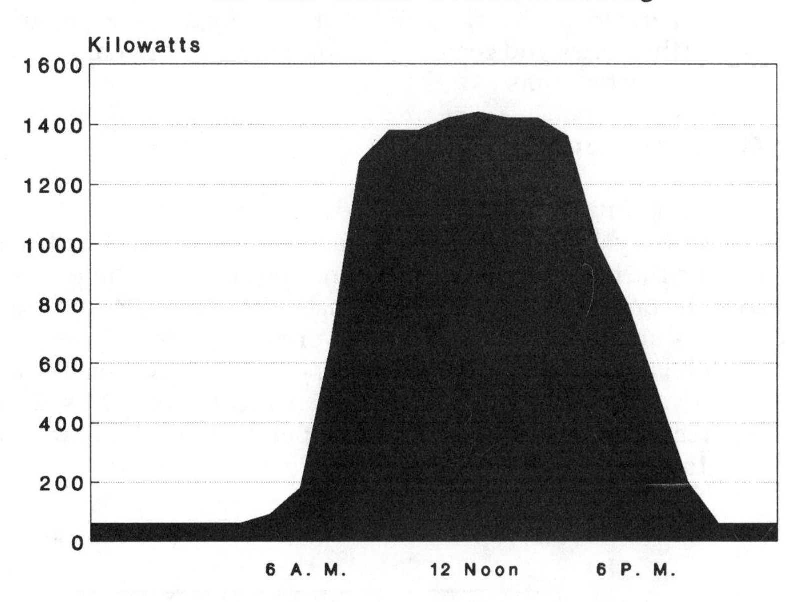

Figure 8-23 A surface graph showing electric load.

Guidelines for Graph Decisions

If you want to show:	Then use:
Change in volume	Surface
Change over time periods	Bar/Line combo
Emphasis on a part of a whole on volume on one of several series	 Pie with cut slice Surface Line or Bar/Line combo
Parts of a Whole at a specific time at 2 different times over a few time periods over many time periods	 Pie or Column 2 Pies or 2 Columns or 2 Stacked Bars Stacked Bars Surface or Stacked Bar
Relationships between 2 series over a few time periods over many time periods	 Bar or Bar/Line combo Line
Trends statistical trends over many time periods over a few time periods	 Line Vertical Bar Line

Figure 8-24 Use this table when you are not sure what type of graph to construct.

Pictographs (pictorial graphs) are graphs that use pictures and images rather than lines and geometric shapes to represent data. Figures 8-25 and 8-26 are pictographs.

8.5.4 Projections

Projecting future trends is an inexact science at best. Whenever you are projecting, you need to make this fact very clear to the reader, and you need to state the basis for your projection; this should include how you came up with the projections and how you made the calculations. Note how the projections are handled in Figure 8-27 prepared by the Federal Reserve Bank of Cleveland for appearance in their periodical *Economic Trends.*

Two different projections are given, highlighting the speculative nature of forecasting the future. The exact point where the graph changes from past to future should be clearly marked on your graph.

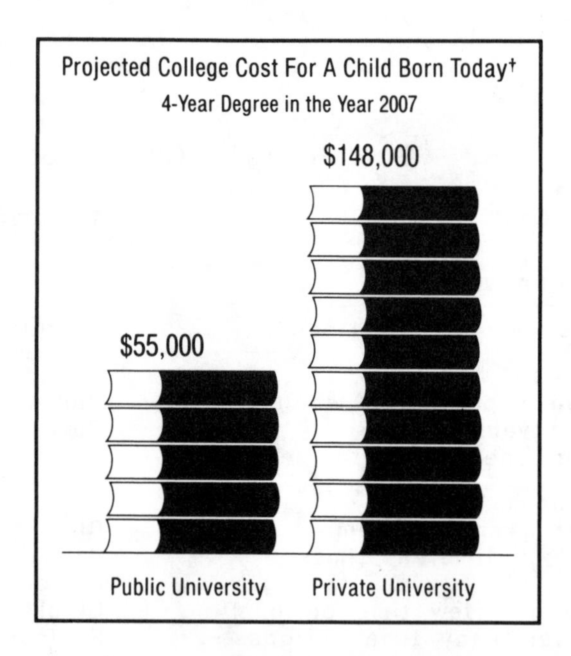

Figure 8-25 This comparison is based on a 6% average annual rate of inflation and average annual expenses for the 1988–1989 school year of $4,445 for a 4-year public college and $11,330 for a 4-year private college. **Reprinted with the permission of The Franklin Group of Mutual Funds.**

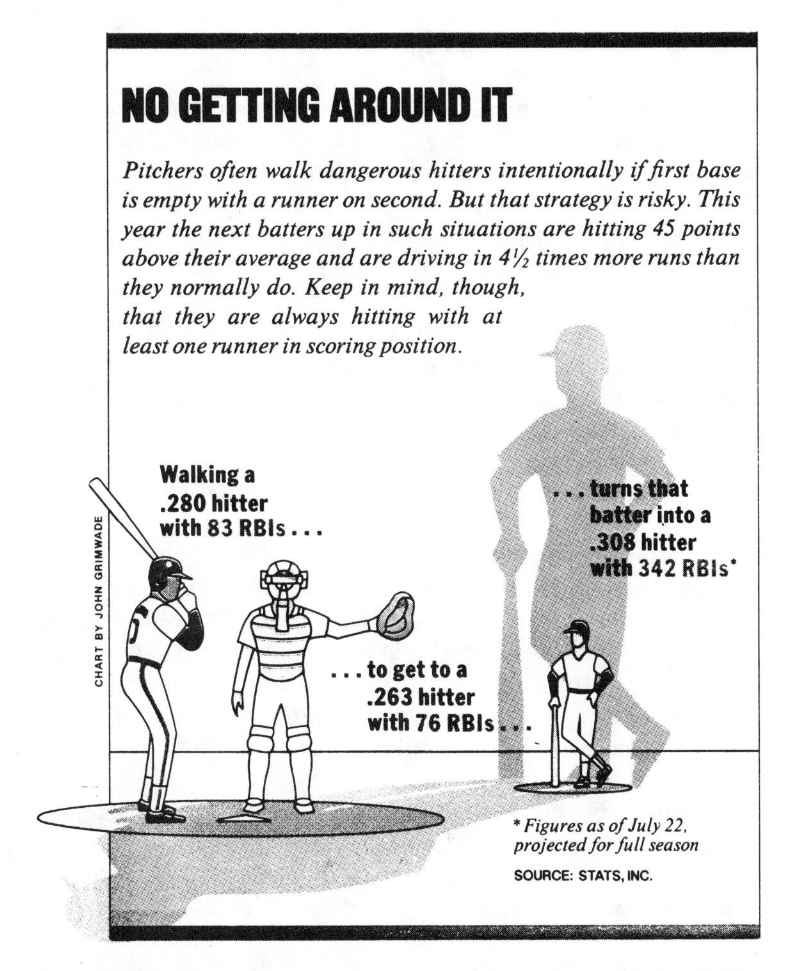

NO GETTING AROUND IT
Pitchers often walk dangerous hitters intentionally if first base is empty with a runner on second. But that strategy is risky. This year the next batters up in such situations are hitting 45 points above their average and are driving in 4½ times more runs than they normally do. Keep in mind, though, that they are always hitting with at least one runner in scoring position.
Walking a .280 hitter with 83 RBIs . . .
. . . to get to a .263 hitter with 76 RBIs . . .
. . . turns that batter into a .308 hitter with 342 RBIs*
CHART BY JOHN GRIMWADE
* Figures as of July 22, projected for full season
SOURCE: STATS, INC.

Figure 8-26 This chart appeared in *Sports Illustrated* (July 31, 1989). **Reprinted with the permission of John Grimwade.**

8.6 DRAWINGS

Drawings, sketches, diagrams, and schematics allow the reader to picture what you are saying in your text. The typical reader can visualize the arrangement of components (*configuration drawing*) or the relationships of the parts of a system (*schematic diagram*) much more easily when the information is presented visually. Whether you work with an artist or draw an illustration yourself, there are a number of things to remember:

- Plan the illustration carefully.
- Construct a rough draft. Evaluate this draft and make the necessary changes.
- Keep the illustration simple.

8.6.1 Freehand Drawings

The ability to draw accurate representations of technical parts, machinery, material, processes and products is a very helpful one. If you have it, by all means use it. You should also realize that not everyone has artistic talent; if you do not, you need to find another way to illustrate your work. Many companies have professional artists on staff for important documents, and you can find freelance artists in the phone book and by inquiring at art supply stores.

The artists will need your technical expertise to guide them in completing the illustrations you need. Make sure you provide specific written instructions that express exactly what you want. Another way is to utilize a computer design program.

8.6.2 Computer-Aided-Design Illustrations

CAD (Computer-Aided-Design) and CADD (Computer-Aided-Design and Drafting) software has revolutionized the design process. Knowledge of a CAD program such as FastCAD, AutoCAD, CADKEY or VersaCAD will enable you to present readable and accurate drawings and diagrams. The programs are too costly for an individual to purchase, and they can take some time to learn, but the results are worth the effort. Some of the programs enable you to do 3-dimensional modeling. Fig. 8-29 was drawn using AutoCAD Release 10.

Knowledge of a CAD program is viewed as a very valuable commodity by technical industries. The knowledge alone may get you a job.

Some paint and draw programs also enable you to create detailed sketches and diagrams.

8.7 OTHER KINDS OF ILLUSTRATIONS

New technology allows us access to a wide spectrum of graphic applications. We can do just about anything we can think of if we have the right equipment. This section suggests some of what you can do to enhance your documents and get your message across:

clip art Many software programs allow you to take a pre-drawn image and place it anywhere in your document. You can size the image to your liking, and with some programs you can change the proportions of the clip art. (Some programs allow you to manipulate the image.) The

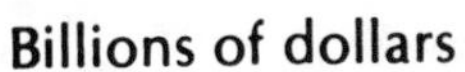

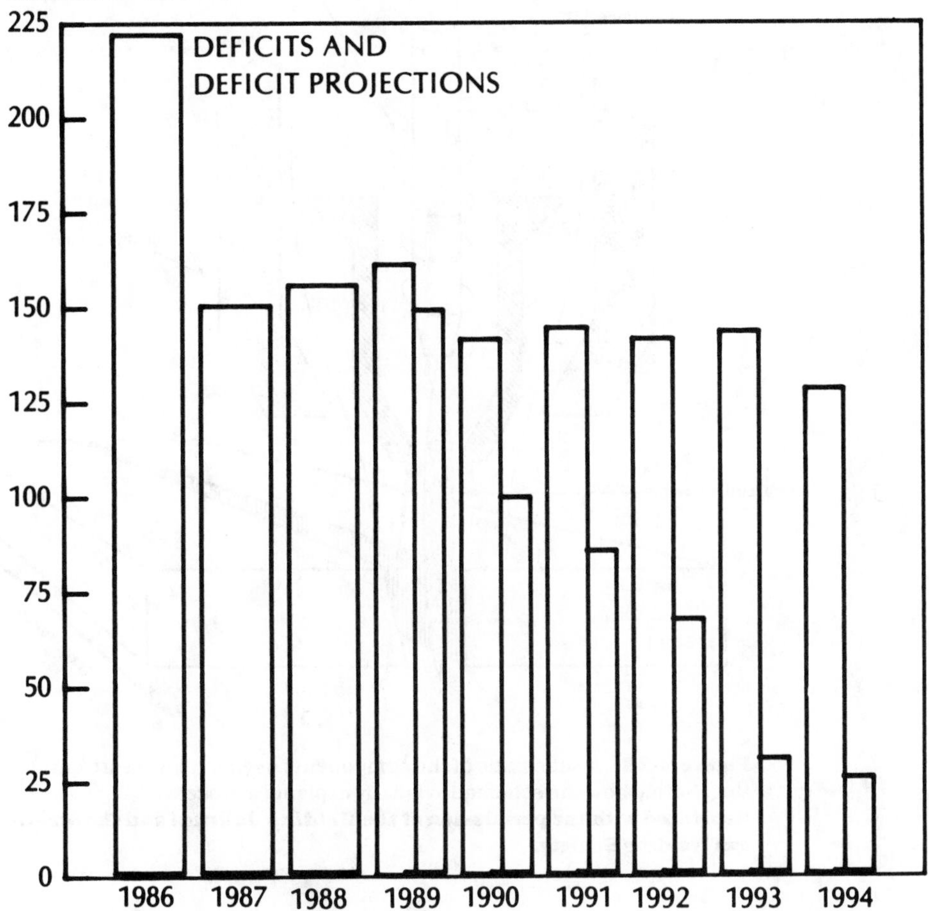

Fig. 8-27 A bar graph depicting the outlook for federal deficits according to the Congressional Budget Office (CBO) and the Office of Management and Budget (OMB).
Source: *Economic Trends*, September 1989, page 7.
Reprinted with permission of the Federal Reserve Bank of Cleveland.

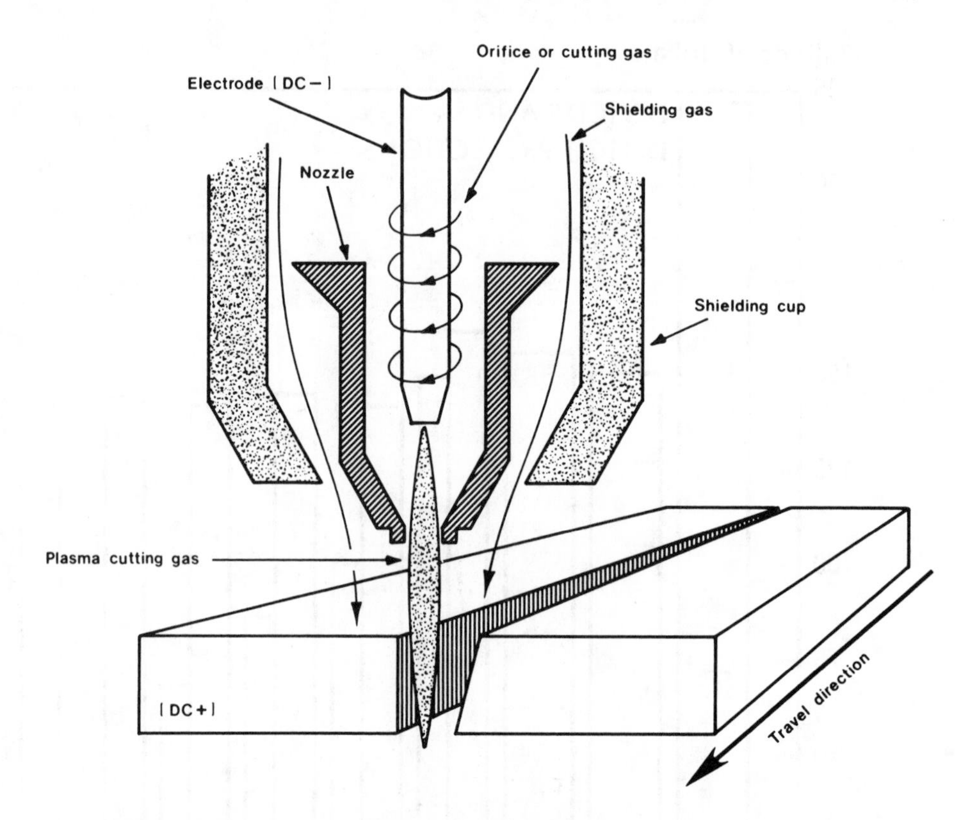

Figure 8-28 A schematic of the components used in plasma arc cutting. Notice how the schematic visually explains a process.
Reprinted with the permission of the *Welding Journal* and the American Welding Society.

images include drawings, symbols, icons, and other art work that may enhance your documents.

maps Don't underestimate the importance of placing your readers within a geographical locale. Maps show relationships of elements in space. Some clip art programs include maps of major cities, countries and continents. Topographical, astronomical, historical, nautical, economic and a whole host of other types of maps are excellent communication aids. If you photocopy a map, make sure that you obtain permission to do so.

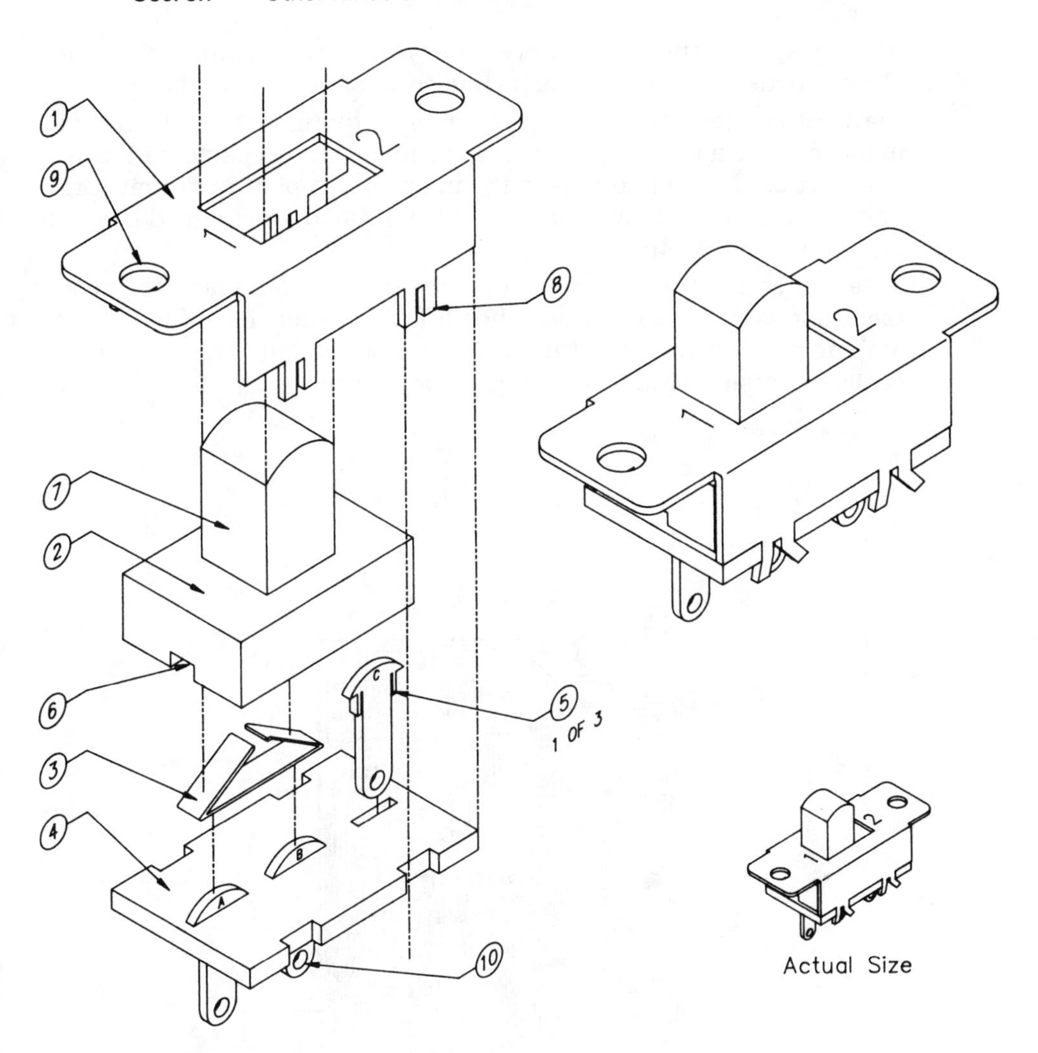

Figure 8-29 A technical illustration of a single-pole double-throw slide switch.
Reprinted with the permission of Mike Richmond.

photographs Although not everyone is a great photographer, any clearly-focused representation is indeed worth a thousand words.

scanned images Until recently, most of us thought that the only way to incorporate a photograph into a document was to paste it in. Scanners digitize a page and import the image into a picture format. Once imported, it can be manipulated, scaled, rotated, or reversed before it is included in a document.

lists You certainly do not need high technology equipment to accomplish one very basic and very helpful illustration for your readers. Making a list of important material in a series puts the items in an easily understood frame of reference for your reader.

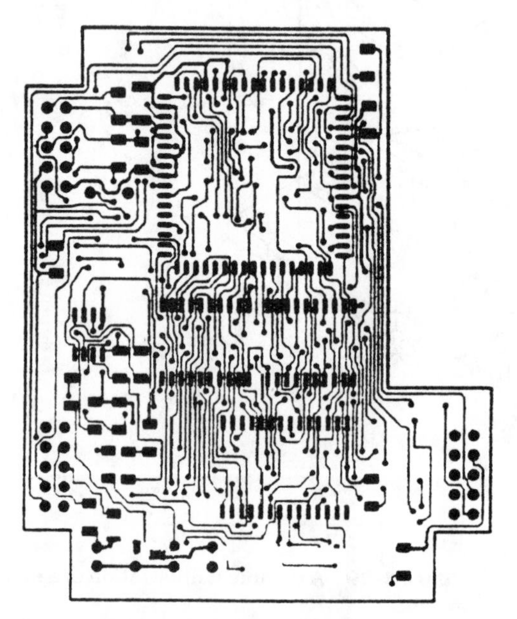

Figure 8-30 A technical illustration of a surface mount board. **Reprinted with the permission of** *Circuit Design*.

8.8 POSITIONING GRAPHICS AND ILLUSTRATIONS

When most people look at a page that has both text and an illustration, their eyes are immediately drawn to the illustration. If you can, you should try to position the illustration within, or surrounded by, the text. If you send a reader to an appendix, it is highly unlikely that the illustration will have a forceful impact. It may not even be seen.

In Chapter 7, Selecting a Format, we encouraged you to treat your document as blocks of information. We wanted you to start thinking of your text in moveable blocks. Illustrations are easily viewed as blocks because they are usually rectangular in nature. Carefully place the illustrations on the page where the readers will see them. Don't force the reader to manipulate the page to see the graphic. Some specific guidelines are offered in Section 7.5 on the placement of graphics. Positioning should be part of your overall design plan.

8.9 USING SPECIAL EFFECTS

Color charts and graphs in texts, and color overheads during presentations, at one time were striking because of their rarity. Special effects are special because they have not yet become something that many people can do. Now people are impressed with moving images on a computer screen, computer

Figure 8-31 A potentiometer.
Reprinted with the permission of Maurey Instruments.

sound effects, and holograms. You may have the opportunity to work with these and other special effects during your career. When you attend meetings and conferences, or when you see exciting graphics and illustrations in books, magazines, and newspapers, try to learn new techniques that can improve your illustrations.

Why Do We Use Visual Aids?

- To present information effectively
- To save comprehension time
- To direct the audience's attention
- To display statistical relationships
- To show comparisons
- To influence the audience visually

Figure 8-32 An example of a bullet (or bulleted) list.

8.10 A FINAL WORD

Once you have mastered the basics of the preparation of visuals, you may become very excited by what you can do. There is a danger here. Don't get carried away and overwhelm your reader with fancy visuals.

We would like to re-emphasize the second and third rules in Figure 8-2: Keep your illustrations simple, and illustrate one main point with each visual. Tell one main story with each visual, and tell that story completely.

As a general rule, assume that your readers have 20 to 30 seconds to grasp the main point of your illustrations. Don't expect them to spend ten minutes deciphering material, no matter how important it is. Most readers will not make the effort. Your task, particularly with technical material, is to simplify.

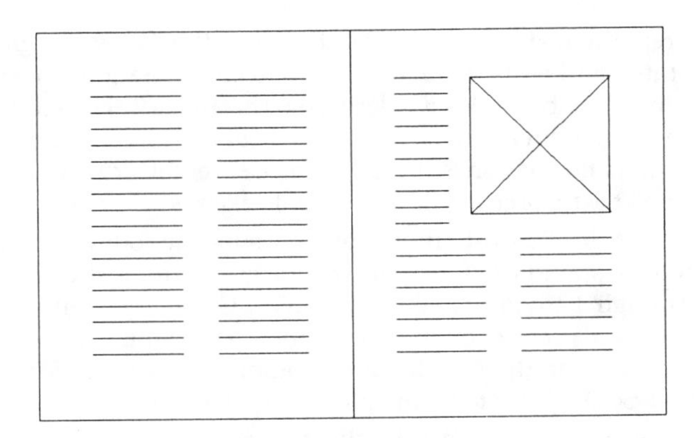

Figure 8-33 Notice how the placement of the illustration draws the reader's attention, particularly because of the way it breaks into the left column of the right-hand page.

ILLUSTRATIONS EXERCISES

1. Make a *table* based on the information that follows:

According to the German Employers Association and the U. S. Bureau of Labor Statistics, of the major industrialized countries, the United States, France, the United Kingdom, West Germany, and Japan, the country with the longest average work week is Japan with an average of 41 hours per week. In comparison, West Germany is the shortest with an average of 38.5 hours in the week. The United States averages 40, while the United Kingdom and France average 39.

In terms of the average number of vacation days, West German workers have the most with 30 days per year. French workers average 25.5, while workers in the United Kingdom receive 25. There is a significant drop-off when you look at the United States and Japan. Workers in the United States average 12 vacation days, while Japanese workers have the least number of days: 10.5.

Japanese workers, however, receive 13 holidays on average, the most holidays of the countries being compared. West German and U. S. workers receive 10 holidays, while French workers receive 9 and British 8 on average.

2. Create a text chart or graphic using the following information: The United States ranked 14th out of 16 industrialized nations in terms of money spent on education for grades Kindergarten through 12. The study calculated the money spent on education as a percentage of the gross domestic product. The source of this information is the UNESCO Center for Education Statistics (1988). Hint: Don't try to force a graph out of this exercise.
3. Create an illustration or graphic using the following information: The United States spends 4.1 percent of its gross domestic product on Kindergarten through 12th grade education; West Germany spends 4.4 percent of its gross domestic product on K-12 education; and Japan spends 4.6 percent of its gross domestic product on K-12 education. The source for this information is the UNESCO Center for Education Statistics (1988).
4. Construct a flow chart that illustrates a process that you know well. The process may be non-technical in nature, but it should have at least seven steps before completion. Remember, you can add text to help explain material presented in the flow chart.
5. Create graphs to illustrate the material presented in the three paragraphs in italics of Exercise #1.
6. According to the Boston Public Works Department and Energy Systems Research Group, the City of Boston generates an estimated 231,702 tons of trash a year. Construct a graph that illustrates the top ten categories of trash.

Category	**Tons per year**
1. Newspapers	21,763
2. Leaves	20,145
3. Food waste	16,999
4. Books/magazines	14,583
5. Grass and brush	14,332
6. Clear glass	13,844
7. Cardboard boxes	13,614
8. Plastic packages & containers	10,421
9. Miscellaneous scrap iron	10,375
10. Nonpackaging plastic	8,520

7. Draw a graph to illustrate the following estimated percentages of trash generated by the City of Boston in a year:

Paper and cardboard	32%
Leaves and yard debris	15%
Glass	12%
Metals	12%
Plastic	8%
Miscellaneous	21%

8. The City of Boston generates the following amounts of glass trash in a year:

Category	**Tons per year**
Clear glass	13,844
Green glass	7,411
Amber glass	4,816
Miscellaneous glass	2,637

(a) Construct a pie graph to illustrate the above information.

(b) Now construct a pie graph that **emphasizes** the estimated amount of green glass generated by the City of Boston in a year.

9. The State University in your state has recently conducted a survey of alumni who graduated between 5 and 10 years ago. The alumni were asked to respond to two questions: (1.) How often do you make oral proposals as part of your current job? and (2.) How often do you make written proposals as part of your job?

frequency of					
oral proposals			written proposals		
4-5	times a week	33%	4-5	times a week	19%
2-3	times a week	28%	2-3	times a week	34%
1	time a week	29%	1	time a week	32%
0	times a week	10%	0	times a week	15%

Draw an illustration that highlights the above information. Your caption should lead the reader to a conclusion about the information: what does the survey show? Draw your own conclusion, but make sure that your illustration emphasizes this conclusion. (Hint: one way of drawing a conclusion is to group frequency categories together.)

10. Graph the following information provided by the American Facsimile Association in 1990: In 1987 12 billion pages were transmitted by Fax machines; in 1988 19 billion pages were transmitted; in 1989 30 billion pages; in 1990 the Association estimates 43 billion pages will be transmitted; and in 1991 58 billion pages. Make sure your graph clearly shows which values are estimates.

11. Look at Figure 8-19. Draw a conclusion about the increase in college tuition from 1980–1987 compared to the increase in the Consumer Price Index for the same time period. Write a memo to your instructor conveying your conclusion, and include a new graph that highlights your conclusion. (You can use estimates based on Fig. 8-19.)

12. List at least four ways that the following graph could be improved. If you are able, restructure the graph based on your recommendations.

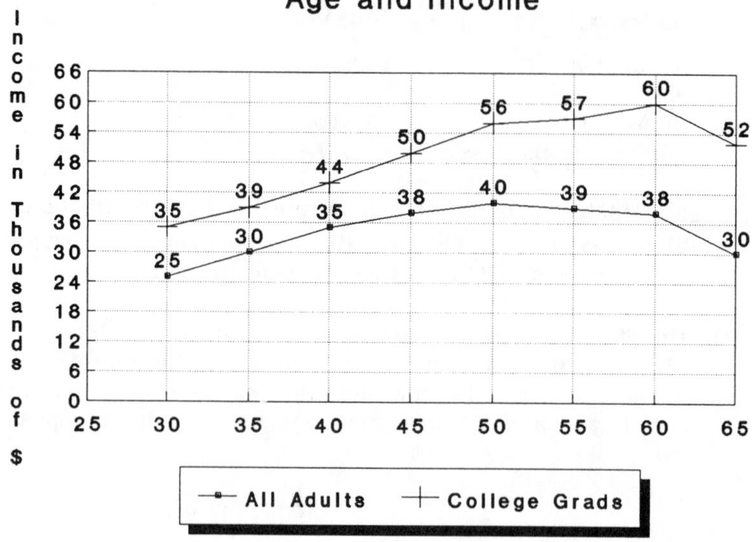

13. Prepare an illustration to depict the following information about the countries emitting the largest amounts of carbon dioxide. Estimated emissions for 1990 in millions of metric tons are:

United States	1,360
Soviet Union	1,020
China	600
Japan	310
West Germany	200
United Kingdom	170

The source of this information is a draft report of the Energy & Industry Subgroup of the Intergovernmental Panel on Climate Change.

14. Use color or some other special effects to construct an illustration highlighting what you consider to be a significant point, one that you want other people to understand and appreciate.

15. Choose an appropriate format to illustrate the following information, all of which involves the typical range of some common sounds.

Sound	Range (in decibels)
Clothes dryer	50-73
Vacuum cleaner	60-84
Automobile	60-89
Train	72-92
Power Lawnmower	80-94
Motorcycle	80-110
Snowmobile	84-109

Include in your illustrations the following information:
Prolonged exposure to noise of 85 decibels causes light hearing loss. Prolonged exposure to noise of 90 decibels causes mild to moderate hearing loss. Prolonged exposure to noise at 95 decibels causes moderate to severe loss. Short exposure to noise at 100 decibels can cause permanent hearing loss.

16. The town of Sureport, Oregon is attempting to monitor its water use more closely. Prepare a series of appropriate illustrations for a presentation based on the last four years of water usage. See the figures below.

	Year			
	1	2	3	4
Total*	76.62	83.59	73.42	96.61
Maximum Day	0.35	0.40	0.35	0.95
Average Day	0.21	0.23	0.20	0.26

*all figures in millions of gallons

9

Designing Effective Memos

Note to Users: **This is a task chapter. The purpose of this chapter is to show you how to design more effective memos. If you want to work on another task, refer to the table of contents.**

9.1 INTRODUCTION

Recent studies have shown that as many as one out of every three memos are written to make clear what was already written in a previous memo. Poorly designed memos create increased document costs and a less productive work force.

This chapter will provide you with effective solutions to reduce the time you spend on writing memos. You will learn strategies and techniques that can be applied to a variety of memo tasks including requests for information, short reports, reminders, and statements of company policy.

In this chapter you will learn how to:

- Use formats to make your message more effective.
- Apply strategies to avoid replies and circular correspondence.
- Consider memo etiquette.
- Decide when to send a memo.
- Design memos that call for action.

9.2 DEFINITION

A *memo* is a document designed to pass information between people and departments **within** an organization. Memos are used to make readers aware of something, to offer instruction, to prompt action, and to serve as a reminder. They are vital to the smooth operation of an organization.

It is important for you to remember that memos are designed as in-house documents. They are written for and to people within a company or organization. They provide a written record and history of company decisions, of alternatives considered, and of responsibility for actions. More frequently they are used to tell persons in the company about a new policy or product or even a new place for the company picnic.

Within organizations which have more than just a few employees, it is impossible for all information to be shared in face-to-face meetings. Even if it were possible, some information is too important to be trusted to memory. It makes more sense to send one written memo to twenty employees than to spend an entire day going to see twenty people.

In business and industry *put it in writing* is the common practice. The memo is second only to the telephone in terms of modern business communication and the sharing of information.

Here, for example, is a typical memo:

June 22, 1992

To: Frank Landy, Maintenance
From: Jack Gore
Subject: Leaking Window in Office

The window in my office, located on the fifth floor, room 7-518, has started to leak. There is water damage to the sill and immediate wall area.

I would appreciate it if you could get to this before Friday because we are having a products meeting here.

Thanks.

This document, like any other, tells a lot about itself. For instance, you notice that the style is less flowery than a letter. There is no *Dear Mr. Landy,* just a simple *To:.* Because this document is intended to communicate within an organization, there is no need for formality. Notice how Jack uses his nickname, something he ordinarily would not use in documents which were intended for persons outside of the company. We also notice that Jack is giving Frank the instructions as to what he wants done and the time by which he expects it completed. This says something about Jack and Frank and their positions within the company hierarchy.

9.3 FORMAT

Many people will judge you on the basis of the memos that you send to them and it is important to be *correct* as well as clear and accurate. The first thing you need to be sure about is your format. Here is a typical memo format:

To:	Date:
From:	
Subject:	

Figure 9-1 A typical memo format.

What distinguishes the memo from the letter, other than its internal audience, is the format. Many companies have preprinted forms for the sending of memos. Sometimes these forms include a space for the reader's response. These are known as *turn-around* memos.

Electronic mail refers to the distribution of messages by computers. Many large organizations have computer systems which allow everyone with access to terminals to communicate directly with each other by leaving computer messages.

Most electronic mail systems automatically format your document as a memo.

All of these formats are simple and direct. Since memos are designed to communicate within a company, they can be less formal than a business

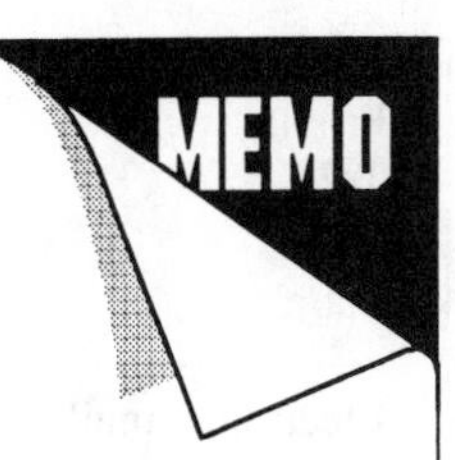

To: Sheila O'Brien
From: Cindy Bauer
Subject: Installation Appointment
Date: November 29, 1992

I would like to install the Postscript emulator in your printer some time early next week. It would be helpful if you were available for about an hour so that I can show you how to use it most effectively. Please let me know what times you have available. Thanks.

Reply

Date:
From:

Figure 9-2 A turn-around memo.

From: ERNIE::GREENEM 18-MAR-1991 11:56:47.71
To: GREENEM
CC:
Subj: Electronic mail

Electronic mail is a computer utility which allows you to send text messages from your microcomputer, local area network, or computer system to another. Your message is stored until the recipient retrieves it. Messages can be read, forwarded, extracted, printed, and edited. You can send your message to one person or a list of persons.

Figure 9-3 An electronic mail message.

letter. Memos can leave out greetings and salutations, and get right to the subject. Clearness and accuracy are expected rather than elaborate politeness.

To: Marty Weiss Date: June 6, 1992
From: Dick Topler
Subject: Installation of Temporary Outlet Board

Please check the power available at our booth site for the Jacksonville Electronics Exhibit. Sales will need a temporary outlet board installed at the Fredericks' Armory for the Robotics display on the first three days of next month.

Because we will be demonstrating the Zona 3 model Industrial Welder we will need a minimum of 15 amps and a maximum capability of 600 volts.

We are located in the same spot as last year (AA12) but the new power requirements may force us to change location.

Let me know by Thursday noon if we have to move to another floor location.

cc: Lois Greer

9.4 THE PARTS OF THE MEMO

If you look at each of the parts of this memo, you will be familiar with the format that is standard in most companies.

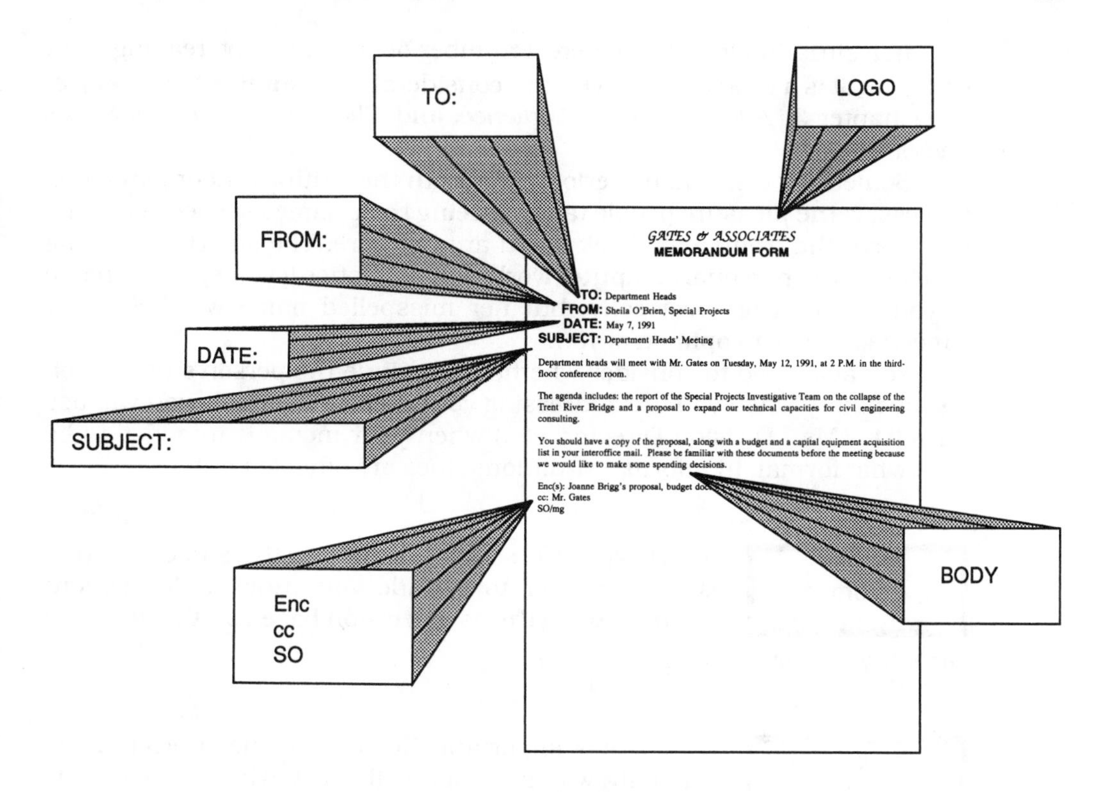

Figure 9-4 An example of a standard memo.

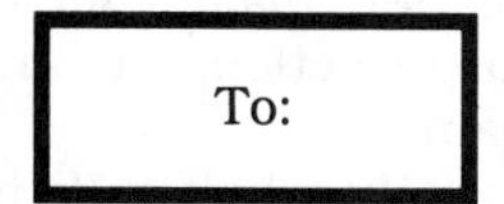

Frequently this is obvious. You know whom you are writing to and you know how this person will respond to what you are saying.

Sometimes though, your memo will have a multiple audience. A number of different people may end up reading what you write.

For example, you send a routine memo reporting your weekly progress on a chemical sorting process. This week you write about a possible method of recycling one of the more expensive chemicals used in the process. Your idea could save the company a good deal of money. It is quite reasonable for you to expect that this memo will be read by others than your usual supervisor. He is going to pass this memo up through the hierarchy and it will be read by persons even further away from the process than your boss. Will these persons have the technical expertise and background data necessary to understand what you are saying?

In a situation like this, where a number of people will be reading what you write, it is a good idea to give extra consideration to **audience** and **style.** *See* Chapter 3, *Addressing the Audience,* and *Chapter 6, Choosing Styles and Strategy.*

Something else, often overlooked in both the writing and proofreading process, is the pleasure people take in seeing their names spelled correctly. It is worth the trouble to check if you are not sure. The smartest spelling checker in the personal computer world will not notice if you spell the name of your supervisor incorrectly, but her misspelled name will help your supervisor form an opinion of you.

You also need to consider whether to include the person's title or job responsibility when you are addressing your memo. Generally you will use titles (Dr., Ms., Division Director, etc.) when your memo is intended to be somewhat formal. In informal situations titles are often left out.

From:

When you are sending a memo, the same question about whether to include your title or department occurs. Again this is a decision based on the degree of formality you are looking to achieve.

Subject:

This one is important. Remember, the subject line is the only way someone will know what the memo is about when they find it in their files seven months after you first wrote it.

Mysterious subject lines like "Important Matter" make your memos difficult to file. If they are meant as a record for future action, they are quite useless since they are unlikely to be looked at again.

Some people are very busy. They spend more than half of their work time attending to documents. They want to know what they are looking at immediately. You need to be specific. The subject line is what will draw the attention of your reader to what you are saying. Generally your memos should only contain one idea, particularly if they are short documents. Try to restate the whole idea of the memo in the subject line.

Subject: Pat Bowen's Promotion to Sales

This subject line makes it clear what the memo is about and makes the memo easy to find in the files. *Note:* Some memo formats use **RE:** (Regarding) instead of Subject.

Date

Putting the date on your memos can be very important. Some persons file memos received by date rather than subject. Sometimes the date of a memo determines its meaning. A memo that is written about safety procedures the day after a fire is less useful than one written on the day before. Electronic mail often has features which provide automatic date stamping.

cc:

This stands for carbon copy. What it means is a copy of this memo has been sent to the person or persons listed after the colon. It is an expected courtesy to let your readers know who else has a copy of the document they are reading.

Enc.

Letting your reader(s) know that they should expect to find something else is always a good idea. An enclosure line alerts the audience to look for what you placed with your message.

9.5 *WHEN TO WRITE A MEMO*

A good time to write a memo is when you want to do one of the things in the following list and you need to have a written record of your having done so.

- CONFIRM
- REQUEST
- ANNOUNCE
- SUGGEST
- EXPLAIN
- REPORT

CONFIRM You can use a memo to confirm the details of a meeting, conversation, or telephone call. Your purpose is to have a written record of decisions which were made or promises that were given.

SUGGEST You can use a memo to recommend solutions to technical problems, to offer the services of yourself or your department, to bring up new ideas or ways of doing things.

REQUEST Use the memo to ask for action or information. This way you have a written record of what you have asked. It is more difficult for your audience to forget or ignore.

EXPLAIN You can use a memo to define clearly for the reader something that is not understood. Your purpose in this case is to make something clear for your reader.

ANNOUNCE Memos are useful for giving formal notice to readers, publicly informing them about new procedures or new products or anything which you want to be publicly known.

REPORT Memos are often used to informally give an account of a project at regular intervals as a way of helping the organization keep track of progress and problems.

Whichever of these purposes you have, it is important for you to document your activities and decisions in writing. Having a good idea doesn't matter if you are unable to communicate it to others within the company. No matter what technical activity you are engaged in, your work and ideas exist as information for other persons working with you. If they do not have this information, they cannot use it in their own work. Much of this necessary information is shared through memos.

When Not to Write a Memo

- When you are in a hurry. If time is very important, use the telephone.
- When the subject is very sensitive or confidential. Many writers have been embarrassed when their memos turned up in unexpected places.

9.6 *HOW TO WRITE A MEMO*

Begin with a brief summary.

The best way to start is to set down in a sentence exactly the point you want to make to your reader or readers. This helps you to be sure of what you are doing. Are you going to explain to your reader how to build a better mousetrap or tell him you need a key for your office? If you are not clear about what it is you want, then how can you expect your reader to be clear about what to do?

One good way to summarize is to complete this sentence:

The purpose of this memo is . . . .

If you have difficulties in completing this sentence, you have a problem that must be resolved. Do not try to skip over this step. Keep working at it. Until you know exactly what point you are trying to make, you are not ready to communicate. You may need to *review Chapter 2, Discovering Your Purpose.*

It's amazing how often ideas that sound good while you are thinking about them lose all of their clarity when you set them down in words. However, you cannot send ideas. You must use words to document what you mean.

9.7 *THE MOST IMPORTANT POINT*

The most important point is always yours when you are the writer. Begin your memo with your point. Get right to the point. Stick to your point. All of these are ways of expressing what ought to be obvious. Your readers expect you to say something, clearly and directly, and they should not have to wonder what it is. A memo is not a mystery story in which the readers search for the writer's clues.

- Start by saying something. Beginnings are strategic. They determine whether memos are read or filed away without any further consideration.
- Secondly, after making your purpose known, give your audience any necessary background information that they will need to understand your point.

For example, you write a memo explaining how greater reliability can be achieved in a balance beam without linking this to the company testing lab where this process occurs. Why would anyone be interested in this information? You must make it clear to the audience how what you are saying affects them and the organization. No person reads memos for entertainment and the usefulness of what you are saying is the key to its value. Make it clear what difference your memo will make.

9.8 *MEMO MANNERS*

Whatever suggestions we give you here, it is still important for you to remember that memos are for use within a company. Each organization sets its own rules and standards for how and when and why memos are to be used; sometimes these are printed in a formal set of guidelines. You have to pay close attention to the environment in which you are working.

One company, for instance, might encourage frequent memos to its top managers, keeping them up-to-date on a variety of projects while another company may discourage written communication to anyone but immediate supervisors. There are some bosses who hate to send or receive memos. Whatever the situation in your company, it is up to you to discover what is expected and to follow that custom.

EDITORIAL

The Fine Art of Memo Writing

Memo writing has developed into an art form. Memos are the communications link that keeps members of the same organization informed about things that are going on or things that they are supposed to do. Because memos go out in batches with a lot of readers, any sender of memos can expect scrutiny from management as well as hostile colleagues. As a result, memo writing has developed into a form where the words and the meaning are not necessarily the same.

For instance; the word "interesting" usually means anything from "questionable" to "stupid," so "your interesting memo of the tenth . . ." really is intended to convey a bit of annoyance without actually saying so. Every industry or organization has a set of key words that the in-crowd understands, but outsiders don't.

Memos that tell somebody what to do are often the subject of stalling maneuvers. A favorite tactic is to switch into government-ese and respond "the implementation of the points specified in recent communications has been deferred pending clarification of procedural questions." Neither friends nor enemies will understand this, but it will buy you a little time. Another responds, "As per instruction, a committee was organized immediately to set up steps for rapid action for compliance with instructions."

When there is no chance to weasel out of an order to do something, a memo stating that you "defer to the experience of John Doe in setting up a plan of action" makes it plain that you are just going along with the idea and it didn't come from you.

Saying that you "cannot decline" to do something really means that you are going to do it. A "conditional recommendation" is no recommendation at all. When you say "no person is better for the position" you are saying the job should be left unfilled. "Unrecognized talent" means just that—nobody has recognized any yet.

With a little thought and practice nearly anyone can become proficient in writing memos with words that mean one thing but have a different hidden message. As I said, memo writing has developed into an art form. Any design engineer can play the game and it seems that most do.

Figure 9-5 An editorial by Lars Soderholm.
Reprinted with the permission of *Design News*, **A Cahners Publication.**

9.9 *THE A, B, C & D OF MEMO DESIGN*

Announce your purpose immediately.
Be sure that you have a point to make.
Conclude by telling readers what to do.
Don't ramble. Stick to your point.

ANNOUNCE YOUR PURPOSE IMMEDIATELY.
It is very important that you state your purpose right away. This is true even if you are conveying bad news. Nothing annoys memo readers more than having to search for the reason the memo was sent.

If you think about this from the viewpoint of the reader, you'll see that this makes sense. At a minimum the reader has a right to know *why* you have sent him this memo to read. And if he doesn't know that, what makes you think he'll be able to figure out the rest of your message?

BE SURE YOU HAVE A POINT TO MAKE.
The first rule of style is to have something to say. It is no courtesy to send a memo which says little or nothing in several paragraphs to a busy person. If your reader needs to contact you to find out what point you were making, you have not written your memo properly.

Stick to making one point in a short memo. If you have to talk about more than one subject in a memo, keep the subjects separate. First say one thing and then the other.

CONCLUDE BY TELLING YOUR READER WHAT TO DO.
Unless you are simply sharing information, you should design your memos to call for action. The best way to do this is to end your memos by setting some sort of deadline for what you want accomplished. If you want someone to do something, then you need to tell them what, where, and when.

DON'T RAMBLE.
This rule is a good one to follow. Far too many memos are too long. Unless your organization uses the memo format for lengthy

reports, then it is a good idea to keep your memos to a single page. Think about what it is that you want to make happen. Say it and leave out irrelevant detail.

You aren't writing a letter to your Aunt Molly who wants you to tell her everything that is going on in your life. Anything that doesn't help the reader to know what, where, when or how to do something can be cut out. Consider the following example:

June 22, 1992

To: Frank Landy, Maintenance
From: Jack Gore
Subject: Leaking Window in Office

The window in my office, located on the fifth floor, room 518, has started to leak. There is water damage to the sill and immediate wall area.

I told one of your men about it but he just yawned and told me that he couldn't get to it for another two weeks. I think you could get better employees for your department.

One of the people who is coming to the products meeting had a similar problem and his company had to replace the window.

I would appreciate it if you could repair this before Friday because we are having a products meeting here.

Thanks.

All of the information in italics does not belong in the memo. It is irrelevant to what the writer wants done which is to get the leaky window fixed. It will probably annoy Frank Landy. Remember, you are writing a memo to get action. Stick to the point and don't get the reader sidetracked.

9.10 A FINAL WORD

The only reason you have for writing a memo is to share information or to get something done. This means that you have to be clear and accurate. In order for you to be clear and accurate you must know what you want. If you are unsure about what you want, review the material on Document Design in *Chapter 1, Readme.1st* and on purpose in *Chapter 2, Discovering Your Purpose.*

MEMO EXERCISES

1. Over the weekend you have read about a Technical Writing seminar which is offered by a nearby college. This seminar, which focuses on the skills necessary to write clear and accurate reports, takes place on a Thursday and Friday three weeks from this week, and is scheduled from nine to five. You believe that taking this course would improve your prospects within Gates and Associates where the written communication is the main product. In order to take this seminar, however, you will need to be absent from work for those two days and you will need to have the tuition paid for by the company.

   Write a memo to your supervisor, Sheila O'Brien, asking if this can be arranged. Be sure to include enough specific information so that Sheila can be able to make a decision without writing back to you. (Be sure to include the cost of the seminar.) You want to avoid the circular response which just wastes time. Take into account her preference for clear and concise writing. (See Section 26.5.)
2. Mr. Gates has asked that any time one of his employees sees an article that has information which might be useful to the company, the employee summarize the information in memo form and forward it to him. Find a recent article in your area of technical expertise which might be of interest to Mr. Gates and send this idea to him in memo form. Keep your report to one page and remember to send a copy (cc:) to Sheila O'Brien, your immediate supervisor, to keep her informed.
3. During the past three months your project team has been assisted by Kim Chen, a coop student (a student working full-time for a semester while enrolled in a degree program) from a nearby college. His work has been very good, going beyond the scope of his job description, and you want to put his work on record and make sure that the company is aware of his fine performance. Address a memo to David Donatelli, Head of Personnel, informing him about Kim's efforts. Be sure to include at least one specific example of Kim's superior performance.
4. You have arrived at your office and the window has leaked again, this time staining the plaster, soaking the carpet, and destroying some computer diskettes with important information on them. You have asked Frank Topler, the Building Superintendent, to fix this problem before, but he has not done anything. You have heard stories in the cafeteria that he does not have enough helpers and you think that he should hire a temporary worker from an agency. You also think that the whole window frame will need to be replaced and recaulked. It would be nice to have a thermal window with triple glazing because the office gets drafty in the winter months. You have also heard rumors that Frank is looking for another job because he is unhappy with the condition of the building. Jorge told you that Frank is having problems at home with his teenage son.

   Write a memo to Frank Topler to address this situation.
5. Several times in the past month you have mentioned to Sheila O'Brien your idea that Gates and Associates should put together an employee manual. This manual would be a complete record of company policies and procedures for everything from vacation schedules to sick leave to reporting formats. Finally Sheila

says to you, "It's a good idea, but put it in writing so that I can bring it to the managers' meeting on Friday and get a decision."

Write a memo in which you put your suggestion forward. Address it to Sheila but remember that it will be read and discussed by all of the department heads within Gates and Associates. See Chapter 26 if you require more information on Gates and Associates.

6. Follow the instructions in Exercise 3, part e in Chapter 26. Pamela Russell needs information about the Hazardous Material Training Center in Pueblo, Colorado. Make sure you inform her if you attach material to your memo.
7. Read 26.7, The Collapse of the Trent River Bridge. You are a member of the Special Projects Investigation Team that will be meeting to determine research priorities, recommendations, and to write a report. You are very enthused about the application of new technologies to this problem and you want Gates and Associates to become involved in testing these applications. You believe this will become a very active priority for consulting firms in the nineties.

   Write a memo to the members of the team asking that your idea be placed on the agenda for the meeting. Give them sufficient information so that they will be prepared to discuss your topic.
8. Apply the A, B, C, and D of memo design formula to Sheila O'Brien's memo.

   Now apply these same standards to a memo which you have written.

GATES & ASSOCIATES

MEMORANDUM FORM

TO: Department Heads
FROM: Sheila O'Brien, Special Projects
DATE: May 7, 1991
SUBJECT: Department Heads' Meeting

Department heads will meet with Mr. Gates on Tuesday, May 12, 1991, at 2 P.M. in the third-floor conference room.

The agenda includes: the report of the Special Projects Investigative Team on the collapse of the Trent River Bridge and a proposal to expand our technical capacities for civil engineering consulting.

You should have a copy of the proposal, along with a budget and a capital equipment acquisition list in your interoffice mail. Please be familiar with these documents before the meeting because we would like to make some spending decisions.

Enc(s): Joanne Brigg's proposal, budget document, acquisition list
cc: Mr. Gates
SO/mg

9. Mr. Gates is planning to travel to Budapest where he hopes to negotiate a contract with a local engineering firm that will act as our East European subsidiary. Sheila O'Brien assigns you to conduct some research and prepare a memo report for Mr. Gates. In particular, she wants you to provide information about the business climate, customs, and negotiating styles which Mr. Gates might expect to encounter.

This memo should be 3 to 4 pages long. Headings and sub-headings should indicate your organization of the information and make it easily accessible to Mr. Gates.

10. In some organizations a good memo is hard to find. Careless writing, inattention to detail, and poor organization, all combine to produce confused communications. If you have the opportunity, collect several bad memos. Show them to the others in your group and discuss what makes these memos negative examples.

10

Writing Business Letters

Note to Users: **This is a task chapter. It is intended to show you ways to design letters that work for you and those you represent.**

10.1 INTRODUCTION

Every day millions of people send and receive business letters. These letters are designed to get things done: claims adjusted, bills paid, products ordered, machinery repaired, proposals accepted. Letters are the written record of commerce between persons and companies. People expect them to be well prepared.

Your letters will represent you and your company when you are not present to explain what you really intended to say. You need to carefully prepare letters that get things accomplished. Your ability to write these kinds of letters will be an important and valuable tool for business success.

In this chapter you will learn how to:

- Create letters that gain attention and follow-through.
- Plan letters that are clear and to the point.

- Customize correspondence for your customers.
- Organize for rapid turn-around and quick responses to your mail.
- Adjust your attitudes to bring about reader approval.

10.2 DEFINITION

A letter is a written message sent to an individual or a group of persons. Generally, it is contained in an envelope and sent through the mails, public and private. Business letters are most frequently addressed to persons outside of the organization or company. Sometimes, for formal occasions, they are used within the company. Accepting and resigning a position, for example, are situations where you should use a letter.

What is most important to remember is that letters are always sent to a person or group of persons. Letters are not mailed to companies but to people. Business correspondence consists of written messages between human beings. It needs to be both professional and friendly at the same time.

10.3 LETTER FORMATS

Format includes the arrangement, shape, size, and general design of your document. You can use these elements to improve both the readability and impact of your letters. Wide margins, short paragraphs, and appropriate emphasis by headings, boldfacing, and underlining can help you to get your message through to your readers.

There are several formats which can be used for business letters. All of these formats, you will notice, provide key information which allows for accurate delivery, storage, and retrieval.

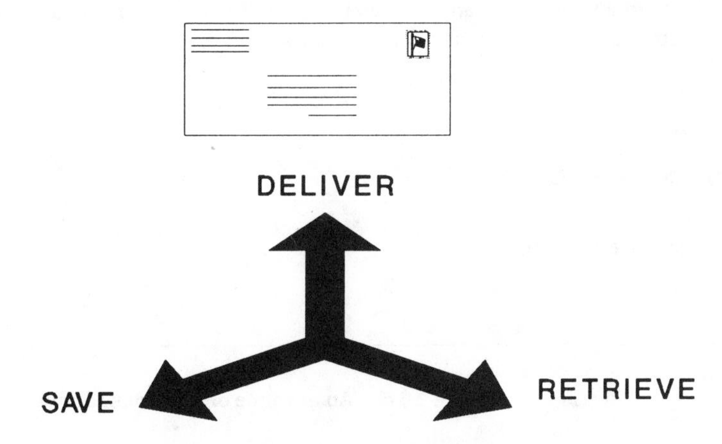

Figure 10-1 The cycle of use of a business letter.

The date, address, and signature are often used for filing and referencing. Attention and subject lines are sometimes used to assist the recipients. Like most technical documents, business letters are meant to be used and your format should be chosen to help your readers.

GATES & ASSOCIATES
4234 North Capitol Street
Washington DC 20002
(202) 617-2345

June 17, 1992

Mr. Simon Cortland
Cortland Manufacturing, Inc.
534 Drysdale Avenue
Leipzig, ND 23232

Subject: FAX Board Eprom Upgrade

Dear Mr. Cortland:

The fourteen Fax boards which we shipped to you on May 29th (Invoice #2862-R) were equipped with the wrong memory chips. The new eproms are being shipped to you by Delivery Express today.

I am sorry that this problem occurred. The difficulty was a small change in the subcontractor's assembly schedule, caused by the upgrade in eproms. Unfortunately, several dozen boards were shipped before the problem was discovered.

Again, I want to apologize for any inconvenience and thank you for your kind understanding. We will keep you informed about future upgrades.

Sincerely yours,

Sheila Lancaster

Sheila Lancaster
Quality Assurance Department

cc: Mr. Gates

Figure 10-2 An example of full block format.

GATES & ASSOCIATES
4234 North Capitol Street
Washington DC 20002
(202) 617-2345

June 6, 1992

Rensselaer Polytechnic Institute
Division of Continuing Education
RPI, Troy, NY 12181

Dear Persons,

I would like to obtain some information about your five-day Technical Writing Institute.

Gates and Associates, where I work in the mechanical engineering department, frequently produces reports and proposals for government agencies and private industry. I am looking for a short term course which will improve my ability to write quickly and with more confidence.

Please let me know if your program has any special prerequisites and if accommodations are available on campus.

Sincerely yours,
Marjorie Amman
Marjorie Amman

Figure 10-3 An example of modified block format.

The most common letter formats are the *full block, modified block,* and *modified block with indented paragraphs.* If your company has a uniform design for its letters, then obviously you should use that one. If your business does not have a preferred format, you can use any of these. (See Figures 10-2, 10-3, and 10-4.) All of them are correct.

What is most important to remember here is that your readers will be influenced by the design and appearance of your letters and technical documents. You are trying to communicate balance, pride, and careful attention to detail. The way your document looks will convey all of these things.

GATES & ASSOCIATES
4234 North Capitol Street
Washington DC 20002
(202) 617-2345

July 7, 1992

Dr. William Manchester
Infovision Animation Studios, Inc.
644 Mateos Boulevard
San Matinas, TX 39485

Dear Dr. Manchester,

Thank you for your kind assistance during the recent Computer Animation Conference. Your quick loan of a working VGA board saved my presentation.

As I told you, this was my first time through a public demonstration of the animation software. The difficulties I was experiencing with my VGA output would have forced me to cancel the session.

Again, I want to thank you very much. Your help was invaluable and your subsequent interest in the presentation caused me to feel much more confident.

Thank you,
Ai-Li Chin

Ai-Li Chin
Marketing Department

Figure 10-4 An example of modified block format with indented paragraphs.

10.4 PARTS OF THE BUSINESS LETTER

Letterhead stationery. Most companies use stationery which has the business name, address, and telephone number printed on it. If you use letterhead stationery, only the date must be added. An unblocked format works well with a centered letterhead. The block form is preferable with a left-side letterhead.

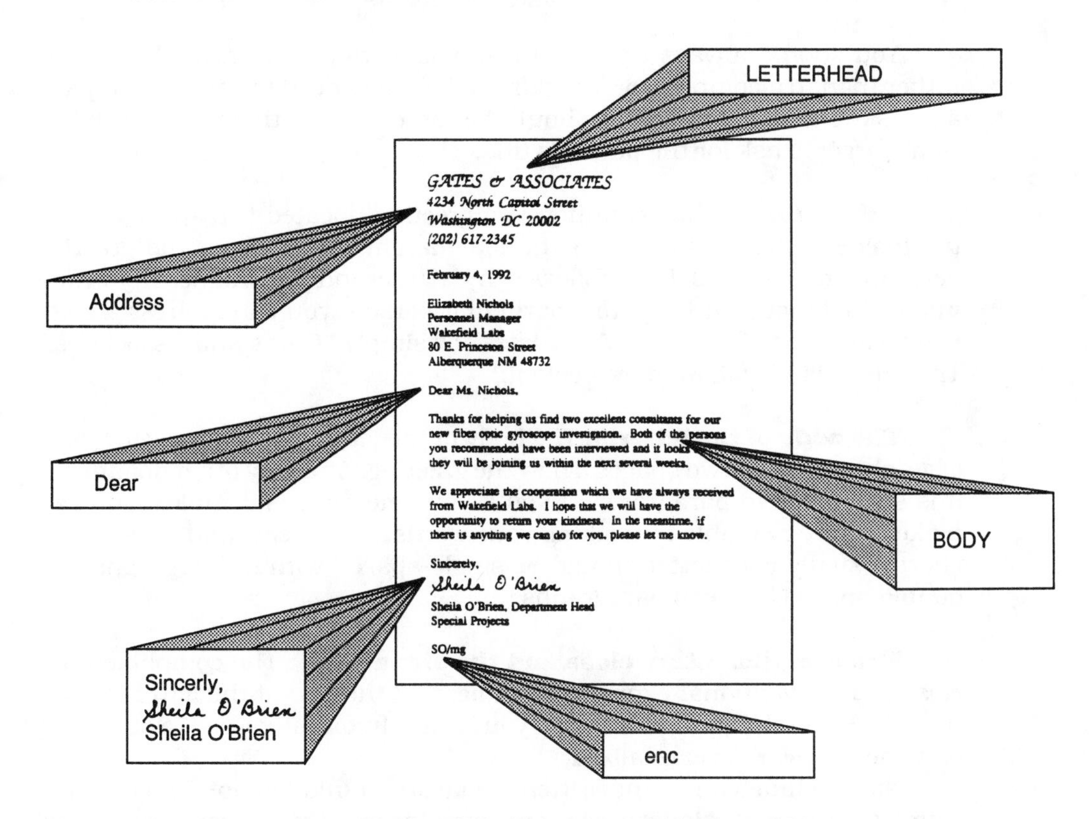

Figure 10-5 The parts of the business letter.

You use the letterhead for the first page alone. Additional pages are on plain paper.

Heading. If you do not have a letterhead, then you need to use a heading. This includes your complete address and the date. You do not include your name in the heading. The heading is located at the top of the page according to the format you select.

Inside address. The inside address is placed even with the left margin, at least two spaces below the heading. You should include the full name of the person or company and the complete mailing address:

Dr. Roy Suarez
Power Equipment Sales
487 Manila Avenue
Leadvale TE 37821

Topliffe Corporation
Marketing Department
95 Russell Road
Rawson ND 58831

You should always try to address your letter to a particular person rather than to a company or an address. You may need to call the company and ask to whom your letter should be directed, but this is worth doing. Don't forget to ask for the person's title.

Salutation. The salutation is a greeting, located two spaces under the inside address, flush with the left margin. Most often, the greeting begins with the word *Dear* followed by a title and last name. If you are unable to direct your letter to a particular person, you can address a company with "Dear Gates and Associates" or simply "Gates and Associates." The salutation is followed by a colon(:).

The body of the letter. The body of the letter, containing your message, should begin two spaces below the greeting. As is any other document, it is structured in paragraphs. Even if your message is complex, you can make it more readable by making your words, sentences, and paragraphs short. Usually your letter should be single-spaced within paragraphs and double-spaced between paragraphs.

The complimentary close and the signature. The complimentary close is a conventional expression, indicating the end of the letter. "Yours truly," "Sincerely," and "Sincerely yours" are frequently used. Notice that only the first word is capitalized.

Now include your handwritten signature in ink. Do not forget to sign your letter; a legible signature is considered necessary and courteous. Type your name and title underneath the signature.

Placement of this part of the letter depends on the format you have chosen.

Optional parts. Sometimes attention and subject lines are used to direct your letter to a particular department or person, or to save time and space.

Attention: Legal Dept. **Subject: VGA Warranty 371a**
Attn: Joan Fontanella **Re: 384K Multifunction Card**

Enclosure lines are used to let your reader know what you have included with your letter. For example, if you include a brochure or a check, you may want to indicate this:

Enclosure: Gates Software License Agreement

When you send a copy of your letter to another person, generally you indicate this with a copy notation:

cc: Ms. Ruth Hibbard

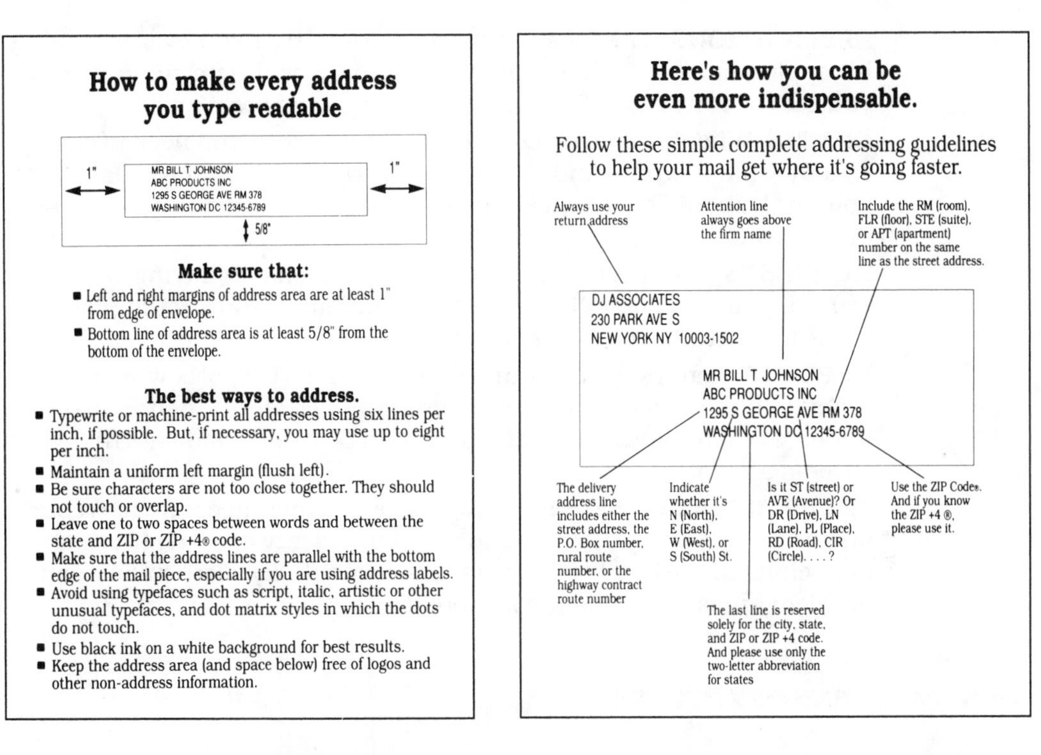

Figure 10-6 An example of a single-spaced envelope.
Reprinted with the permission of the U. S. Postal Service.

Envelope. The envelope has two regular parts: the outside address and the return address. Needless to say, both need to be correct. The outside address should be exactly the same as the inside address. Usually, business stationery includes preprinted envelopes. According to the U. S. Postal Service, single spacing allows for electronic sorting, providing you with quicker and more reliable deliveries.

10.5 *WHEN TO WRITE A LETTER*

People write letters to do all sorts of things: send greetings and news, exchange information, or just keep in touch. Business letters are written to get something done.

If you need to write a business letter, you need to begin with your purpose. Here are some common reasons for writing business letters:

ADJUSTMENTS Letters are frequently used to deal with customer problems. Acknowledge the difficulty and describe clearly what steps you will take to resolve it. If you do not intend to do what your customer requests, explain why not in detail.

APPLICATIONS The most important letters that you ever write may be applications for employment. Your letter will determine whether you get the interview. (*See Chapter 11, Applying For Work.*)

CLAIMS When you have a complaint or a problem, you need to explain exactly what is wrong and then state clearly and definitely what action you think should be taken. Be polite. You need to decide between anger and action.

REQUESTS You need to be precise and specific in saying what you want. If you are requesting more than one thing, you should place the items in a list. Remember to say thanks.

REPORTS Letters are sometimes used for short reports which do not include illustrations or extensive graphics. Headings can be used to improve readability. A subject line should be used to indicate that your letter is a report.

TRANSMITTALS These are cover letters which accompany reports and proposals. They allow you to make necessary comments which do not belong in the body of your main document. Your letter can identify problem areas or emphasize key points.

10.6 PATTERNS OF LETTERS

For all of the variety in types of letters, most letters come down to a simple pattern: open, body, close. This is a three stage process in which you state, elaborate, and summarize your message. A good formula for letter-writing: "Say what you are going to say, say it, and then remind people of what you have just said."

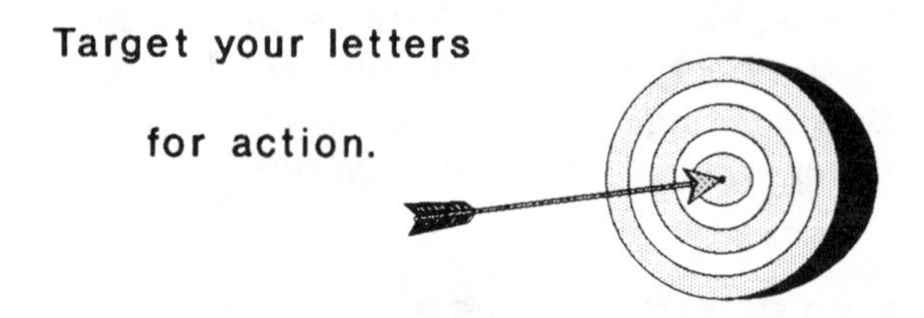
Target your letters
for action.

Open Here's what I'm going to say.

Body Now I'm saying it.

Close Here's what I just said.

You begin by telling your reader why you are writing, your purpose, what you want to happen. A good way to open is to say, "The purpose of this letter is to. . . ." Then tell the reader whether you are writing to inform, request, complain, adjust, support a particular decision, or something else.

What do you want? Why did you write the letter and why should someone finish reading it?

In the body of your letter, you exercise your judgment. You have to decide how much detail is sufficient, what to include and what to leave out. In general, if you are conveying bad news, a refusal for instance, then it is courteous to explain in detail. Good news requires less supporting information.

Technical explanations should be handled in as simple and clear a manner as you can devise. You have to consider what your audience already knows while you are deciding on the level of detail. Try to include enough information so that a second letter or explanation will not be necessary.

Something that few letter writers consider is the value of an enclosure. If you have brochures or samples, something that can be seen and examined physically, send this with your letter. Always include a letter with an enclosure. The letter is evidence of your interest and attention.

Your close has to tell your reader exactly what you want. You should summarize and restate your purpose. Everything in the letter has led up to the conclusion. You need to be direct at this point. You cannot always rely on your reader to understand you. What you want needs to be clear, definite, and specific.

A good letter is like a boomerang. It returns to its starting point. Like a circle, you need to finish by returning to your beginning. Summarize your purpose and directly state what you want to happen.

10.7 HOW TO WRITE A BUSINESS LETTER

Imagine. The first step in writing a business letter is to try to imagine your reader. You start by studying your audience. This is what a coach does when she analyzes the opposing team, or what a business woman does when she evaluates her competition.

Begin by visualizing your audience. This will make the writing more personal, friendly, and understandable. Remember, you write letters to people, not to companies.

Begin with the ending. When faced with difficult subjects, begin at the end. If you write the endings to your letters first, the beginnings will come much easier. In a letter everything leads up to the conclusion. The best way to think about how to begin your letter is to think about the ending. What do you want to happen? Business letters are written to get something done.

Make it readable. The best way to make your letters readable is to keep your words, sentences, and paragraphs short, even if the letter is long. Be sure to repeat your major ideas.

Include keep-on-reading signs. Your letters should provide reasons to keep reading. Most people will be interested if you are writing about their problems, their benefits, or their questions. These are topics which will sharpen your readers' focus. The more you remind your readers of the benefits to them, the closer their attention to your message.

Stay on the trail. When you are writing a letter for business reasons you need to stick to your subject. Don't mix things together. Separate matters should be taken up in separate letters. The sudden introduction of new subjects will confuse your readers, slow their response, and perhaps cause them to stop reading. Don't let every thought that occurs to you become a detour from your path. Stay on the trail.

Stick to the facts. Writing is a way of conversing with someone who is not there with you. Your written words cannot respond to your reader's confusion or disbelief. You need to anticipate and answer objections before they actually occur. You are better off to write only what you can support with evidence. If you can demonstrate something, it is a fair and factual claim.

10.8 YOU COMES BEFORE I

When you spell the word "business" *u* comes before *i*. When you are writing letters which are meant to do business, YOU comes before I. This is a very important point to remember when you are writing letters which are intended to get something done.

If you need to refer to yourself, of course you can. You should feel free to use the word "I" but use it no more than once in a sentence, and not in every sentence. The pronouns you use in your business letters signal your attitude to the persons who read them. Don't fill the message with references to yourself and your problems.

Instead, keep your focus on the reader and try to write letters from the other person's perspective. This means understanding and responding to problems from your reader's point of view. Leave yourself out and put the reader in when you are writing a business letter.

Make an effort to imagine why someone is upset or angry about something. When someone takes the trouble to write a letter of complaint, then they are certainly convinced that they are correct. You should show attention and concern.

Whether you convey concern or interest, you need to understand if you want to be understood. Try to imagine your audience and put yourself in the place of the reader. That person may have different problems, needs, and responsibilities than you. You need to recognize these differences in order to communicate effectively.

10.9 BUILDING CREDIBILITY

Building credibility with your readers and listeners is a step-by-step process. Each step has to be correct. Perception is everything here. Many readers see a grammar mistake, a misspelled word, or even a typographical error as evidence that your whole document is careless and wrong. Even if their perception is unfair, it is a fact of life.

Letters need to be prepared very carefully because they are read very carefully, sometimes again and again. They have to withstand this kind of scrutiny.

If your readers distrust what you write, your letter cannot respond to their disbelief. It cannot explain names that are misspelled, statements that are exaggerated, nor factual errors. Mistakes cost credibility.

Your goal should be to create letters that build confidence in your ability to get a job done, to get a task finished. Here are some effective techniques which will work to create confidence in your readers:

Fast answer. A quick response is one of the most effective techniques a writer can use. The faster you respond to a letter, request, or telephone call, the closer you respond to what motivated the person who tried to reach you. Even if your message is negative, a rapid response will be appreciated as an indication that you are trying hard.

Good paper. It is difficult to underestimate the importance of a good letterhead and good quality rag-content bond paper. The letter represents you. Envelopes and stationery should match. Your letter should have a professional appearance.

Direct language. You want to use a low-key, reasonable approach to the readers' interests. Be straightforward. Tell facts simply and directly. This will build credibility with your readers.

In business, written communications are more formal than when we are talking. "No problem!" is an acceptable response when you are speaking on the phone, but in a letter may be inappropriate and even upsetting. A respectful professional style is the correct style for business letters to persons, especially when you are not friendly with them already.

Tone. Remembering that you are a person and that you are talking to people may be the most effective technique for building trust. If you are friendly when you write to people, even when you have to convey bad news, they will recognize your attitude and credit you for it. If you want people to feel friendly toward you, then you need to feel friendly toward them.

Brevity. "I've made this letter longer than usual," Pascal once apologized, "because I lacked the time to make it short." It takes work and effort to design a letter which is brief and to the point. It is easier to put extra information in or leave things out than to write a short letter.

Two things you should remember: busy people get lots of letters; and, busy people are very busy. They expect you to be direct and to-the-point.

10.10 SAYING NO AND SAYING YES

One of the most difficult letter-writing tasks is having to say no. Whether you are refusing to honor a warranty claim or rejecting a proposal, you want to make particular efforts to avoid offending your audience. You don't want to blame or embarrass the person who writes to complain or ask for something you will not do.

If someone takes the trouble to write, they are sure they are correct. If you have to disagree with them, your letter should convey your sympathetic and helpful attitude. You want to explain your refusal from your reader's viewpoint. The best way to convey bad news is with this simple formula: Thanks, Sorry, Because, Thanks. Surround your negative message with courtesy and polite attention.

THANKS: "Thank you for the opportunity to explain this situation." This is the type of positive attitude you want to communicate.

If you are responding to an angry complaint, an effective technique is to restate the complaint and then offer a complete explanation. By repeating the complaint accurately, you demonstrate that you understand the situation and are giving it your full attention. You are not just sending a form letter back to an address.

SORRY: There is nothing wrong with sending an apology. You can simply say you are sorry and move directly to the bad news. If you need to apologize, do so.

BECAUSE: You should always explain a refusal in detail. Your effort to explain gives evidence of your sympathy and concern. A brief refusal is seen as abrupt and dismissive. A no letter requires explanation. Give reasons, details, specifics, and support for your decision. The more trouble you take to explain, the more your reader will appreciate your attitude.

THANKS: You should end your letter with courtesy, and appreciation. Say again that you are sorry for the circumstances and offer to do whatever you can. This will be remembered along with your negative message.

Then there is the business of saying yes. When you are agreeing to a customer's request, say so politely. Many people think that if they are saying yes, they can be as curt and abrupt as possible. Instead, think of this as an opportunity to gather points with your readers.

Politeness Quiz: Which should go before c? b or a?

A: "You claim your electronic ignition failed."

or

B: "I understand that your electronic ignition failed."

before

C: "I agree to repair it under the warranty."

In other words, don't be afraid to be a pleasant human being just because you have to write a letter. A negative attitude can destroy the effect of the kindest agreement. Remember, the tone you use for letters should represent your company's attitude toward its clients, not your mood while writing. Every letter is a public relations effort.

Crossing the international dateline can cost more time than you realize. In the U.S. 5/7/90 will be read as May 7, 1990, but in Europe and Latin America, it will be understood as July 5, 1990. For international mail, write the date out.

10.11 FORM LETTERS

Occasionally you will see advertising for pre-written business letters. These ads suggest that you will never have to write another business letter. You can just copy your letters from diskette or book. You simply make minor changes like names and addresses.

1~ 2~ 3~
4~
5~, 6~ 7~

Dear 1~ 2~:

Gates & Associates will be holding our annual information meeting for subcontractors on Friday morning, November 6, 10 A.M. We will be placing particular emphasis pollution control projects which we anticipate during the coming year.

We hope you will be able to attend.

Cordially,

Sheila O'Brien
Special Projects

Mr. James Duggan
80 Princeton Street
Jackson, VA 47376

Dear Mr. Duggan:

Gates & Associates will be holding our annual information meeting for subcontractors on Friday morning, November 6, 10 A.M. We will be placing particular emphasis pollution control projects which we anticipate during the coming year.

We hope you will be able to attend.

Cordially,

Sheila O'Brien

Sheila O'Brien
Special Projects

Sorry about the form letter. I have to notify everyone. Will call you next week.

Sheila.

Figure 10-7 A form letter with a personalized note.

If you consider this product seriously, you can see how unlikely it is a solution to your need for effective correspondence. Real business letters have to be designed for particular situations and persons.

Yet, some people treat letters as if they were simple commodities, to be taken from the shelf, dusted off, stuffed in an envelope, and sent on their way. You do not want to be represented by the equivalent of a mimeographed letter. Writing letters to people is not a recitation where you repeat rote phrases. Every letter is an opportunity to make a good impression. People who receive your letters will make judgments about you and your abilities.

This does not mean that you should never use form letters. Sometimes you need to communicate the same message to lots of different people or you find yourself writing the same letter on a frequent basis. The merge and file functions of most word processors allow you to save and re-use effective letters.

You need to remember though that you are writing to people. You should sound like a human being, not a phrase-making computer program. A personal letter is more effective than a form letter, but a form letter can be personalized.

10.12 CHOOSING AN ATTITUDE

Think before, not while you are writing your letter. The best way to write a good letter is to know what you want. You need to decide between feelings and action. If you are writing a letter to show that you are upset, there is little reason to expect it will get the job done.

You can't be angry with people you want to cooperate with you. Instead, you need to give your readers the opportunity to respond in a positive way. As we said before in this chapter: if you want people to feel friendly toward you, then you have to feel friendly toward them.

Be definite. Say what you want done. When you are writing a letter, discuss when your reader will do something rather than if your reader will do something. You want to focus your reader's attention on what you want done in a pleasant and positive way.

10.13 A FINAL WORD

One type of letter that is not mentioned or practiced as frequently as it should be is the letter of appreciation. You can write a letter to thank people, or to praise them for something well done, or to show your interest in what they are doing. People will appreciate and remember you.

Businesses which receive a high volume of mail tend to value positive letters more highly than negative ones. It takes a high degree of motivation to sit down and write a letter to praise some excellent service, quality, or attention to detail.

Too often we write more complaints than compliments. It is worth considering the pleasure we get from other people recognizing that we are doing a good job. Honest praise and sincere compliments will be valued.

BUSINESS LETTER EXERCISES

1. You receive a letter from Gates and Associates which says they have narrowed down their candidate search to three persons. They are impressed with your resume and your performance during the two interviews you had.

   The hiring process has reached a snag, however, because Mr. Gates has raised a question about communication skills. He wants a letter from each candidate briefly describing their job goals and where they want to be in five years.

   Obviously, this is more than a simple task. You need to choose a style and strategy. You want to sound energetic, lively, practical, and competent. The head of the company will be reviewing your work and you will be creating a long-term impression.
2. Write a letter requesting this Request for Proposal:

> The Commonwealth Bay Transportation Authority invites interested parties to respond to a Request for Proposals for the establishment of an Authorized Licensing Agent for Commercial Use of CBTA designs (logos). The Request for Proposals may be obtained during normal business hours hours commencing on October 27, 199_ from: CBTA- Department of Real Estate Management 10 Broadway Room 2520 Boston MA 02117. Responses to the Request for Proposals will be accepted until 4:00 PM on December 6, 199_ . Interested parties may attend a pre-bid conference on November 8, 199_ at 10: 00 AM in Conference Room 6 on the second floor of the Transportation Building, 10 Broadway, Boston. For further information please contact Herman Wells, Assistant Director, Real Estate Management, 10 Broadway, Boston MA.

3. Find the name of a professional society or trade association in your area. Choose a letter format and write an inquiry letter asking about the requirements and costs for membership. Ask if there are special rates for students or new members.

4. At a recent corporate meeting, Gates and Associates agreed to change the policy regarding the lending of company-developed software. The issue is expert systems, databases and knowledge-based software programs designed to quickly organize information resources in disaster situations.

The company has had a generous policy of loaning these programs for temporary use to government agencies and commercial clients. We have spent a lot of money developing these expert systems. The software code represents a detailed blueprint of our company's approach to certain types of environmental and engineering problems. This information would be very valuable to company competitors. Its general availability would harm the company's chances of successfully bidding on certain contracts.

When a company engineer discovered an unauthorized copy of one of these programs had been installed on the mainframe system of a government agency we have been working with, we decided to have a strict software lending policy:

> "Proprietary (company-developed) software will not be made available to anyone outside the company unless it is copy-protected."

Unfortunately, this policy creates a serious problem for one of our most important clients, Kern Cranes, Inc. Copy-protected software is difficult and expensive to customize for clients and adapt to specific problems.

For the last eight years the Kern Cranes, Inc. has used Gates' software to analyze construction crane failures. They employ three full-time persons who were hired to adapt our programs for this purpose. Our new policy will cause them considerable inconvenience in terms of time and expense. We may be forcing them to lay off three employees as well.

Patti McDevitt, General Manager of Kern Cranes, has sent a letter to Mr. Gates which states in part:

> "Although we highly value our continuing relationship with Gates and Associates, this new copy-protection scheme is unworkable for our needs, and we will have to turn to another software provider."

While he is sympathetic to their problem, Mr. Gates does not feel that we can make any exceptions to the new policy for at least a few months. He asks you to write a letter, for his signature, explaining this position to Ms. McDevitt.

5. List some of the things which are wrong with the format and appearance of the following letter. What impression does it create?

Ron Chimer
7 Seaver Str.
Boxwood, Md

Dear Mr Gaits,
My father suggested I write to you. He met you at some kind of enginering conferenc e last year. I am lookig for a job which would suit me over the summer months. I am in college but I don;t take classes in the summer because that is the time to relax a little. The kind of job I am looking for would involve me meeting with people. My teachers say I am looking f good at this and working with people is my ambition.
If your interested in hiring me for your company,
please let me know as soon as possible as I need to make plans. I am a hard worker and very eager to get a weekly paycheck.

Yours very Truly,

Ronald Chimer

PS My dads name is William.

6. A friend of yours who works at another company tells you about a post office program to update company mailing lists and avoid delivery problems. Contact your local post office, find out about this program, and write a memo to Mr. Gates suggesting that we use this free postal service to improve the efficiency of our mailings.
7. Using your own word processor, investigate the mail/merge functions. Most word processors allow you to write one letter and combine it with a list of names and addresses to create many letters. Can you write a letter to five different persons which still sounds personal and specific? Can you write a form letter which will not be recognized as a form letter?
8. Write a complimentary thank-you letter to a person who has been helpful to you. Thank them for their assistance and let them know that you appreciate what they have done.
9. Sheila O'Brien asks you to design a form letter which will be sent out by Mr. Gates to all of our clients, informing them that the company will be opening a branch office in Budapest, Hungary. We are going to aggressively pursue pollution control contracts throughout Eastern Europe and we expect to be involved in some large scale projects. The office will be located at 17 Tad Zols Square, Budapest, Hungary.
10. What circumstances would make you decide on a letter rather than a telephone call? Make a list of possible circumstances, then change your list into a set of procedures for general use within Gates and Associates.

11. You have been offered a very attractive position in a technical company in your spouse's hometown. Although you are regretful, you decide this is too good an opportunity to let pass. Both you and your spouse would like to live near your relatives and you prefer country life to living in Washington, D.C.

Write a letter to Gates and Associates resigning your position. Try to make sure that your tone expresses your sincere regret, along with your determination to do what thing is right for your family and yourself. Leave a good impression behind your exit.

12. This may be your most difficult letter writing assignment. Write a letter to your (potential) child, or a child you want to influence, explaining why you would recommend or not recommend a career in technology. Imagine that this child is ten years old and that you sincerely want to influence a decision.

11

Applying for Work

Note to Users: **This is a task chapter. It is designed to help you create effective application letters and a resume which will support your job search.**

11.1 INTRODUCTION

In today's technical job market, an attractive, well-designed resume and application letters are essential tools for your job search. They have one simple purpose: to get you an interview. Very few persons are hired on the simple basis of a letter or resume, but your demonstrated communication skills in these documents can gain you a face-to-face meeting where you can make your own case for the job.

The job search is a competitive process with many applicants for every position. By one estimate the average employment manager reviews almost one hundred resumes for each hire.

An initial review will screen out underqualified candidates, so you need to be realistic with your applications. This means careful self-assessment and investigation of the available job market. Preparing your resume, and following through with letters of application, interviews and follow-up letters are activities which deserve your closest attention.

In this chapter you will learn how to:

- Inventory your qualifications.
- Investigate the employment market.
- Choose the right resume.
- Prepare for the job interview.
- Keep credentials up-to-date.

11.2 DEFINITION

A systematic job search, with effective documentation, involves a series of steps:

- **self-assessment**
- **investigation of the job market**
- **resume design**
- **letters of application**
- **preparation for the interview**
- **follow-up**

Each of these steps is important and will affect whether you achieve the position, salary, and location you are seeking.

You need to begin your job search with a realistic inventory of your skills, abilities and interests, an honest **self-assessment.** You have to know your own abilities and accomplishments. You need to understand what type of job you want and whether you have the right qualifications. Be realistic and know what employers are looking for.

A thorough job search involves an **investigation** of the job market. This means using all available sources of information. Study the want ads.

Talk to professionals. Visit the library and do some research. The more you know about what employers want, the more you will be able to match your abilities to their needs.

The two most common job application documents are resumes and cover letters. Design both to showcase your skills and abilities and to get you the interview. Busy managers will spend less than thirty seconds examining your credentials. Format is crucial. These documents must have a professional appearance because they are used to estimate your ability to do the job.

Your **resume** is many different things: an ad for your services, a summary of your experience and qualifications, a description of your skills, and a background sheet. Your resume should be a persuasive document. It is designed to sell you and your skills and services to the reader. Since the resume supplies the reader with a one- or two-page reference for you, it must be attractive and error-free. Mistakes, clutter, or misrepresentations tell the reader your document is not credible.

Letters of application are sales letters. You are marketing yourself. You are trying to persuade the person who screens the resumes that you are qualified for serious consideration. Focus on what you can do for the organization to which you are applying. The purpose of the letter is to create enough interest to get you the interview. Your purpose is to make someone want to meet with you.

Keep your application letters short. Express your interest in the position, highlight any particular strengths on your resume, and ask for the opportunity of an interview. This is all that the letter has to do. Three paragraphs, each with its own task, will accomplish your purpose.

Here, for example, is a want ad and a letter of application.

FIELD SERVICE ENGINEER
DC firm seeks Field Service Engineer w/PC & Network experience. Sal., benefits commensurate with background and ability. Engineering consultant support. Written applications only to Sheila O'Brien, Gates & Associates, 4234 North Capitol Street, Washington DC 20002

21 Hillside Road
Lower Fields, MD, 55555
June 17, 1992

Ms. Sheila O'Brien
Gates & Associates
4234 North Capitol Street
Washington DC 20002

Dear Ms. O'Brien:

The May issue of Women in Engineering magazine carries your advertisement for a Field Service Engineer. I would like to apply for that position.

Enclosed you will find my resume which details my work and educational background. You will notice that I have worked with field engineers and a variety of commercial software packages. I am confident that I can learn new applications quickly and reliably.

I live close to the Washington area and am available for an interview at your convenience. I would be pleased to bring samples of my written reports and CAD drawings. Thank you for your consideration.

Sincerely yours,

Susan Kehoe

Susan Kehoe

Preparation for the interview means learning as much as you can about your prospective employers. This involves two activities: researching and anticipating. You need to know as much as you can about the company you are applying to. You need to expect certain types of questions and be ready with thoughtful answers.

Do your homework. Know what the company does, what it produces, and what is the competition. Interest and curiosity are a sincere form of flattery and you should not go to an interview if you are not sure what the company does.

Send the interviewer a **follow-up** letter two or three days after your interview. Express your thanks for the interview and provide any further information which may have been requested. These letters are not only courteous but practical. They bring your name and qualifications to the attention of the interviewer for a second time.

11.3 SELF-ASSESSMENT

Before beginning your job search, you need to conduct a rigorous and realistic inventory of your skills, abilities, and qualifications. A careful analysis of what you are qualified to do and what you want to do will help you with every stage of your career planning from resume preparation through the interview process. Knowing what you want is very important. Too many persons make the mistake of trying to channel their own objectives into whatever vacancies happen to exist at the moment. You may have to take temporary jobs, but you should try to decide what you would enjoy doing, what you would like for a career. Then you need to determine if you have the credentials and qualifications.

This means comparing your work experience, education and training, and other skills to the requirements of the jobs you want. Unless you have persuasive substitutions, there is little sense in applying for jobs which require credentials or experience you simply do not have.

11.4 INVESTIGATION

You need to be familiar with the current job market. This means establishing an information network which will keep you up-to-date on job qualifications, employment trends, and sudden opportunities. When you are aware of what employers require, you can focus your course work, work experience, and credentials to meet their needs.

Use all of the resources available to you. Browse through the *help wanted pages* in your Sunday newspaper and become familiar with local salary levels and job qualifications. Use the *library* at your school, company, or hometown to help in your job search. *Occupational handbooks, professional journals,* and *government publications* all provide information about work opportunities.

Another important source of information for your job search is expert advice. Join the local chapter of the *professional organization* in your field of interest. Attend their meetings and talk to people who are doing what you would like to do. Practical advice and inside job leads are only two of the benefits you will receive. Membership in these groups looks good on your resume.

AMERICAN SOCIETY OF MECHANICAL ENGINEERS (ASME)
INSTITUTE OF ELECTRICAL AND ELECTRONICS ENGINEERS (IEEE)
SOCIETY OF WOMEN ENGINEERS (SWE)
AMERICAN INSTITUTE OF ARCHITECTURE STUDENTS (AIAS)
ASSOCIATED BUILDERS AND CONTRACTORS (ABC)
ASSOCIATION FOR COMPUTING MACHINERY (ACM)
AMERICAN SOCIETY OF CIVIL ENGINEERS (ASCE)
SOCIETY OF MANUFACTURING ENGINEERS (SME)

Don't forget to use the placement services at your school or the *Human Resources Department* where you work. Placement offices can provide useful counseling and suggestions about the job search process. Campus interviews provide an opportunity to meet company representatives in a comfortable setting. Human resources personnel have a strong interest in promoting and training individuals from within the organization, so take advantage of their help.

11.5 TYPES OF APPLICATION LETTERS

Letters of application are always sales letters. What you are selling is your ability to do a job. The letter is your sales representative. It speaks for you and represents you. If the letter catches the reader's interest and the resume supports this initial response, you will get the interview.

Remember, potential employers will examine your letter carefully to see whether you can communicate effectively in writing. The quality of your application will indicate your ability to do the job.

Prospecting Applications

A prospective employer is someone you have targeted to ask for employment. Prospecting is a description used for unsolicited letters, where you send out an inquiry without being asked in the hopes that a job might be available. Large firms receive many applications like this every day, so your letter and resume need to be perfect for you to have a chance of success. You are advertising for an interview.

You need to start out with your central selling point, whether it is experience or education, and quickly explain how your abilities can benefit the company. Make it clear that you are looking for a specific job, not just work. Substantiate your abilities by pointing to accomplishments which can be verified. In other words, prove yourself.

One way to make a good impression is to demonstrate that effort and research have gone into your letter. Address your letter to the correct person within the company if you can obtain this information. Show that you are aware of the company's operations, products, services, and competitors. Your letter should highlight information in your resume, not provide details. The letter is a first step, pointing to the resume, not replacing it.

Invited Applications

When you are invited to apply for a job, whether through an ad or personal request, you should begin by referring to the source of your invitation. Be sure to mention the specific job you are applying for. This benefits the reader who may be responsible for filling many different positions and who is interested in knowing how you heard of the opening.

Avoid specifics about salary until the interview. If the job description asks for salary information, use a vague phrase like "your usual rate" to postpone this topic for the interview. Instead, use your letter to connect your education and experience to the job. Readers are less interested in your general background than your ability to help them with a particular problem. Try to focus on the benefits your employment will bring to the reader. You want the reader to want to meet with you.

Use the letter to explain briefly how your interests, abilities, and experience match the job or career you are seeking. Make yourself stand out from your competition by focusing attention on your resume and a specific skill or ability. Connect your capabilities with the employer's needs.

Follow-Up Letters

Following the interview, you should compose a letter thanking the interviewer and indicating your continued interest in the job. Your follow-up letter should refer to the interview or some common interest. Briefly refer to the benefits your employer will receive if you are hired. The simple courtesy of a thank you will help you stand out from other candidates for the same job.

11.6 TYPES OF RESUMES

Your resume is a sales document which needs to be designed for a quick sale. On average, the reader will spend less than thirty seconds before deciding whether you will be interviewed. Some person or persons will narrow the selection of applicants to those who meet specific qualifications. Your resume needs to persuade this limited audience that you should be considered. Keep this audience in mind.

These busy readers want information quickly, so use conventional formats. Your audience is familiar with these patterns and can go right to what they want to find. Almost all resumes use the same parts, but the arrangement and focus are shifted to emphasize your strong points. In general, there are three types of resumes: the chronological, the functional, and the targeted resume.

LOURDES SANTIAGO
410 Brown Street
Butler Springs, GA 55555
(555) 434-7897

OBJECTIVE: Seeking a challenging position in the field of electrical engineering/computer science where I can use my skills and experience.

EDUCATION: **SOUTHERN INSTITUTE OF TECHNOLOGY, MARIETTA, GA.**
Candidate for a Bachelor of Science Degree in Electrical Engineering Technology to be awarded in May, 1992. Dean's List; G.P.A. 3.1

Courses included instruction and hands-on experience in the following areas:

Linear Circuit Analysis	Fiber Optics
Instrumentation	Semiconductor Theory
Microwaves	C Language
Robotics	Automated Processes

WORK EXPERIENCE: **Diprete Research Associates**
Loudon, GA June 1991 - August 1992 (part-time)
Research Technician -- Worked with project team to develop software controls for automated textile cutters. Responsibilities included software testing and debugging as well as hardware modifications.

Marchionda Controls, Inc.
Marietta, GA June 1990 -- September 1990
Instrumentation Technician -- Tested, calibrated and maintained instruments for 17 person research project. Duties included lab manual preparation and client presentations. Responsible for all graphics in final project report.

Radio Shack
Marietta, GA April 1989 -- June 1990
Salesclerk -- Assisted customers in choosing various components for home electronic systems.

AFFILIATIONS: IEEE (student member)
Macintosh Club, President (senior year)
Society of Women Engineers (SWE) 3 years

- REFERENCES AVAILABLE UPON REQUEST -

Figure 11-1 A traditional resume.

WILLIAM GIBSON
52 Waquene Street
Sierra Vista, AZ 85635
(602) 555-2864

PROFESSIONAL WORK EXPERIENCE

FEBRUARY 1989 -- PRESENT

Wakefield Labs, 893 Paseo San Marco, Sierra Vista, AZ 85635
Field Service Engineer, Site Manager
Responsible for the set-up, operation, and repair of 20 ruggedized Micro-Vax II computers used for testing the Mobile Subscriber Equipment communication system for the U.S.Army at Ford Hood, TX and GTE, Taunton, MA. Troubleshoot and correct system problems as they occur. Supervise instrumentation technicians. Responsible for all administrative and logistic functions associated with site operation. Maintain spares inventory.

JULY 1987 -- JANUARY 1989

Communications Learning Systems, 5 Lee Road, Canton, MA 02021
Computer Test Technician
Responsible for testing and repair of microprocessor-based barcode data collection terminals. Troubleshoot to component level and tested PCB's using a Zhentel automated test machine. Responsible for meeting customer orders.

JUNE 1985 -- JUNE 1987

Bramhall Corporation, 80 Main Street, Nashua NH
System test and integration technician
Assembled, integrated, and tested digital imaging subsystems consisting of a DEC Micro-Vax II, Dec terminals, and a modified backplane enclosure for use in scientific field experiments and research.

EDUCATION

Lowell Institute School, MIT campus, Cambridge, MA
January - May 1986 Printed Circuit Design
September - December 1986 Digital Electronics

GTE Sylvania Technical School, Waltham, MA
Graduated June 1985 with a Certificate in Computer Electronics
Perfect attendance and 92 percent grade average

REFERENCES

Available on request

Figure 11-2 A chronological resume.

Susan Kehoe
21 Hillside Road
Lower Fields, MD 55555
(843) 555-6394

OBJECTIVES

Entry level position with Engineering Consultants firm where I can gain experience and continue my management studies towards an M.B.A.

EDUCATION

B.S., Technical Management — Wentworth Institute of Technology, Boston, June, 1991.
G.P.A. 3.68/4.0. Concentration: Computer Studies

QUALIFICATIONS

* Worked extensively with Wordperfect, Lotus 1-2-3, Harvard Graphics, and current releases of Autocad
* Experienced with the operation of plotters, scanners, and fax machines
* Excellent communication skills, written and oral
* Prepared and delivered career information seminars for the 2nd and 3rd annual Women in Technology Conference

EXPERIENCE

Engineering Support Technician, Weston Labs, Massachusetts (two semester co-ops)
Worked with field consultant team to develop installation and training procedures for new production line automated systems. Maintained computer communications with home plant engineers; responsible for quality control reporting, client presentations, and all stages of technical documentation.

ACTIVITIES

Student Investment Society (Vice-President) senior year
Society of Women Engineers (SWE) active member three years

INTERESTS

Sailing, horseback competitions, computer bulletin boards

REFERENCES

Furnished on request

Figure 11-3 A functional resume.

Mark Nguyen
33 Woodchester Drive
Milton, Oregon 97601 Telephone 803-555-1234

Objective
Seeking an entry-level position in engineering consulting firm, with opportunity for advancement. Willing to relocate.

Education
Oregon Institute of Technology, Klamath Falls, OR
Bachelor of Science in Industrial Design Technology with a minor in International Studies, June 1992

Engineering Field Support Experience

Mechanical Engineering Department (work study, Spring 1992)
* Wrote efficiency/cost analysis for proposed telephone system
* Managed department data collection, storage and retrieval
* Coordinated presentation for Industrial Cooperation Day
* Conducted preliminary study of video/computer projection systems
* Presented study findings to Engineering Department Heads

St. Elizabeth's Hospital, Rehabilitation Unit (Summer 1991)
* Fabricated mechanical devices to facilitate computer use
* Designed software to scale monitor text (Pascal language)
* Collaborated on writing teams for three NIH proposals
* Presented the Mechanical Assistance Project to business groups

Pioneer Telecommunications, Inc. (Summer 1990)
* Wrote technical proposals for telecommunication upgrades
* Prepared Cad schematics for fiber optic ISDN networks
* Presented company proposals at three qualification meetings
* Designed on-screen documentation for graphics package

Awards and Activities
* President, Vietnamese Students Club, 1991-92
* Calculus Tutor in Learning Center, four hours per week
* Dean's List, all semesters
* First Place, Annual Egg Drop Competition, 1990

References
Furnished on request

Figure 11-4 A targeted resume.

Chronological or **traditional** resumes list your education and work experience beginning with your present situation and moving backwards. Time is the organizing principle. Many reviewers prefer this style because they can go right to the information they need.

Functional resumes arrange work experience by function rather than time. In other words, you list the functions you have performed rather than the order or length of time you have done them. This is a useful style for persons who want to focus on what they can do, rather than where they have worked.

Targeted resumes incorporate elements from both chronological and functional resumes. Using a traditional organization pattern, you target a specific job by focusing on skills and abilities related to that job. All of the information you provide is aimed to this end. Work experience, education, and personal activities are all used to support your interest in a particular position.

11.7 PARTS OF THE RESUME

Whichever type of resume you select to present yourself to employers, you will include the same categories of information:

- **name and address**
- **career objective**
- **educational background**
- **work experience**
- **personal interests, awards, and special skills**

The heading should contain your **name** and **address.** This needs to be complete and accurate. Provide your zip code. Include a telephone number where you can be reached, along with the area code and extension, if needed. Make sure that possible employers can reach you easily. Center the heading. Put your name in all caps and use boldface so that it will stand out. Don't center any other parts of the resume because centered text is more difficult to read.

BONNIE J. KERN
237 Los Almadeiros Highway
San Pedro, CA 90731
(000) 000-0000

Your **career objective** should show a clear sense of purpose and a long-range focus. Your objective provides organization and coherence to the rest of the resume. State your immediate objective and then what you would like to be doing in the next five to ten years. Make your objective broad enough so you can be considered for more than one job, but specific enough to indicate your interests and aptitudes. One effective strategy is to state two goals: short-term and long-range. This can indicate thoughtfulness, ambition, and a defined career path.

> **Career Objectives:** Initial employment in the field of mechanical design and fabrication; long-term goal: supervisory management.

It is your **educational background** which will qualify you for many technical jobs. List your education and training in reverse order, beginning with the most recent. Name the school or workshop; the city where it is located; the degree, certificate, or credits you earned; and the date you participated or graduated. First-time applicants often include their high school and the dates attended.

Work experience, particularly if it relates to the job you are seeking, will provide strong support for your application. List the jobs, employers, concise descriptions of your responsibilities and accomplishments, and the dates of employment. The resume is a brief reference sheet so you need to focus attention on your ability to get things done. Use action words, verbs which emphasize the characteristics, skills, and abilities called for in the job description.

If you are designing a functional resume, list the activities you have performed instead of listing the jobs. Use categories like Management, Sales, and Training to group skills and functions for easy reference. This is a good arrangement if you want to emphasize your versatility and ability to do many different tasks.

ETHICAL CONSIDERATIONS

When people are looking for work they frequently exaggerate their qualifications. There is a difference between enthusiasm and dishonesty. You should describe your experience and background in positive terms, but you need to stick with the truth. It is unethical to misrepresent yourself.

It is a serious mistake to make up credentials. Even if you are not discovered, you are vulnerable later when attention is focused on your performance or records. If you resort to false details and misleading information, you are violating the essential expectations of honesty which provide the basis for technical interactions.

Listing awards, achievements and special skills is an effective way to provide employers with a quick profile of your **personal interests.** Often your outside activities reveal management and leadership abilities. Use this section to include anything which will relate positively to the job you are seeking: professional involvements, additional skills, and community participation. Don't pad your resume, but if you have relevant material include it.

Don't include your **references.** Always state that these are available on request. This way an employer will have to contact you to reach your references and you will be able to let them expect a call or letter. Obviously, you cannot use someone as a reference without obtaining their permission. You want the persons who recommend you to say specific things, not make general comments. Provide your references with a list of your skills and achievements.

11.8 FORMAT

Your resume must be attractive and readable. Highlight your main sections with headings, boldface, margins, and white space. Flush left and ragged right margins are the easiest to read. Single-space your text; double-space between sections. Readers should immediately understand your outline. Visually communicate your organization.

- **Boldface important terms.**
- **Be consistent. Use a similar style throughout.**
- **Bullet items for emphasis.**

At the very least, your resume must be neatly typed, on good quality paper, and free of spelling, mechanical, and grammatical errors. This document represents you during the job search and it must be perfect. Technology has raised standards so you might want to have your resume typeset. With desktop publishing, this is easy and inexpensive to do.

If you are using your resume to apply for different types of jobs, use a word processor and a laser printer. This allows you to change your resume and still have perfect copies. Many service bureaus provide laser output and advice on how to construct multiple versions of your resume.

11.9 PREPARING FOR THE INTERVIEW

Preparation for the interview involves two activities: research and anticipation. Begin by finding out everything you can about the organization. What does it do? What does it make? Who is the competition? The more you know about the organization, the more you will be able to relate your experience and skills to the needs of the organization. You also prepare by learning to anticipate. This means exercising your powers of imagination. Start with some questions that change your focus and point of view from your own perspective to that of the interviewer. What would you ask if you were in the position of hiring someone to do that job?

You can also prepare by anticipating the general types of questions that are common to many job interviews:

- **What are your career goals?**
- **Why do you want to work for this company?**
- **How does your experience relate to this job?**

Preparation means rehearsing by thinking about what you will say and imagining the response. If you are asked about your long-range plans or ability to work with a project team, your response should show that you have thought about the answer. The interviewer will recognize if this is the first time you have ever considered your answer. Don't go to the other extreme and memorize your answer; people expect to interview a person, not a stage performer.

Finally, there is practice. Every interview builds your experience. Go to interviews even when you do not expect to get the position. You can be more relaxed if you do not expect success. Every interview is good practice if you are attentive and try to learn from your growing experience.

11.10 THE INTERVIEW

Employers interview you because they want to meet you face-to-face, to see what you are like and how you behave and interact with them. If credentials were enough, your resume and references would provide enough information to make a decision. Instead, your resume serves as a focal point for a personal meeting, discussion, and possible negotiation.

Your interviewer will expect you to be somewhat nervous. After all, you are aware of being compared to other candidates. You know that you are being judged on how you present yourself. An interview is a stressful occasion. For a short time you are required to perform at a very intense level. The best preparation is being ready.

According to Mary Scott Welsh, every job applicant should be prepared for six tough questions:

- Tell me about yourself.
- Why do you want to work here?
- What do you expect to be doing five years from now?
- Do you plan to have (more) children?
- What was your previous salary?
- Have we covered everything?

You should be prepared to answer these types of questions. (The fourth question is one that is illegal to ask, but you may be asked, and so you need a response.) The more you prepare yourself, the more confident and reassured you will be during the interview.

You also want to make a good impression. This means a firm handshake, a pleasant smile, and a confident attitude. You are responsible for providing your interviewer with information about your skills and abilities which correspond to the organization's needs. Research the organization and prepare some intelligent questions before the interview. Be prepared to discuss personnel policies and future directions. The knowledge you demonstrate will show your interest in the position.

Remember, when a company or organization is hiring a new employee it is making an investment in time and money. Naturally the people who do the hiring are going to be careful. They need to discover what sort of person they are hiring. Interviewers will want to know how your various experiences and achievements relate and how you will perform in the job. They will expect you to provide the relationships between your abilities and their needs.

LEGAL CONSIDERATIONS

State and Federal laws provide strict guidelines to interviewers. You should not be questioned about spouses, children, age, or other personal information. Discriminatory questions are illegal.

11.11 A FINAL WORD

Whether you are in school or happily employed in a job you enjoy, it is a good idea for you to stay in touch with the job application process. Scan the want ads in the Sunday papers. Read the trade publications and journals in your field. Keep aware of employment trends, current salaries, and job descriptions.

Maintain your employment credentials by keeping them current and easily available. This means reviewing your resume on a regular basis and adding accomplishments or changing the emphasis. Don't let your resume get more than six months old and keep some fresh copies at home and work. Be sure that you have up-to-date references who can provide specifics about what you are currently doing.

Your ability to respond quickly to an immediate employment opportunity may be the difference in whether or not you get a special job. Even if you are satisfied with your present position, awareness of the current job market is valuable information.

JOB SEARCH EXERCISES

1. Make a list of your interests, abilities, and experience. Now write down three types of jobs you would like to have. Compare the two lists. Do your interests, abilities, and experience match with the jobs you would like? What do you need to do to create a match between the two lists?
2. The primary exercise for this chapter is for you to create an effective resume and prospecting letter for yourself. Show your job search documents to other persons in your class or group and get their feedback. (See Exercise # 3.)
3. This exercise requires the formation of small groups. One member of the group is designated as the Personnel Director. The other group members should give their resumes to the Personnel Director and ask this person to rank the candidates seeking the job. Discuss the rankings, keeping in mind that the Personnel Director has a difficult task to do.

   Now each of you can decide whether to make changes in your resume.
4. This exercise also requires the formation of groups. Look through the help wanted section of your Sunday newspaper. Circle any positions you think you might reasonably apply for. Choose one of the ads and write a letter applying for the job. Show the letter to your group for feedback.
5. In this exercise you will role-play a job interview. Two members of the class or group are needed: one to serve as the interviewer, a Personnel Manager for a large manufacturer of household products, and one to serve as the interviewee applying for a specific job. (The interviewee should be allowed some time to work out the details of the job.)

   When the interview begins, the Personnel Manager should explain that one of the members of the secretarial pool accidentally misplaced all of the paperwork. The Personnel Manager will need to find out what job the interviewee is applying for. The interviewee should make up a job based on his or her career goals. The six-minute interview will be held in the office of the Personnel Manager.

After the interview, members of the class or group should discuss their observations with both of the participants. Remember, you may very well find yourself in each of these roles at some time during your career, so both are important to do well. You might consider having the participants reverse roles and try to see themselves in the other position.

6. Gates and Associates is looking to hire a field-site technician who will be able to work with a variety of temporary project teams. The firm is looking for an individual with a wide range of interests and abilities, someone who is flexible and able to contribute to different types of projects.

   Look carefully at each of the four resumes in Section 11.6 and decide which of these persons you would recommend for the position. Write a memo to Sheila O'Brien in which you rank the four candidates. Provide brief explanations for your decisions.

7. Carefully review Lourdes Santiago's resume in Section 11.1. Do you think this document would be more effective if it were typeset? Ragged right? Targeted for a specific job? What suggestions can you come up with to improve it? Make a list of your suggestions and submit the list to your instructor.

8. Using the six most commonly asked job interview questions from Section 11.10, role-play a job interview with someone else in the class. The interviewee needs to remember to bring a resume to the interview. The interviewer should be sure to ask all six of these questions in some form during the course of the interview.

   If you have the opportunity, videotape the interview and play it back for discussion and analysis. How well did the interviewee handle the tough questions? What strategies can you suggest to make the interview go more smoothly?

9. One effective way to practice for something as stressful as a job interview is to engage in a negative rehearsal. In other words, work up the worst possible interview you can imagine, one in which everything goes wrong. Both the interviewer and the interviewee should plan some strategies beforehand. Discuss with your partner the types of mistakes which create the worst impressions.

10. Write an unsolicited letter which seeks to get Mr. F. Robert Gates of Gates and Associates to read your resume and consider you for a position. Make sure you focus Gates'attention on solutions to problems he faces. See Chapter 26 for more information about F. Robert Gates and Gates and Associates. On a separate sheet of paper, make a list of the strategies you employ in your letter. Submit both the letter and the list to your instructor.

12

Reading Technical Documents

Note to Users: **This is a task chapter. It is designed to help you read technical documents more quickly and efficiently. As a technical professional you will need to read with speed and accuracy.**

12.1 INTRODUCTION

Anyone who is serious about developing a career as a technical professional should practice the skills of reading technical materials. Reading will be a regular part of your job. People in professional positions depend on reading as a way of keeping up with key developments and trends in their fields. They also rely on reading to keep them in touch with the activities of their project team and the rest of their organization.

Reading technical documents can be difficult work. You have to extract information and organize it so it will be useful. You need to use different strategies for different reading tasks. You have to shift your attention between numbers and words. Sometimes you will have to puzzle your way through a poorly written document, trying to figure out what the writer

Figure 12-1 Data shifting, confused writing and the clock are all obstacles to efficient reading habits.

intended. What is even more difficult, you need to get through lots of technical reading but you will never have enough time.

Reading technical materials effectively, with speed and comprehension, requires a wide range of skills and techniques which can be mastered only by practice and use. The further you progress in your technical profession, the more you will be asked to read and evaluate recommendations, reports, and technical proposals.

In this chapter you will learn how to:

- Extract and organize technical data.
- Utilize effective reading techniques.
- Develop your technical vocabulary.
- Use illustrations to support your reading.

12.2 DEFINITION

The activity of reading technical materials means taking in the sense of the words and illustrations you are looking at, understanding what these documents mean, not simply looking at them. Comprehension means making sense of the various voices you are decoding and listening to in your mind. Reading is one of the most common and frequent tasks performed by technical professionals, yet it is seldom considered in employment interviews or referred to in job descriptions. It is assumed, often without justification, that technical professionals will know how to read efficiently.

More than half of the United States work force is involved in communicating information: collecting, storing, retrieving, handling, and distributing technical writing and other types of business documentation. What you need to remember is that people have to read most of this information, extract the data they need, and organize it in some useful fashion.

Different people will read the same document in different ways, because they have different needs, different levels of expertise, and different responsibilities. Some of these persons will be reading for overall meaning while others will be searching for specific information. All of these persons will need to grasp, understand, and remember what they are reading.

Reading for work is different than reading for pleasure, diversion, or entertainment. The suggestions we make in this chapter are not designed for reading the sports or hobby pages of your favorite magazine or newspaper. They are designed to help you approach technical reading with systematic, logical techniques which will increase your ability to grasp and use what you read.

12.3 CRITICAL READING

When you read critically, you are involved in making active judgments about what you are reading. Critical reading considers the evidence and the arguments. You can't take what you read at its face value; instead you need to consider whether what you are reading is *relevant, accurate,* and *useful.* This involves several tasks.

First, you need to evaluate the source of what you read. There is a big difference between information from the Surgeon General of the United States and the Tobacco Institute of America. Technical information draws a good deal of its credibility from its point of origin. The best evidence comes from expert witnesses, specialists whose knowledge, credibility, and authority have been demonstrated in the past and are widely recognized.

The second thing you need to be clear about is exactly what is being said. What is the main idea? In other words, know the point of what you are reading. This means careful reading, where you try to distinguish important ideas from information that is supporting the central points.

You need to consider whether the underlying assumptions are correct. What evidence is used to support the positions the writer takes? Is there sufficient detail to provide proof that the ideas make sense? Are statements and claims backed up in a way that is convincing? Critical reading means that you are looking for flaws and misstatements in what you are reviewing. You are inspecting the document for logic, clarity, and completeness. You accept only what you decide meets these standards.

Finally, you need to consider the implications of what you read. Sometimes, a simple piece of information can have significant effects in surprising areas. An article on mechanical miniaturization, for example, might provide answers to a problem in vehicle design. Serendipity, the chance discovery of important information, is one very positive side effect of reading. When you read regularly, you create an information network where ideas can cross-fertilize each other and lead to unanticipated rewards.

12.4 USEFUL TECHNIQUES

Pre-Reading. Begin with a conceptual mapping strategy that helps you to understand the underlying structure of the text. These are big words to describe the process of pre-reading a document. Pre-reading is a technique where you use every resource of the text to help you efficiently process the information it contains. By becoming familiar with the organization, key words, and general plan of the document, you increase your chances of understanding and retaining what you read. Pre-reading involves a general review of the text before you read from beginning to end.

Start with the table of contents. Read this section carefully and attentively. The table of contents is a detailed map of the document which locates all of the key ideas and lets you know what information is important. Headings and subheadings indicate relationships between and among different topics.

Pre-reading

✓ Cover

✓ Table of contents

✓ Preface

✓ Introduction

✓ Table of figures

✓ Index

✓ Glossary

Read the preface and any other introductory material, and then scan the index looking for familiar terms, distribution of topics, and key terms. Flip through the text and quickly look at the illustrations, reading the captions. What you are trying to do is to build a mental map of the text so as you read specifics you will have a place to locate them.

As a reader you must be concerned with the same problems as the writer: what is the purpose?, who is the audience?, and what are the key details? If writing is an assembly process where you put a document together, then reading is the dis-assembly where you take the document apart to get what you need from it. The more you can map the structure, the more you will retain and be able to use.

Questions. Avoid mechanical, monotonous reading by focusing your attention with questions. If you are looking for specific answers, you will find it easier to maintain your interest in what you are reading. If you know what you are trying to find, it will be more recognizable when you encounter it.

Begin the reading task by defining your purpose. It helps to know why you are reading something. What do you want to know or do after reading the material? Who wrote the message or sent it? A Message from the Committee to Restore the Use of Slide Rules is less significant than one from your supervisor.

Reading Plan Know Your Purpose	
Questions	Notes
What do you need to know?	
What do you need to do?	
What questions do you have?	

Clumping. A very important skill you should work to acquire is the ability to grasp information from groups of words rather than from one word at a time. Two suggestions:

- Train yourself to look at more than one or two words at a glance. Look for chunks of words, blocks of text.
- Practice reading for sense rather than sound. Grasp for the idea, not the details. Go back for the details later.

Notes. Make notes as you read. Your notes can include key words and phrases, headings and sub-headings, concise summaries, simple diagrams or sketches, whatever will help you organize and make use of the information. Many people make the notes on the document itself, marking it up and highlighting key information.

The effort of making the notes will reinforce your memory of what you are reading, and it will help you to make discriminations between what is important and what is not. The notes will help you review and refresh your knowledge of the material later.

Definitions. Definitions are basic to all technical writing. You need to understand exactly what you are reading. Careful reading means using the specific, precise definitions of technical terms. While you are reading pay attention to definitions; mark them, and think about what they mean. This will help you to understand and remember what you read.

Patterns. Another useful technique is to read for patterns. Try to remember this number quickly:

1491625364964

This is a lot easier to do when you recognize the series as the squares of one through eight ($1^2 - 8^2$). Information is easier to remember and understand when you know how it is organized.

Frequently a technical subject can be broken down into a list of details and sub-categories, but only when you realize how the material is classified and organized. Use chapter titles, headers and footers, headlines, and highlights to look for patterns in the information.

Specifics. It is important to consider how much of the information you actually need. If you are looking for specific information, can you go directly to it? The index and table of contents may be able to guide you directly to what you want to find. Ask yourself if you can you get what you need from reading selections instead of the whole text. By refining your search to the specific information you need, you can save time and effort.

12.5 SPEEDREADING

If you want to read more quickly, you have to read more quickly. This sentence isn't a mistake. It represents the simple truth about speedreading. You are the one who is doing the reading. Only you can speed up your reading. In order to read more quickly, you have to read faster. The first and most important step is to begin to make the effort to read faster. Try to read faster, and you will read faster.

You need to remember that different materials are read at different rates of speed. A long-term lease deserves more careful examination than the instructions for installing the paper cups in the distribution device. You don't want to speedread when you are reviewing the applicant qualifications for a major contract. On the other hand, you don't want to dawdle over routine, repetitious information.

The more familiar you are with a topic, the quicker you will be able to read information about that topic. Familiarity supplies the context that we need for the easy acquisition of technical data. When you are reading technical material, you want to discover the essential message, the main idea. You also want to judge, reliably and quickly, whether this idea is supported sufficiently to be credible. The more familiar you are with the topics discussed, the more easily you will accomplish this task.

12.6 KEEPING INFORMED

If you want to keep current with technical developments in your own and related areas, find out what new products and services are offered, and get an overview, a context for your work, develop a reading list of the important magazines, newsletters, and journals in your field.

Regular reading is an important way for technical professionals to keep up with current developments in their fields. Many professional organizations provide periodicals and journals of special interest to their members. There are also many specialized technical magazines and newsletters, available for free to professionals within the areas they write about. Trade publications target a very specific audience—persons employed in a particular technical field or business. These magazines keep you informed about people, products, services, and trends in the industry in which you are employed.

We strongly recommend that you subscribe to several professional journals, even if you are still in school. You will get the opportunity to read the words of your fellow technical professionals. You will be developing a sense of context, building your sense of standards and of what is appropriate, and responding to the written voices of your peers. If you actively read and pursue the ideas and suggestions in what you read, you will become a better employee and your job will stay a challenge instead of a chore.

12.7 READING ILLUSTRATIONS

Some people are quite comfortable reading text, but they avoid the illustrations because they find them confusing and frustrating to interpret and decipher. This is unfortunate, because, if they are used correctly, graphics can add a great deal to the understanding of a written explanation.

One of the chief difficulties of reading illustrations is the effort of switching between reading numbers and reading text. Read the text first, including all labels and captions. This will help you to make sense of the graphics and numerical information. A systematic, problem-solving approach will enable you to figure out what each illustration means.

12.8 READING SHORT DOCUMENTS

Frequently you will be called on to read reports, journal articles, technical directions, and other short documents. Begin by skimming the text. Read the first and last paragraphs. The introduction and conclusion provide valuable summaries. Then quickly read the first sentence of every paragraph. This will help you get oriented to the material and provide a framework for the information you are reading.

New Entrants Into Labor Force 1988-2000

Total = 42,832,000

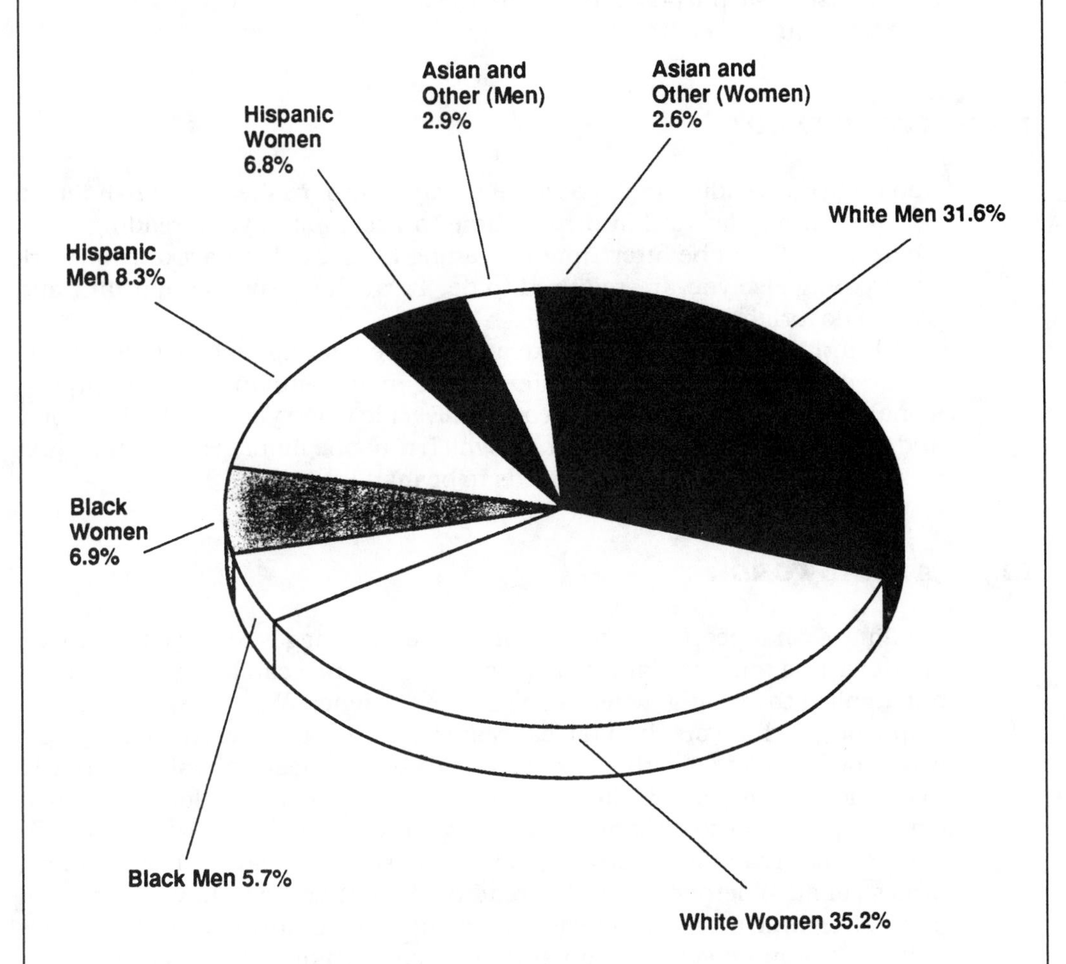

Our pool of talent for new scientists and engineers is predominantly female or minority or disabled—the very segments of our population we have not attracted to science and engineering careers in the past.

Source: U.S. Bureau of Labor Statistics

Figure 12-2 What hints and clues can you find in this illustration to help you understand it?
Source: The Final Report of The Task Force on Women, Minorities, and the Handicapped in Science and Technology.

Focus your attention by determining in advance what information you are looking for. This means knowing your purpose. You are not reading for pleasure; what do you need to know or do? How will this document help you accomplish that purpose? It is difficult to concentrate when you are not sure why you are reading.

12.9 *TIME TO READ*

Don't let your reading tasks become a gray blur of rushed responsibilities. Set aside a regular and sufficient time to accomplish your reading, time where you will not be interrupted. Reading is as much a part of your work as anything else you are required to do. Treat this task as important and plan to do it right.

Prioritize your reading. Make a list of the reading tasks which you do on a regular basis. Don't forget to include mail, company memos, professional journals, and trade publications. Decide when you will do this work, and in what order. Reading is a task which must be managed carefully, just as any other technical activity needs to be managed.

12.10 *A FINAL WORD*

Reading is what people do with writing. All of the suggestions we have made about writing and speaking clearly to your audience can be turned around and applied to reading, where you are the audience. What we have tried to emphasize is that you should treat reading technical materials in the same ways you would deal with other technical communication tasks. You need to consider what you are doing, what you are trying to get done, and then devise a plan to accomplish your reading more efficiently.

Regular reading will be an important part of your career as a technical professional. Whether you enjoy reading or not, you will have to read all sorts of technical and business documents, some of them not very well-written. The more you develop and practice the skills of efficient reading, the more valuable your work performance will become. Reading skills are developed and practiced by only one method: reading.

READING EXERCISES

1. Bring three magazine articles to class. Choose one that is below your reading level, one that is above, and one that you find appropriate. Examine the three articles and explain what criteria you used to place them at a particular level of difficulty.
2. Read the technical briefing from The Office of Technology Assessment (OTA):

March 1990

Coping With An Oiled Sea:
An Analysis of Oil Spill Response Technologies

Cleaning up millions of gallons of oil at sea under even moderate environmental conditions is an extraordinary problem. Current national capabilities to respond effectively to such an accident are marginal at best. OTA's analysis shows that improvements could be made in our oil spill response capabilities. Those offering the greatest benefits would not require technological breakthroughs—just good engineering design and testing, skilled maintenance and training, timely access to and availability of the most appropriate and substantial systems, and the means to make rapid, informed, decisions. **Even the best national response system, however, will have practical limitations that will hinder spill response efforts for catastrophic events—sometimes to a major extent. For that reason it is important to pay at least as much attention to preventive measures as to response systems. In this area, the proverbial ounce of prevention is worth many, many pounds of cure.**

The March 24, 1989, *Exxon Valdez* oil spill in Prince William Sound, Alaska, dramatically illuminated the gap between the assumed and actual capability of industry and government to respond to catastrophic oil spills. There are many reasons why this gap wasn't better appreciated before March 24: elaborate oil spill contingency plans had been prepared and approved; oil spill equipment had been developed and stocked; major damaging spills had occurred infrequently, and almost never in the United States; and a nebulous faith had existed that technology and American corporate management and know-how could prevent and/or significantly mitigate the worst disasters.

It is now understood that the U.S. industry has concentrated on developing technology to fight the numerous small spills in harbors and protected waters, and it oversold its ability to fight major spills. The government and the public uncritically accepted industries' assurances, and relied on their capabilities. Expectations about what can be accomplished once a major spill has occurred have clearly been too high.

OTA's study of oil spill response capabilities focuses on technological promises and limitations. The technology now available for oil spill cleanup in the United States and overseas has resulted in little actual cleanup for almost all past major ocean spills. OTA obtained data from several open-ocean, large-tanker spills that show that less than 10 percent of the oil that was discharged into the sea was recovered—usually much less. Probably between 6 and 8 percent of the oil spilled by the *Exxon Valdez* was recovered at sea, although, as of this writing, Exxon is still in the process of developing a recovery estimate. **Some sources claim that the most oil that can be recovered after a major spill is 10 to 15 percent.**

Improvements in mechanical recovery technologies that can be expected in the future are unlikely to result in *dramatic* increases in oil recovered from a catastrophic spill. However, some improvements are possible and those that are likely to offer increased effectiveness for large offshore spills involve larger, more costly equipment, strategically located for quick response.

Techniques other than mechanical recovery can also be used to attempt to mitigate the effects of a large offshore oil spill without actually picking up the oil—for example, burning the oil or using dispersants. In fact these other techniques have seldom been used successfully.

Despite the shortcomings of all existing countermeasure approaches, each may have applications in certain situations. Many technologies may be very effective in certain applications but completely inappropriate in others. **Regardless of the technique(s) employed, the effectiveness of the response will be greatly enhanced if there is a *rapid* response by a professional response team that understands which techniques are best under which conditions.**

Increased R&D on oil spill response technologies will likely yield incremental benefits. Important problems can be better understood, but technological breakthroughs that would result in major improvements in mechanical cleanup capabilities are unlikely. The most important problems have to do with: 1) providing technical backup for decisions on use of techniques such as dispersants and burning, 2) developing technical standards based on full-scale tests of specific equipment, and 3) sound engineering design of capable and reliable systems.

Copies of the OTA report, "Coping With An Oiled Sea: An Analysis of Oil Spill Response Technologies—Background Paper," are available from the Superintendent of Documents, U.S. Government Printing Office, Washington, DC 20402-9325; (202) 783-3238. The GPO stock number is 052-003-01183-2; the price is $3.75. Copies of the report for ***congressional*** *use are available by calling 4-9241.*

Non-congressional *requests for the report can be ordered from the U.S. Government Printing Office or for further information, contact OTA's Publications Office. Address: OTA, U.S. Congress, Washington, DC 20510-8025; (202) 224-8996.*

The Office of Technology Assessment (OTA) is an analytical arm of the U.S. Congress. OTA's basic function is to help legislators anticipate and plan for the positive and negative impacts of technological changes.
John H. Gibbons, *Director*

(a) Make a list of the text devices the designers of this document use to help the readers.

(b) Make three bullet list charts which communicate the key information in this briefing. Show them to other persons in your class and take questions on the information.

(c) Write an executive summary of this briefing for Sheila O'Brien with a copy to Mr. Gates. Keep it to less than a full page.

(d) Write a letter requesting the full OTA report.

(e) Write a memo to let your group know whether or not they can find the full report in a local library.

3. Figure out how much reading you need to do on a regular basis. Write down a brief description of the strategies you use (or intend to use) to accomplish this work effectively.
4. Choose an article from a professional journal and read only the first and last paragraphs, and then the first sentence of every paragraph. Did you get a good sense of what the article had to say? Experiment with this technique on different types of reading.
5. Examine the illustration on page 199 carefully. Then write a memo to Sheila O'Brien explaining what effects this information will have on future recruitment, hiring, and personnel policies here at Gates and Associates. What are some of the implications of this data for our company over the next ten years?
6. Measure your own reading speed. Use a stop watch to time yourself reading for sixty seconds and then count up how many words you have read. Try yourself on different types of reading material. Does your reading rate change with the material? What is your average? What steps can you take to improve?
7. Choose a technical book which you have not read. Spend ten minutes by the clock to review the items on the pre-reading checklist.

Pre-reading

✓ Cover

✓ Table of contents

✓ Preface

✓ Introduction

✓ Table of figures

✓ Index

✓ Glossary

Science and Engineering Doctorates

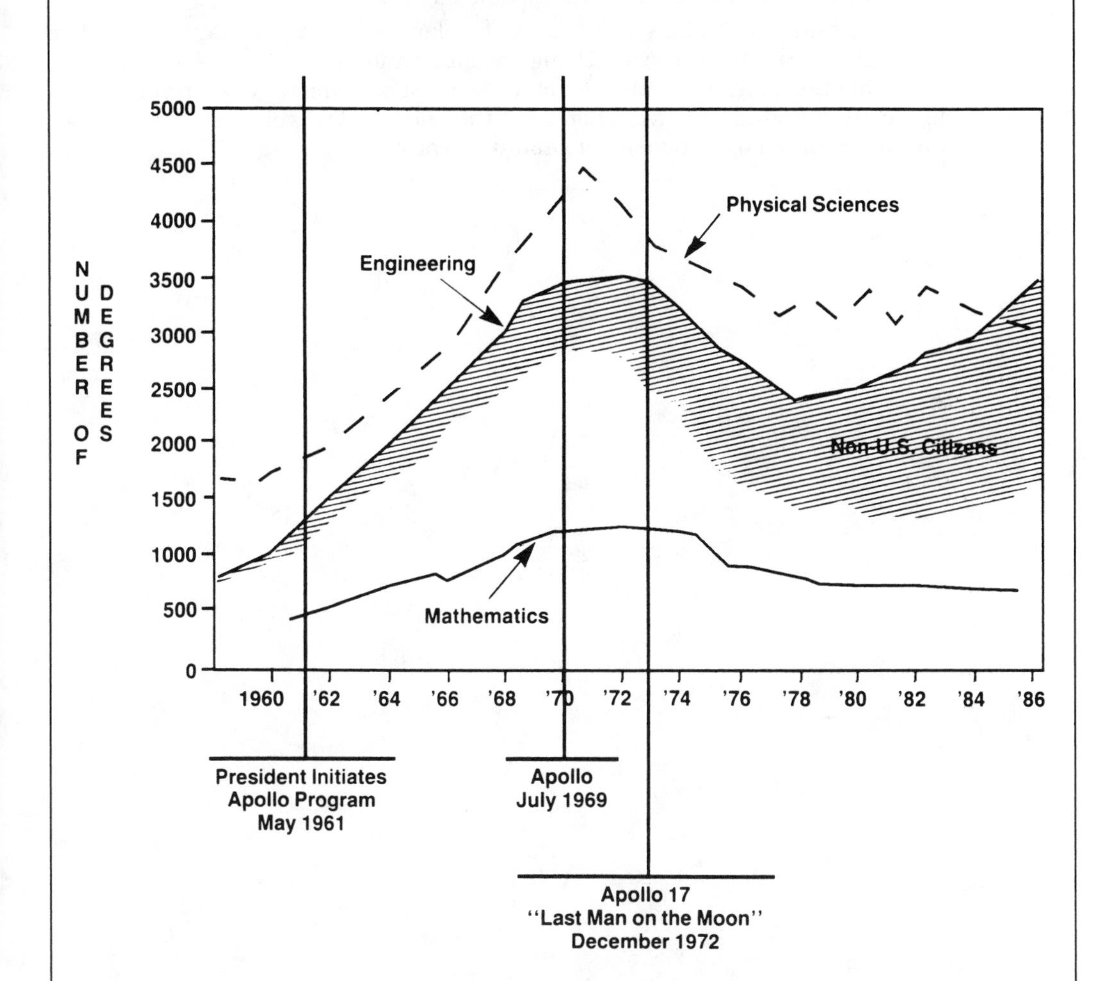

Students' choice of science and engineering careers is clearly related to the technological challenges on our national agenda, as this chart showing the history of the space program indicates.

Source: National Aeronautics and Space Administration

Are you better prepared to read this book now? What did you discover about the organization of the book and its contents? Did you learn anything from the index? Could you go directly to specific information which you needed? Did you find the pre-reading activity worth the time it took?

8. Reading is an activity. This means it can be planned. Choose a chapter from this text which you intend to read. Using the illustration in Section 12.4 on asking the right questions, make up a list of three questions which will keep your reading focused and productive. Then read the chapter. Did your list of questions improve your ability to remember what you read?

13

Communicating by Telephone

Note to Users: **This is a task chapter. It is intended to show you ways to use telephone systems and devices more productively.**

13.1 INTRODUCTION

In the last few years the telephone and switchboard have evolved into a complex electronic messaging center. The copper lines which used to carry human words and voices are crowded with bits and bytes, data, information and machine talk—*machines talking to each other.* Even the copper wires are being replaced—by fibers!

With these new technologies—voice data systems, fax machines, teleconferencing and alpha-pagers—human voices are almost as likely to be transmitted electronically as face-to-face. This chapter will provide you with an overview of these new technologies as well as some practical suggestions to make your calling time more productive.

In this chapter you will learn how to:

- Decide when to use the phone.
- Apply strategies to avoid telephone tag and circle calls.
- Mind your electronic message manners.
- Use telephone techniques to improve your everyday communications.
- Design conversations that get things done.
- Discover new telephone technologies that work for you.

13.2 *DEFINITION*

Defining telephone communication in the 1990's is a matter of choosing among telephone devices. Alexander Graham Bell's "*electrographic voice transmitter*" has become a telecommunications system. The telephone outlet is a terminal through which you can access and send information to different people in many different ways.

You can telephone your voice, numbers, drawings, and data to almost anywhere in the world. Your information, whether it is voice or picture, is modulated, demodulated, microwaved, and bounced off satellites which swing round the earth. So much for the simple telephone!

While new devices are changing the ways we communicate by phone, most business telephone calls are still voice-to-voice communications between individuals. More than fifty billion business calls were placed in the United States in 1990. The telephone system provides an immediate form of communication necessary to technical businesses.

What you need to remember is that *all of these telephone calls are made for the same purpose, to pass information between people.* The secret for successful communication remains courtesy. Whether you are using a fax, modem, or telephone handset, your object is the same: to communicate your information. Consideration for other people requires accuracy, clarity, and politeness.

13.3 *TELEPHONE DEVICES*

The communication of voice, data and images over telephone circuits and systems involves a variety of equipment and transmission facilities from the telephone itself to the computer and fax machine. Here is a partial list:

ANSWERING MACHINES Devices which answer your telephone. The caller hears your recorded message and is able to leave a message for you.

CELLULAR PHONES Portable phones which send and receive signals on a cellular network instead of telephone lines. These are generally found in vehicles such as cars and planes. They can be used with modems and fax machines to send text and images as well as voice.

CENTREX Local phone companies are providing centralized services as an option to hardware solutions. Some of these services include selective call forwarding, priority ringing, call return and calling-party identification.

ELECTRONIC MAIL Computer utilities which allow text messages to be sent from one terminal or computer system to another. The message is stored until the recipient retrieves the mail by logging on to the system.

FACSIMILE/FAX Equipment which sends and receives copies of documents over regular telephone lines. Any image which is on paper—sketches, handwritten notes, drawings, text, photos—can be sent and received with facsimile machines. Fax boards can be installed in personal computers allowing users to send and receive whatever they can place on their screens.

MODEM Short for modulator/demodulator. This device changes digital signals to analog and the reverse so that computers can exchange information, text and graphics, across telephone networks.

PAGERS/ALPHA PAGERS Devices which signal persons that they have messages or are needed. Alpha Pagers can transmit brief text messages as well as numbers and signals.

SATELLITE COMMUNICATION Several organizations including NASA provide satellite-based, high-capacity digital services for voice, image, and data communications. Satellite link-ups allow for direct communications with remote projects and locations.

VOICE MAIL A voice mailbox using computer-phone systems allows you to pick up and leave voice messages using touchtone phones. This same technology can provide information and data by providing telephone access to text files and databases. This is what happens when you combine the computer with the answering machine.

VIDEOCONFERENCE An online meeting between two locations linked with video and sound. Facilities available in many large cities avoid the need for direct travel.

13.4 DESIGNING A CONVERSATION

A telephone conversation requires the same type of planning that goes into a written memo. You will need to consider your audience—the person or persons who will be listening to what you say. You will need to know what you want to learn or what you want to be done. Your listener will be wondering too, and it is up to you to provide the answers.

Whenever you are on the telephone, you need to remember that you are in direct and immediate contact with someone who cannot see your expressions or the chart you are pointing to. You need to be prepared to give your listeners the verbal assistance which will help them to meet your needs.

The telephone sometimes gives us the sense that we are face-to-face with a person. Instead, we are voice-to-voice with each other, and neither can see. When we are explaining technical details we need to be particularly careful that the information is passed on accurately.

A Few Simple Rules

- **IDENTIFY YOURSELF.** You should always start out a phone conversation by telling the listener who you are and why you are calling. There are not many people who can be helpful while they are wondering who you are.

  Identification can be many different things. Some computer systems require elaborate passwords when calling by modem. For most person-to-person conversations, however, simply stating your name and company is enough. Then say what you want.

  Answering the phone requires the same type of directness. Unless your company has a particular phone protocol, you should answer by stating the company's name and your project group, if necessary. Your name is optional. You do not have to give an elaborate response that includes your employment qualifications.
- **SAY WHAT YOU WANT.** Your listener needs to know this. Before engaging in any telephone communication you should always be prepared to respond to the direct question, "What do you want?". Very few people will ask this, but, if you do not make it clear, all of them will be wondering.
- **KEEP IT BRIEF.** The phone is most effective for short communications. For one thing, busy people don't have lots of time for conversations, particularly ones they didn't schedule. Technical information may not be available when you call and complex technical details are easily confused on the telephone.
- **WRITE IT DOWN.** Don't rely on memory. If it is important enough to remember, then it is important enough to write down. We are trying to shout this rule. So Listen:

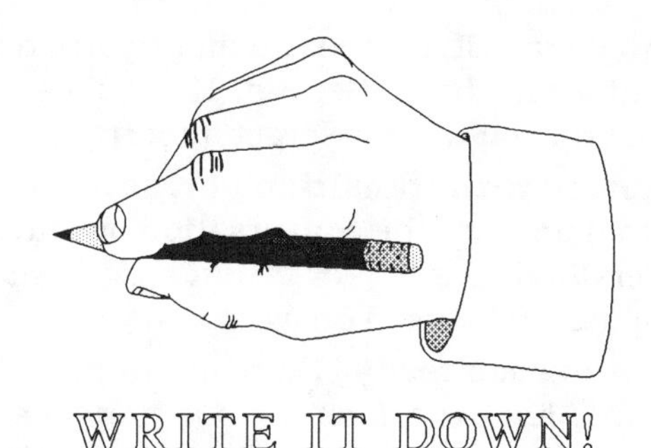

13.5 WHEN TO MAKE A TELEPHONE CALL

A telephone call may often seem like the easiest way to deal with a situation. You should remember, however, that telephone calls have advantages and disadvantages.

The phone is immediate. You don't have to wait for a decision or agreement. You can get the information you want when you need it. On the other hand, there will be no written record of that information or agreement. Complex technical details can easily get confused. Busy people do not like to be interrupted. Some material is too sensitive to be discussed on the phone.

You want to be sure the message is appropriate for the telephone. In general, phone conversations are useful for transmitting small amounts of information that is relatively simple.

13.6 GUIDING THE CONVERSATION

One reason that many persons dislike the telephone is that they consider it an interruption, an unplanned interference with their schedule. When you initiate a call, you are responsible for seeing that things get done, that the purpose of the call is accomplished, and that you do not waste your listener's time. If someone you call is reluctant to talk to you, ask for a convenient time to call back.

- **Take control of the discussion.** You do this with your voice, tone, and confident delivery. This means preparation, knowing what you want and what you are prepared to do. When you have planned your call, you will sound prepared and confident.

- **Outline your call.** Before you make your important calls take the time to fill out a call sheet. (See Fig. 13-1.) This will keep your conversation focused and you will not forget important matters.
- **Be prepared with transition phrases.** You should have a few phrases ready to help keep the conversation on track. Simple expressions like *"I understand, but. . ."* or *"That's interesting, however. . ."* can keep the discussion focused on what you want to talk about.
- **Have an excuse ready.** Be ready to get out of a conversation if you want to. *"I have a call on the other line. Let me get back to you."* is a polite way for you to get further information or check with others in your organization. If you have to leave the phone to get information, offer to call back.

You can help others listen to you by using a pleasant voice and lively tone. Organize what you have to say ahead of time and let your listener know how you have organized your message. For instance, you might describe the steps to do something in sequence, or the parts of a device from back to front. Let your listeners know so they can follow along. Even more importantly, follow through. If you agree to return a call or to do something, make sure that you do so.

13.7 THE CALL SHEET

A call sheet is a simple form which you can use to focus your attention during important phone calls. Write down why the call is important and what you want to accomplish. Then use the form to note key phrases.

13.8 AVOIDING TELEPHONE TAG

Some researchers have estimated that as many as 70% of all business calls do not reach their intended audience. The person you should be most prepared to speak to when you make a call is a person who is not there. Be ready, at least, to do your best in reaching this person. Be prepared. Tell the person that you do reach the individual you want to talk to, the time and date of your call, and why you are calling: **WHO, WHEN** and **WHY.** Ask the person who answers to record this information.

Then be prepared to receive a return call. There is little sense in asking a person to call back if they will not be able to reach you. If you are going to be away from your phone, leave instructions for how to reach you. It takes two sides to play a really good game of telephone tag.

If you do not receive an answer within a reasonable time, two days for instance, then send your message in writing. Don't forget to keep a copy for yourself.

Gates & Associates **Call Focus Sheet**

1. This call is important because:

2. By the end of this call I want to accomplish this result(s):

KEY PHRASES	PROMISES
	I agree to:
	They agree to:

Figure 13-1 A Call Sheet.

These suggestions, of course, apply in reverse. You should return calls as quickly as possible, certainly within two days if you are going to return them at all. You should also be prepared to collect full information (WHO, WHEN and WHY) for fellow workers who cannot take a call.

13.9 LEARNING TO LISTEN

The reason we were given two ears and one mouth, says a very old proverb, *is so we can listen twice as much as we speak.* Unfortunately there are many persons who have not heard this saying, perhaps because they are too busy talking. The ability to listen effectively is one of the prime abilities of successful communicators.

Listening is not the same as hearing. When you listen, you should be engaged in an active process of trying to make sense of what you are hearing. You should be trying to organize what you are hearing and relating it to your needs and purposes. Depending on the context (speaker, subject, and situation) you need to listen in different ways.

Sometimes you need to listen critically, carefully noting the implications of what the speaker is saying. At other times you need to listen supportively, listening sympathetically for the feelings that are being expressed.

You can learn to be a better listener by using a variety of techniques. These include **asking, waiting** and **writing:**

Ask questions. This keeps you involved and participating in the conversation.
Wait to prepare answers. Give the other person a chance to finish speaking. Until then, think about what they are saying, not what you are going to say.
Write it down. Jot down key words from the other person's conversation to keep clearly in mind the main points of what they are saying.

Listening on the phone is difficult because it is easy to get distracted. Maintain your attention by becoming involved in the conversation. It is your responsibility to listen well, even when you are not interested. To do this you have to avoid negative listening habits like selective listening, daydreaming, and inattention.

An effective listener is able to repeat back what was said, to the satisfaction of the other person. Even a computer checks to see if a message is being correctly received. Occasionally you should repeat back what you are receiving from the message to make sure that you are listening correctly.

13.10 PHONE IT IN WRITING

Put it in writing is still the first rule of business and industry. If you need to remember something, put it in writing. If you want someone else to remember, put it in writing too. Written words reduce the possibilities of garbled and confused messages. Today the telephone system can be used to communicate in writing.

The facsimile machine is a wonderful piece of phone equipment which can be used to send and receive documents—sketches, text, any kind of visual information. A fax unit combines the benefits of the telephone and the copy machine.

Fax output consists of documents which are virtually error-free. There can be no misunderstanding about what was said or suggested. Even better, you can fax drawings, charts, maps, blueprints, and photographs. Fax machines have followed the computer into offices and workshops everywhere, putting people into written contact with each other. Fax machines are now available in many U. S. Post Offices.

Electronic mail is another technology which is increasingly available. Regional, national, and international networks are being created which will let industry, schools, and government rapidly communicate with each other by electronic mail and file transfer. Networking is a tool which allows people to exchange ideas and information quickly and effectively.

Networking has changed the ways in which people work together. Collaboration is no longer limited by physical locations. Through the networks, people have immediate access to distributed resources from distributed locations. Technical projects can involve persons, instruments, and software that are contained in many different places, even continents apart.

13.11 ELECTRONIC MESSAGE MANNERS

The speed with which messages are transmitted by telephone, facsimiles, and electronic mail has been blamed for an increase in poor manners. Rehearsal time is eroded and replies are requested without sufficient time for reflection and thought. Even so, courtesy remains a key component in any successful communication.

If you are using a fax or a modem to send images or text, there are some simple rules of etiquette to follow. These are very similar to telephoning by voice. You need to say who is sending the message, who it is addressed to, and how many pages or bytes of information will follow. A cover sheet is often the most important consideration in whether your message reaches its intended audience.

The conference call represents another instance where special courtesy is required. When multiple parties are attempting to talk by phone from various locations, a moderator is necessary. People need to be invited into the discussion and encouraged to participate by the moderator who clearly sets up and maintains the format of the conversation since the visible signals to do so are absent.

13.12 TECHNICAL SUPPORT

Many companies that sell sophisticated technological products provide technical support on the telephone. This is a difficult task, since it requires both technical knowledge and communication skills. If you do technical support, you may find your job requiring skills roughly similar to an air traffic controller; you will deal with individuals who need immediate attention and precise advice.

Too often, the documentation that accompanies the products is not what it should be, and customers must phone the company to resolve problems. Many products are designed for ease of manufacture, not ease of service. And while design engineers do not create products with support in mind, sales representatives may sometimes not know enough about a product or may misrepresent the capabilities of a product in order to generate sales. Despite these problems, you never want to embarrass your own people. All of you work for the same company, and you have an obligation to them.

Preparation is a key to good technical support. You should have user manuals, other documentation, schematics, and the product itself readily accessible before you pick up the phone. How well the call goes depends on how you treat the customer when you pick up the phone. Remember, you will only receive a call when someone has a problem.

Telephone Support Guidelines

- Know about the product(s).
- Be courteous. Treat the caller with respect.
- Let the person talk. The customer initiated the call, and therefore has the right to the first few minutes of the call.
- Build trust. Ask questions, based on your knowledge, to pinpoint the problem. Work with the caller. Try to visualize all of the steps the caller took before initiating the phone call.
- Know the potential problems with the product. Think about the most basic errors that can be made. Learn from experience.
- Empathize with the caller. Your caller may be working in a cramped space under adverse weather conditions and time pressures. Remember, customers don't forget how they are treated. The future of your company may be at stake with each phone call.
- If your caller begins to be frustrated or irritated, regroup and start over again. If this fails, allow the caller to speak to someone else.
- It's okay to say "I don't know, but I will look into this" to a customer. You may need to take the problem to an engineer and get back to the customer.
- Log in your notes on a database. The customer may call back, and similar problems may arise with other customers.

As with other kinds of communication, the key to successful telephone support is empathy. Try to picture the conditions the caller is working under, and be patient. You may need to be particularly patient when there are language barriers between you and your caller. Slow down. Keep repeating and rephrasing your statements. You may need to find someone in your company who speaks the language of your caller.

13.13 A FINAL WORD

Whatever telephone device you use to communicate, your purpose is the same. The common denominator to all of these choices is that you are sending a message. For the past ten years new products and innovations have been coming into the telecommunications marketplace at a rate that exceeds many people's ability to use them. What this means is that equipment is out there which is not being used because people don't know how to use it.

You can improve your abilities as an effective communicator by learning as much as you can about the telecommunications system used by your school or business. Whether you send your message by voice, image, or text, you are still trying to deliver a message to another person accurately and clearly.

The telephone is useful for a number of purposes, to keep in contact with people, to gather information, to stop and talk something over, to add feeling and emotion to your message. The business phone should be thought of as a tool which can improve your communications by increasing their speed, and making them more personal and dependable.

TELECOMMUNICATIONS EXERCISES

1. You receive a telephone call from Mr. Gates who has a project for you and your team. (*See Chapter 26, Communicating in an Organizational Environment,* for more information on Gates and Associates.) What he wants you to do is to investigate available telecommunications technologies and to recommend an appropriate system for Gates and Associates.

As you review your notes of the conversation with Mr. Gates, you realize that he has some specific concerns. In particular he wants to match the phone system to the needs of our business and our clients.

We are a consulting firm whose very existence depends upon the need for customers. Like every other technical business we need every inquiry we can get. Many potential customers will not call back after hearing a busy signal or getting cut off.

The system needs to handle a high volume of calls with only a few persons available to answer the phones. The system must be cost-efficient and reliable. Reliability is particularly important. Without this, our other equipment, facsimile machines, modems, and voice mail, will be useless.

Gates has heard horror stories from other companies who have been cut off from service for as long as a week. Reliability is very important to him. We will need to use this system for at least five years and cannot afford to make a mistake.

We have ninety-five employees, many of whom are traveling and away from the office. These folks need to be able to send and receive messages from wherever they are. We also have a local area network which connects twenty-seven personal computers. About half of these are equipped with modems and three have PC-fax boards added in. Gates wants to make maximum effective use of this equipment.

Some of the items which should be included in your decision matrix (a ratings table comparing products against specified criteria) include cost, features, warranty, and service. Gates is not sure what features we should look for but he has expressed an interest in automatic callback, speed dialing, and speaker phones for conference calls. You should evaluate the need for these features.

He also wants you to include a separate item in the budget which discusses installation costs, upgrade costs, and training costs.

Gates wants your project group to prepare a fifteen-minute presentation where you will make your recommendations for the next board meeting. You should include a written summary of your recommendations so the board can discuss them after your presentation.

2. Choose either a hardware or software product you are familiar with, for example, an answering machine or word processing program. Assume that it is not working properly and you are planning to call an 800 technical support number. List the preparations you would make before placing your call.

3. Your state has a right-to-know law requiring that all companies which use toxic chemicals post a list of the chemicals they use and their hazards for the benefit of their employees.

Gates and Associates has been hired to find out how well the law is working. (See Chapter 26 for more information.) Sheila O'Brien assigns you to design a telephone survey which will question employees and employers about the level of compliance in their workplaces. Your first task is to prepare a script for the persons who will conduct the survey.

Remember, if the data which is gathered is going to be useful information it needs to be comparable and easy to understand. Your script should include a form for the interviewers to enter the data they collect.

4. Since the court decision which broke up the telephone company monopoly, all telephone service has been divided into three parts: local service, equipment, and long distance carriers. Each of these parts has numerous vendors, selling a wide range of equipment and services. The sales efforts are marked by confusing offers and hard-to-measure technical claims. Sheila asks you to follow-up your presentation on a company telecommunications system by recommending a long-distance carrier. Again, reliability and cost are the two most important criteria.

Write a memo to Sheila O'Brien recommending a long-distance carrier for Gates and Associates. Remember to provide supporting reasons for your recommedation(s). *See Chapter 26, Communicating in an Organizational Environment* for more information about Sheila O'Brien.

5. One way to improve your telephone skills is to practice them. Two persons can sit back to back and conduct a simulated telephone conversation. One advantage of role-playing is that you can reverse roles with the other person and get to experience both sides of the conversation.

You might want a third person to observe, take notes, and provide feedback to you both. Try role-playing some of these conversations.

(a) You are very upset with Ken Bolton, an electronics supplier whose delivery did not arrive by Overnight Delivery as he promised it would. The result is that your project is on hold with the deadline fast approaching. You absolutely need these parts by tomorrow and it is too late to look elsewhere for the parts.

(b) A Mr. Yelpir Noj calls from Moldavia where he represents the Moldavian Mechanical Research Collective. At least you think so. Mr. Noj's accent makes his strained English difficult to follow. You are not sure of what he wants. Arrange for a return call with Mr. Noj and then describe how you would prepare to make this call.

(c) Choose a device with which you are familiar, an overhead projector, for example. Give it to another person. Sitting back to back with that person, try to answer the person's questions without looking at the device they have in front of them. Describe the results.

6. Use the call sheet to prepare for a conversation with Mr. Gates. You recently requested company support for your attending a week-long conference in Montreal, Canada on trade relations between the two countries. Sheila O'Brien, your supervisor, has expressed concern for the cost, $1400, and she suggested that you talk to Mr. Gates.

Gates & Associates

Call Focus Sheet

1. This call is important because:

2. By the end of this call I want to accomplish this result(s):

KEY PHRASES

PROMISES

I agree to:

They agree to:

7. Make a list of 20 common American English idioms. Idioms are phrases and combinations of words that have a unique or unusual meaning. "Take a break" has a meaning different than the literal meaning of the three words that make up the idiom; "break a leg" means good luck when used before a performance; "went broke" also suggests something quite different than the usual meaning of *to break*.

   Now imagine receiving a phone call from someone who has learned English from textbooks. Will this person know the idioms on your list? What strategies would you employ to make sure that you could communicate clearly and accurately with this person?

14

Participating in Meetings

Note to Users: **This is a task chapter. It is intended to discuss ways to lead and participate in meetings.**

14.1 INTRODUCTION

Meetings are a crucial part of any career in technology, and for that reason it is particularly surprising that so little time is spent in classrooms or in board rooms discussing ways to make meetings more productive. As your career takes you into a variety of levels and functions within a company or companies, you will participate in daily, weekly, monthly and impromptu meetings, interviews and briefings. In certain careers, as much as fifty percent of your time can be spent in meetings.

In this chapter you will learn how to:

- Facilitate productive discussions.
- Create an effective agenda.
- Design room arrangements to suit your meetings.
- Use closure to complete tasks.

14.2 DEFINITION

Meetings can be defined as any event where two or more people get together to discuss a subject or group of subjects. One basic difference between the meeting and a presentation is that, more than likely, a meeting asks for input from the participants, whereas a presentation communicates a specific amount of information to an audience. The emphasis in a meeting is on the group process, not the presenter. (The audience at a presentation may have a chance to ask questions, but their primary function is to absorb information.) Meetings also tend to be less formal than presentations. Three workers can meet in a hall near the water cooler and make important decisions about the future of a company.

There is no one standard for a meeting. A meeting may be held to assign projects and tasks, present timetables, summarize completed work, provide briefings on new products, or inform a group about advances in a specific field, among other purposes. Whether the meeting is a daily ritual that begins the activities of the day or is a rare event within your company, there are five essential concepts to remember about meetings:

Leader
Agenda
Minutes
Closure
Location

We will look at each of these concepts in turn. Afterwards, we will discuss the role of the audience and the importance of note-taking. At the end of the chapter, we will look at one specific type of meeting: the interview.

14.3 THE ROLE OF THE LEADER

Every meeting requires a leader, even if that leader has little power or authority. The leader's task is to facilitate discussion and decision-making.

If you are to be the leader, you need to **prepare.** The leader needs to do homework:

- Locate all the necessary background information.
- List your objectives: *What should the meeting accomplish?*
- Decide beforehand what course of action should follow the meeting. This may change during the meeting, but it is important to know what you want the next step to be.
- Select an appropriate site.
- Prepare the site. *See the section on room arrangement, Section 14.7.*

The leader must decide whether a meeting is the best way to use the time. Sometimes, a memo is far more efficient. The memo, or a series of memos, may detail six or seven key points that need very little elaboration or discussion. The leader must ask the question: **Do we really need to meet to discuss this?** Should x number of people sit in this room for two hours? Or is their time more valuably used in some other capacity?

For example
1 meeting = 10 people × 2 hours = 20 work-hours
20 work-hours × $25.00 per hour* = $500.00 a meeting

*(salary plus overhead)

Too often meetings become lectures. One key to a successful meeting is interaction. People like to feel that they have some input into a decision, particularly one that may have significant impact on their lives. It is the leader's responsibility to get everyone involved, to make sure that all of the participants know how decisions will be made, to monitor the progress of the meeting, and to cut off inappropriate behavior.

ETHICAL CONSIDERATIONS

Meetings reflect the values of a company or organization. The leader needs to resolve conflicts and ensure everyone of the opportunity to participate. Every employee, regardless of rank, needs to be treated fairly.

14.3.1 *The Phenomenon of Filling the Time*

A meeting expands to fill the available time. If you allow 30 minutes for a meeting, the meeting will take 30 minutes. If you allow 2 hours for the same meeting, the meeting will take 2 hours. Always err on the side of a shorter meeting.

14.3.2 *Awareness of Group Dynamics*

The group leader needs to know his audience. The past histories of those in attendance, including their personal relationships, have a strong influence on the performance of a group. Past successes and failures, friction because of social relationships, malice bred of mistrust, and a host of other problems may ruin what should otherwise go smoothly.

It is the leader's responsibility to analyze each participant not only for work skills, but also for how he or she gets along in a group situation. When an individual becomes disruptive to the group process, the leader must speak with that person privately and address the problem. The leader must also ensure that participants work together by being ready to deal with problems as they arise. Avoiding conflict will cause more serious long-term problems.

Example:

When Fred Warren was selected to succeed Bill Wilmington as the Department Head of Section Eight of Global Electronics, little did he realize what he was getting himself into. Warren had been the section leader at Global's Delaware branch office before he was promoted three months ago.

Now he must contend with an unhappy department. Every one of the 13 employees in the department had fully expected Gloria Monahan to get the job. She had been in the department for 11 years, longer than anyone else at her level. Her evaluations were always glowing with praise, and she had the respect of all her co-workers. She had been interviewed for the job, so it came as a complete surprise to everyone in Section Eight when Fred Warren, a relative outsider, was given the task of running the section.

The women workers believed that the situation was a clear case of discrimination, and everyone in the section feels that it was wrong to go outside: that this shows that loyal service to the company isn't rewarded, and therefore there is no reason to work hard. (There are a number of women who hold the position of Department Head, but no one understands why Gloria wasn't promoted.)

Fred was aware of the hard feelings in the beginning, but thought they would all go away. He may have hurt things when he increased the workload to get the General Data Info project done on time. Now meetings are a farce. No one listens, and absenteeism and tardiness are rampant.

To Fred, the situation is hard to fathom. He feels that he has done nothing to warrant such treatment, and now he is beginning to feel resentment towards his staff. The frustration is beginning to show in the way he conducts the meetings.

Analysis:

Fred Warren may rightly consider himself a victim of a situation that was not of his own doing. He needs, however, to address the situation directly and impartially. He should call a meeting with no other purpose than to clear the air, to get everyone talking about ways to solve the problem. Warren must be straightforward, open-minded, and clear about his belief that the employees are mis-directing their feelings of anger. If he doesn't deal with the situation now, it will only get worse.

To attempt to coddle his staff, to win their favor by being nice, may work short-term, but ultimately he will have to make unpopular decisions, and when he does, all of the tension will resurface.

14.3.3 Non-Verbal Communication

We frequently communicate more by our non-verbal behavior than by the words that come out of our mouths. Posture, movement around the room, hand gestures, facial expressions, and the way we look at other people may mean much more to people than a promise we make or a compromise we suggest. When people hear what we have to say, they evaluate our words by all of the signals we transmit as we say the words. They trust us more for the way we speak than for what we say.

You may believe that your non-verbal methods of communication are seen in a positive light, but sometimes we are fooled by our desire to be liked by our audience. The best way to learn about this is to videotape yourself speaking in front of a group. You may be surprised by what you discover.

A mirror can also be helpful, although you would only be able to use a mirror in a rehearsal. You may also want to find some people whom you trust to offer critiques of how you present yourself to others. You need to approach such critiques with an open mind, willing to be receptive to ways to improve your presentation style. If you don't hear the criticism, or if you reject it before you understand it, you won't learn anything of value.

Experienced speakers will scan an audience, gauging their reactions, making sure that they are following the meaning and showing the desired emotional reaction. These speakers are capable of making revisions in their presentation based on the clues they pick up from the audience, for the audience reacts in non-verbal ways also. Every schoolteacher is aware of the approach of the end of the class period without looking at the clock. The audience—the class—begins to fidget, to shift in their seats, to pack up their books and belongings. At least some members of the audience stop looking at the instructor in the same attentive way because they are looking forward to the end of the class.

As you become more comfortable with leading meetings, you will develop confidence in your ability to analyze the reactions of the participants and to adjust accordingly. The experienced leader of groups knows that to be heard a message must be phrased to fit an audience. The experienced speaker is acutely aware of his or her own methods of non-verbal communication and is continually looking for the non-verbal responses of the audience.

14.4 THE IMPORTANCE OF AN AGENDA

For a meeting to be successful, the participants have to know what it's about beforehand. The agenda, usually distributed in a memo, provides appropriate advance notice of the topics that will be covered. Frequently, having the agenda beforehand allows participants time to prepare and thus saves time. Without an agenda, those in attendance at a meeting can legitimately feel that something has been sprung upon them unfairly. People are much more receptive to ideas when they have had time to organize their thoughts.

The agenda should be relatively simple. Two or three key items are all that can reasonably be covered in a meeting. An agenda that is a list of ten items is overwhelming, and no progress may be made at the meeting.

Finally, and obviously, the agenda provides structure to the meeting.

A Sample Agenda Memo

Nov. 20, 1991

To: All staff
From: Stan Cohen, Project Coordinator
Re: Meeting on Friday

There will be a meeting of the staff this Friday at 12:00 in Room C-23. Please bring your files concerning the Wing assembly. We need to discuss:

1. The revised deadline for initial testing.
2. The design changes recommended by Engineering.

14.5 FOR THE RECORD: WHY WE KEEP THE MINUTES

The human mind is prone to forget. What can seem to be the most important business decision you have ever made can quickly fade from your memory. We all need to be reminded of what we have said and done, and we need to keep a record for other people. It's possible that of five people in a room making a key decision, not one will be with the company in five years.

The leader needs to assign some competent individual the task of recording the minutes of a meeting. The recorder need not write down every word that is said, but offer a summary of the key ideas and decisions made during the meeting. The leader must make sure that everyone receives a readable copy of the minutes.

The tasks of the leader do not end here, because the minutes need to be filed into some logical and accessible storage situation (whether in a filing cabinet or onto a computer disk) for future reference. A similar situation may arise five years down the road, or you may need to answer a question about company policy six months after the meeting. Having the written record provides you with a hard copy of what happened, not just some hazy rememberings. You may need the minutes of the meeting as a legal document, or someone else may need to know information long after you have left the company. Minutes are legal documents only if they are approved.

14.6 THE CONCEPT OF CLOSURE

All of us know the value of completing tasks, of getting things done. We all need to see where we've been and the value of the time we have spent. Meetings, particularly those that are scheduled on a regular basis, can become frustrating experiences when participants begin to feel that nothing gets done. The leader is the one who is responsible for closure so that the participants know that the meeting has ended successfully.

The leader should summarize what has been discussed, paying careful attention to the variety of viewpoints that have been expressed and to the conflicts that these viewpoints have generated. Whenever possible, the leader should blend viewpoints to attempt to generate compromise. The leader needs to finalize plans for future action.

At the very least, the leader should reserve the final five minutes of the meeting to answer the two questions that appear in the shaded box.

Two questions to ask at the end of every meeting

- What have we done?
- What happens next?

The best time to schedule a follow-up meeting or to assign an individual a specific task to accomplish is right now, at the end of the meeting. What is said at the end tends to stay with us the longest, and it is what we, the participants, will take back to our offices.

14.7 THE ARRANGEMENT OF THE ROOM

The leader of the meeting needs to ponder and prepare the location of the meeting well in advance. You need to consider your objectives and which room of the ones you have access to will best help you meet your objectives. You may want to think about the following:

- **intimacy** How many people will attend? How close will they sit? Do these people get along? Do you want to encourage or discourage discussion?
- **decor** Will the success of the meeting be influenced by aesthetic considerations? Will your audience perform better in a particular setting? Adequate lighting and wall-to-wall carpeting can influence the way people feel about your ideas.
- **air-conditioning** If one of your potential rooms has it, opt for air-conditioning during hot weather.
- **equipment needs** Plan beforehand to ensure that the slide projector, the overhead projector or whatever other equipment is in the room, works, and can be seen by everyone in attendance. Go into the room before the meeting and test the equipment yourself to make sure you

know how it operates. Don't allow yourself to be in a situation where things can go wrong. *See Section 15.9 in Chapter 15, Designing Presentations*, for more discussion.

If you have options, compare the structures of the different rooms keeping in mind what you want the meeting to accomplish.

14.7.1 Room Structures

Look at the following room designs. Which ones lend themselves to particular situations?

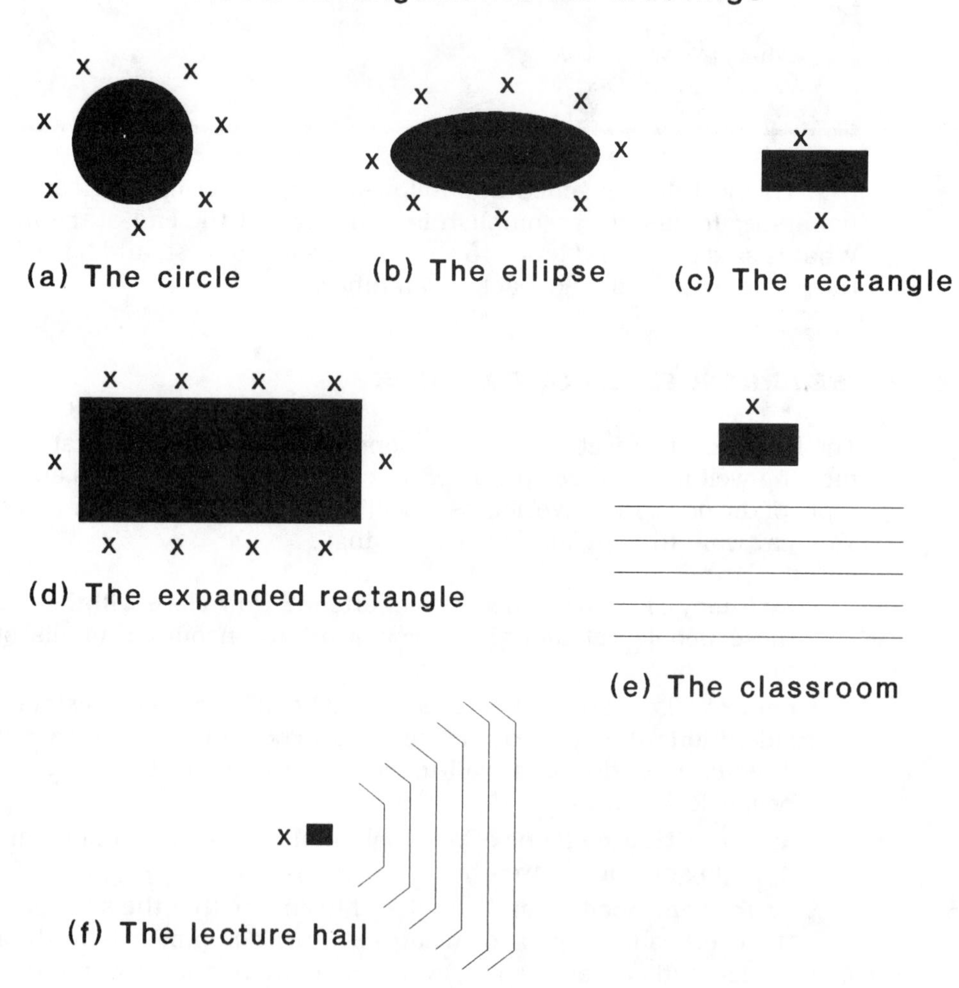

The following is a chart showing some of the advantages and disadvantages of the room arrangements presented above. Clearly, the right room for a meeting varies according to its purpose.

Table 14-1 THE ADVANTAGES AND DISADVANTAGES OF SPECIFIC ROOM ARRANGEMENTS

	Advantages	Disadvantages
The circle	Great for intimacy and discussion: everyone's equal.	Beyond 12–15 people, the circle becomes too large. People are too far away.
The ellipse	Good for discussion and intimacy. The leader can assume a position of control.	Same as above.
The rectangle (intimate)	Perfect for one-on-one meetings, especially interviews.	The one who is behind the desk has control. (This can be seen as an advantage.)
The rectangle (expanded) or Board Room	Efficient space use; people across table are close together. Leader in command of meeting.	People can hide in the corners. Focus on the leader, not the group.
The classroom	Allows the leader to control discussion; good for one-way communication; focuses attention to front.	Discourages discussion; audience can't interact with each other.
The lecture hall or conference room	Great for large groups particularly if there are tiers; great for large screen use.	Can seem empty with a small group; not intimate; speaker must project.

The state-of-the-art in meetings technology is the Electronic Meeting Room. At the University of Arizona, the Decision and Planning Laboratory allows meeting participants sitting at individual workstations to brainstorm and enter ideas that are compiled on large screens. The ideas are analyzed and arranged by priority by the group, enabling decisions to be made and policies formulated.

14.7.2 Booking Space in a Hotel or Conference Center

Most likely, the task of booking space in a meeting place outside of your company will be the responsibility of someone in the Personnel Department. Personnel officers handle approximately seventy-five percent of all such bookings. That means there is another twenty-five percent where employees in other areas of a company decide on an outside meeting site. In the event that the task falls upon your shoulders, or if Personnel handles the bookings but the success of the meeting is your ultimate responsibility, you should know some simple guidelines to follow.

> Visit the site first.
> Visit another site for comparison.
> Sit in the back of the room and imagine the meeting. Will those in the back row be able to see and hear you and the other speakers? Will the audience need writing space?
> Explain your equipment needs before you book the space.
> Work out and sign the contract well beforehand, making sure that the contract spells out all of your requirements. (You will probably be asked to make a deposit.)
> - food and beverage
> - microphone, podium, overhead projector, etc.
> - seating
> - all costs and surcharges
>
> Think about access for the handicapped. Can wheelchairs fit through the doorways?

Even if the negotiations will be handled by the Personnel Officer, you need to take an active role to insure that the meeting is a success.

14.8 THE ROLE OF THE AUDIENCE

Far more of the time you will spend in meetings during your lifetime will be spent as a member of the audience, not as the leader. Your listening skills are critical now. Far too many people assume that if they are in attendance they do not need to listen **actively,** that somehow the message will enter their ears and stay. But to retain information, we need to listen with a focus on the issues, not hearing just the words but their intent and their ramifications. The leader must communicate, but we must be receptive to the message. Attention at a meeting is a professional courtesy, but it is more than that: it is a way of improving job performance. Be aware of your nonverbal communications with the leader and with the other group members.

When the leader distributes the agenda before the meeting, he or she may ask you to add items. Here is your opportunity to participate in the decision-making process of the meeting, and hence the future course of your department or organization. You need to give careful thought to your contribution, keeping in mind the audience and its history.

During the meeting, whether or not you have supplied an agenda item, you may be asked to speak by the leader or you may have an idea to volunteer. Remember to speak clearly and direct your comments to all of the audience, not just the leader. Try to gauge how specific and technical you need to be by paying close attention to non-verbal cues. If you are paying attention, you will know just when to stop.

Finally, those in attendance should come prepared to take notes at meetings. Many college students learn the importance of taking clear and accurate notes during their first test in their Freshman year. Frequently, these same students develop their note-taking skills throughout their college careers, and then promptly forget all they have learned once they get into a work environment. It is difficult to take notes if no one else in the room is doing so. Your notes, however, even if they are brief jottings made at the conclusion of the meeting, can provide you with an invaluable written record of the major items of importance. So take notes. It takes discipline, but it is worthwhile.

Some consulting companies require all their employees to sit down for five minutes after a meeting and write up their notes, which are forwarded to the secretarial pool for typing. The practice supplies the employees with a record of what was accomplished at the meeting and a reminder of what tasks they need to complete.

Here are some suggestions for your notes

- List the key items of the discussion.
- List any decisions that have been made.
- List important dates, especially any deadlines that pertain to you.

14.9 FOLLOW-THROUGH

A successful meeting is one where the leader and the other participants do what they agree to do during closure. The leader must carry out the decisions of the meeting, and must distribute the minutes to everyone involved, including those who did not attend the meeting, within two or three days. The participants must keep their end of the bargain. If they have agreed to or have been assigned a task, they must carry it out. If there is to be another meeting, they must come prepared.

14.10 A SPECIAL KIND OF MEETING: THE INTERVIEW

Interviews are conducted to allow someone or some group to exchange information with another individual or group by asking and answering questions. Although we tend to think of interviews as a "one-way street" with the interviewer picking the brain of the interviewee, it is important to emphasize that a good interview is two-way communication. It is the interchange of ideas that characterizes a successful interview.

Although there are many types of interviews, some of the more common are exit interviews, transition interviews, debriefings, media interviews, and employment interviews.

Exit interviews are conducted by an objective, unbiased individual at the end of an employee's term of service to determine suggestions for improving processes and routines, and for improving inter-company communication. The risk here is that the ex-employee may express only gripes and complaints about personality conflicts.

A **transition interview** is used during times of change, upheaval or intense stress to ease the strain on employees. When departments are realigned, when individuals take on added responsibilities, and when there are significant layoffs, companies employ transition interviews as a way to solve present and potential problems.

A **debriefing** occurs when employees return to a company after working on special projects. A wise company interviews the returning employees to learn what happened: what went right, what went wrong, what improvements can be made on the next project, and why.

A **media interview** is perhaps the form of interview that we know the most about because we see, read, and hear so many of them, but it is also the most difficult to handle well. Only someone well-versed in handling the media should attempt to represent a company because there are so many pitfalls: even a "no comment" can backfire, and a seemingly harmless remark can blow up in your face. The potential for misrepresentation and distortion is very real, but when used well the media interview can have tremendous benefits for a company or enterprise.

The **employment interview** is conducted by the Personnel Department or by the decision-makers in a particular section of the organization. The employment interview is one of the final steps in the process of hiring new employees. Probably the most important interview you will ever attend is the one that lands you your first career job.

Employment interviews can be intimidating. You can feel that you are on the defensive, justifying your life to the interviewer. What you need to do is change your perspective about the interview; that is, you are evaluating the company as much as you are being evaluated. The employment interview is covered in much more depth in *Chapter 11, Applying For Work.* If you would like to explore this topic in more detail, we invite you to read Section 11.9, Preparing For The Interview, and Section 11.10, The Interview.

14.11 A FINAL WORD

We would like to emphasize one final point about meetings: if done well, meetings can have a significant and long-lasting positive effect on those who attend. You can improve productivity, boost morale, resolve conflicts, and inspire the participants. Meetings can work.

Refer to the following checklist before the next meeting you lead.

A Checklist for Meetings

Answer each of the following questions before you hold a meeting. As you answer, check off each question.

- [ ] Do you really need to meet about this?
- [ ] What should the meeting accomplish?
- [ ] What should occur after the meeting?
- [ ] Who should attend the meeting?
- [ ] How long should the meeting be?
- [ ] Where should we meet?
- [ ] How can we prepare the site?
- [ ] What special equipment is needed?
- [ ] What is the agenda?
- [ ] When should the agenda be distributed?
- [ ] Who will keep the minutes?

MEETINGS EXERCISES

Please note: Exercises 1–6 are interconnected and require the participation of the entire class. The entire process may require about 2 hours of class time to complete.

1. Divide the class into groups of five. The class as a whole should decide how to divide into these smaller groups. (You may want to use an arbitrary distinction: for example, the color of the shoes that people are wearing.)
2. Each group should elect a spokesperson. Write down the process used to decide upon the spokesperson.
3. Each group should decide on three low-cost ways of improving the student experience at your school. Make sure you discuss the rationale behind each of your concepts.
4. Each group will contribute three ideas to a meeting of the spokespersons. The agenda for the meeting will have only one item: to decide to implement **one** low-cost way of improving the student experience at your school. Only the spokespersons will be allowed to talk at the meeting. The other members of the class will watch but will not contribute at this point. (They will contribute in Exercise 5, so note-taking should be considered.)
5. Once the spokespersons have come to a final agreement on the one way of improving the experience of students, each member of the audience will write down their analysis of how the decision was made. You may want to look at the

process of the negotiation, the steps that led to a compromise, and the force of individual personalities in determining the final decision. Be particularly aware of non-verbal cues. Discuss two non-verbal communications you witnessed.

6. Discuss the written analyses as a group. The spokespersons should have the opportunity to react to the analyses. (The spokespersons may have been so involved in trying to influence the decision that they were unaware of many of the subtleties of the process.)

Please note: Exercises 7–13 involve conducting interviews. The best way to improve your interview skills is to watch a videotape of yourself with an open mind. Even if you do not have the capability of filming or videotaping your interview, you can learn quite a lot by conducting the interview and getting feedback from the rest of the group. These exercises work best when the interviewer is given a time limit. We suggest a time limit of 4 minutes.

7. Find out as much as you can about the technical expertise of the individual you are interviewing.
8. Discover the ideal career path of your interviewee. What would he or she like to be doing in 10 years? Work out some routes to that objective.
9. Find out what was the most difficult course your interviewee has taken at your school. What made the course so difficult? How could the course be improved?
10. Elicit a problem your interviewee is having at work. Try to find ways to solve this problem. (If your interviewee is not working, revise the exercise to read: elicit a problem your interviewee is having at school. Try to find ways to solve this problem.)
11. Who has been the most significant individual in the interviewee's life? Why?
12. Discover the most difficult individual in the interviewee's life. Elicit ways of improving the relationship with this individual.
13. Find out what the interviewee feels is the most pressing problem about your school. Elicit ways to solve this problem.

---

14. Read Section 26.8. Your task is to call a meeting to discuss how to develop a plan to solve the problem of the lack of information that firefighters face when toxic fires occur. Send a memo to the other members of the Special Projects Team informing them of the meeting and the agenda for the meeting.
15. Refer to the example in Section 14.3.2. Write an analysis of ways for Fred Warren to solve his leadership problem. What role should Gloria Monahan play at the meeting? Should an agenda be distributed? If not, why not? If so, what items should be listed on the agenda? How should the room be arranged? How should Fred Warren begin the meeting?

15

Designing Presentations

Note to Users: **This is a task chapter. It is intended to improve the way you think about and give presentations.**

15.1 INTRODUCTION

Designing presentations is an appropriate title for this chapter because what we would like you to do is use the presentation design process explained in Chapter 1 to design, literally, your presentation.

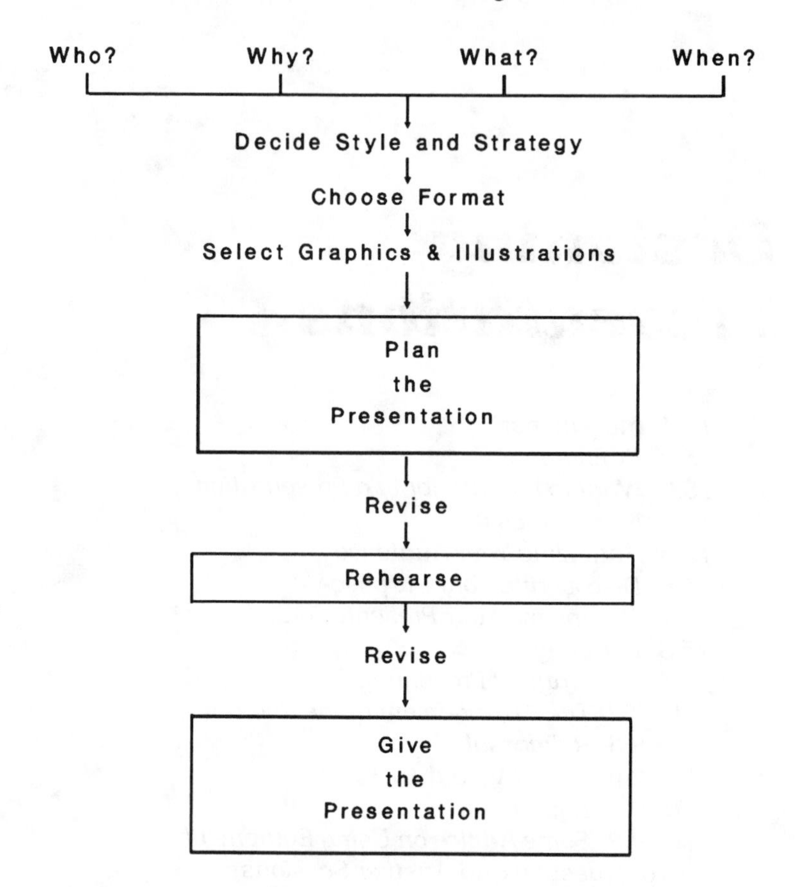

The more details you can prepare beforehand, the better your presentation will be. Imagine how you want the presentation to go. Imagine the reaction of the audience. Visualize the presentation being a complete success.

In this chapter you will learn how to:

- Organize and prepare presentations that persuade.
- Overcome your nervousness.

- Produce effective audio and visual aids.
- Respond to questions from the audience.

15.2 DEFINITION

In Chapter 14 we looked at meetings, and most of the ideas we wrote about in that chapter apply equally to presentations. For our purposes, we will define a presentation as any event where your primary purpose is to speak in front of an audience. In truth, there is very little to distinguish a meeting from a presentation, and there are times when the two words are used interchangeably. As we stated in Chapter 14, the focus of a presentation is on the presenter, while the focus of a meeting is on the group.

Perhaps a second key distinction has to do with the audience: the leader of a meeting usually has control over who will and who will not attend, but a speaker in a presentation rarely has complete control over who attends. For a presentation, a general announcement is made within an organization or to the general public and interested parties become the audience.

Given that you have little control over who attends, presentations tend to be more formal than meetings. In a positive sense, you are putting on a show for your audience, who have come to hear what you have to say; in other words, your task is to communicate a certain amount of information. You may be curious about what the audience has to say in response, but primarily you want them to absorb your message. In contrast, the leader of a meeting is frequently very concerned with getting feedback from the audience; the leader may encourage participation, sometimes to the point of generating a discussion and then standing back and letting the participants determine the direction of the discussion.

15.3 WHAT TO PREPARE FOR IN A PRESENTATION

Presenting information before an audience demands that you consider how your audience will receive your message, and that means that you need to think about the ways that a speech is different from a written document. Look at the Presentation Evaluation Form for a list of some of the factors that make a presentation successful.

Presentation Evaluation Form

1 ▪ Needs improvement
2 ▪ Satisfactory
3 ▪ Good
4 ▪ Excellent ✓

		1	2	3	4
Content	Knowledge of Topic				
	Organization				
	Clarity				
	Awareness of Audience				
	Transitions				
Delivery	Appearance				
	Voice Projection				
	Eye Contact				
	Posture and Gestures				
Visual Aids	Overhead Projections				
	Slides				
	Films				
	Handouts				
	Tables, graphs, charts				
	Posters				
	Other				
Q & A	Content of Responses				
	Rapport with Audience				
Overall Evaluation					

Notes and Comments:

The evaluation form certainly does not take into account every factor that is important in a presentation, but it should give you some ideas about areas to concentrate on. As you give more presentations, your sense of presence before an audience improves. You project more, you remember to make eye contact with the audience, you feel more comfortable being on stage. With the added confidence, your posture improves, your gestures become more natural, and your transitions become smoother. This is why it is so important to get experience by making presentations, and learning what works and what does not work for you. (If you are serious about gaining experience and have few opportunities, investigate joining a Toastmasters' club or a campus organization.)

The most important advice we can give you is to conceptualize the presentation beforehand. Try to imagine your audience, and give the presentation aloud. The shaded box gives some more advice.

Speak to the back row.
Look at your audience.
Pick out three different individuals in the audience and speak to them.
Practice different postures and speaking positions.
Work out the timing of the different parts of your presentation.
Write down the transitions between sections.
Test the equipment you will be using.
Imagine the questions from the audience.

15.4 NERVOUSNESS

First and foremost, you must come to grips with this one basic proposition about your presentation: **You may be nervous before your speech.** Fear of public speaking ranks right up there with fear of snakes and fear of going to the dentist as the top three fears of the American public. So it is perfectly possible that your pulse rate will increase, you will sweat more than usual, your mouth will get dry, and your memory will become hazy.

If this happens, **accept** it. It is normal, and the odds are that it won't affect your presentation. As soon as you get by the first minute, you will be fine, and you will build momentum as it becomes obvious that your worst fears haven't come true. Avoid reacting to your nervousness negatively: "I expected to be nervous and now I am. It's going to ruin my speech. Now I'm even more nervous. I can't do it." Instead, channel your nervousness. Realize that being nervous also means that you have more adrenaline to channel into your presentation. Make the added energy work for you.

From our experience—we have seen hundreds of individual and group student presentations—presentations that do not have a positive energy do not go over well with the audience no matter how knowledgeable the presenters are. A poor presentation actually drains an audience and makes the audience feel uncomfortable. On the other hand, when the presenters are enthusiastic, the energy is transferred to the audience. Therefore, being slightly nervous can have positive consequences.

15.5 ANALYZING YOUR AUDIENCE

As we mentioned before, you may not be able to control who attends your presentation. You can, however, still prepare for your audience. Know as much as you can about their point of view, know their positions within their company or organization, know their responsibilities, and know their history. If you have entered the text at this chapter, you should look back at the questions in *Chapter 3, Analyzing Your Audience.*

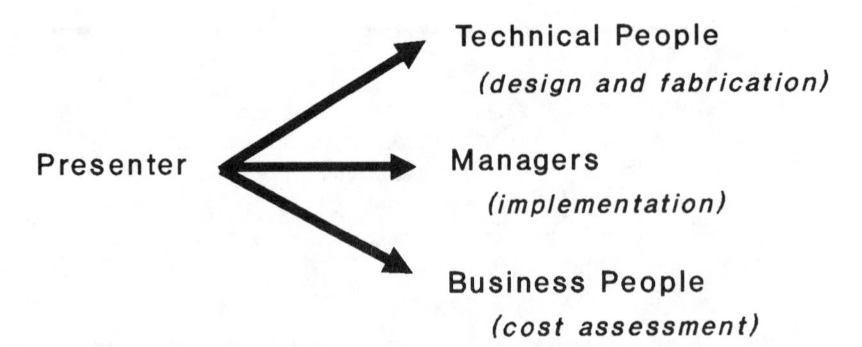

A presenter may be speaking to 3 different groups, each of which may focus on separate issues.

Knowing as much as you can about your audience will help you determine the best ways to bring your point across. You will decide how much detail to include and how sophisticated you can make your explanations.

- What will get the audience's attention?
- Are you influencing a decision? How?

15.6 DISCOVERING YOUR PURPOSE

As we have tried to stress above, the nature of your audience influences your purpose. In The Document Design Process, we placed knowing your purpose before knowing your audience. Here, however, because your audience determines to a great extent your message, once you have analyzed your audience you can begin to understand your purpose. In truth, audience and purpose are so inter-related as to be united. It is difficult to think about one without the other.

If you entered the book at the beginning of this chapter, it would be worthwhile to look at *Chapter 2, Discovering Your Purpose.* Your presentation will have, generally, one of a few basic purposes: to propose, to report, to motivate, to inform, to instruct, or to persuade. You may, however, want to accomplish a combination of the purposes listed above.

You need to decide your primary purpose before you begin. As a way of testing yourself on your purpose, complete the following sentences:

> The purpose of my presentation is to ______________________________
> my audience to (or, if your purpose is to inform, substitute about) ________
> ____
> ____.
> At the conclusion of my presentation, I would like the audience to ________
> ____
> ____.

Since your audience will have to receive information at your pace and they will not have the opportunity to backtrack to clarify important points as they could if they were reading your writing, it is a good idea to state your purpose a number of times in your presentation. Remember, many people do not retain information presented to their ears as well as they retain information presented visually. So follow the advice presented in the shaded box as a way of making certain that your audience understands your message:

> - State your purpose early and clearly.
> - In the middle, review your purpose.
> - Return to your purpose in your closing.

If you follow this advice, you have already completed the first step towards organizing your presentation.

15.7 ORGANIZING YOUR PRESENTATION

This is truly simple advice, essentially the Queen's admonishment to Alice while in Wonderland: Start at the beginning, go to the end and then stop. What could be simpler? All you need to do is to figure out what should be in the beginning, the middle and the end. Unfortunately, sometimes it isn't quite so easy.

Here is a sample rough outline of some areas covered in many presentations. You need to work out a logical order for the material you want to present.

Sample Outline

1.0 Title
2.0 Introduction
3.0 Overview
4.0 Background and History
5.0 The Main Issue or Issues
6.0 The Alternatives
7.0 Your Recommendation
8.0 Cost Analysis
9.0 Closing
10.0 Question and Answer Session

The above organization is not intended as a one-time, all-purpose organizational plan. Let your structure fit your presentation. The important thing is to have a clear and logical sequence to the ideas you present.

Here are some sequences for you to consider:

problem → solution Use this pattern when your task is to solve a problem.

theory → plan of action Similar in nature to the problem → solution approach, this pattern emphasizes an abstract concept that you or your team has evolved.

cause → effect Use this pattern to analyze the possible outcomes of a strategy.

first event → last event (chronological order) This pattern works particularly well when you are analyzing the history of a process. This pattern is often employed when a company is attempting to determine what went wrong.

More patterns of development are presented in Section 2.7 in the chapter on Discovering Your Purpose.

Whatever pattern of development you choose, the end of your presentation should always emphasize **closure,** a note of finality, a sense that you have successfully covered your main points.

15.8 PLANNING

A main aspect of planning for a presentation involves considering how much new information you need. Once you know this, you can figure out where you need to go to get it, and how much advance time this will require. Time to get the information can be costly if you have failed to plan ahead.

15.8.1 Ways of Presenting

People who are uncomfortable in front of an audience frequently read from a prepared text. While this reduces anxiety for the presenter, it is rarely a very rewarding format for the audience. What usually works best is an experienced, dynamic yet relaxed presenter who speaks extemporaneously, or gives the illusion of speaking extemporaneously. Experienced speakers have often planned their talks thoroughly, yet they are relaxed enough that it seems as if they are engaging in a spontaneous conversation with the audience.

Somewhere between reading a speech and speaking "off the cuff" lies a middle ground that we recommend for presentations. (If you are a relaxed extemporaneous speaker, you don't need to read this section.) We suggest that you write down your first three sentences on a notecard and read these sentences. This insures that you will get off to a good start. Now you can switch to using an outline of the main sections of your speech. Obviously, you may want to write down particularly complex ideas or statistics and read these directly, and you may want to read your conclusion from a prepared notecard.

There are some strategies you can employ to reduce the risk of forgetting information or becoming tongue-tied. One is to prepare visual aids to emphasize key elements of your presentation. Visual aids (*discussed in Section 15.9.*) allow you to deflect attention away from yourself and direct the audience to a chart or graph or bullet list. You can point out the relevance of specific items in your visual aids, and you can elaborate on other items. If you are speaking to a large audience, you can hide notes to yourself that the audience members are unable to see between the lines of a chart or poster. (Write these lightly in pencil.) If you are speaking from a podium, you can have a prepared text ready in case you need help. If you do not need the assistance, the audience need never know that it was there. Finally, you can use a mixture of media (for example, slides and overhead projections) with each one serving as a prompt for you to discuss another aspect of your main approach.

As a general rule, the more times you speak in front of an audience, the more comfortable you will become. You will be able to determine what works best for you. And, by all means, learn from others. If you see a presenter employ a technique that works well, adapt it to your own presentations.

15.8.2 The Arrangement of the Room

Do not overlook the importance of familiarizing yourself with the location of your presentation. See Section 14.7 in the chapter on meetings for a more detailed look at this issue. What applies to a meeting also applies to a presentation.

Room Arrangements for Presentations

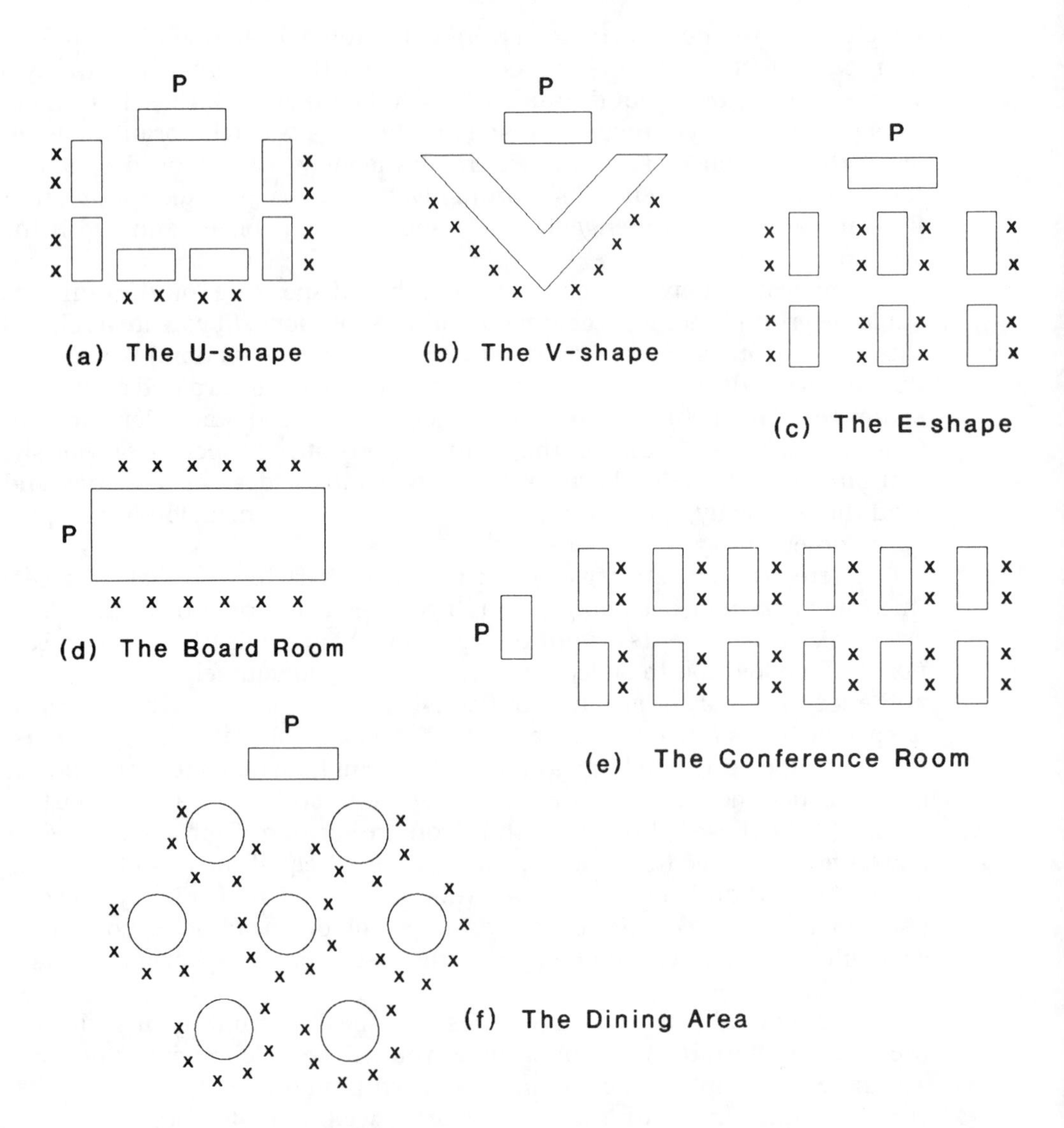

Whenever possible, visit the site before your presentation. Locate the outlets and the light switches. Set up your equipment and arrange the room to **your** needs. This gives you a sense of control and allows you to create a particular atmosphere. For example, if you can move the chairs in the room, you can create intimacy or distance between you and the audience.

The final thing you should do before the audience arrives at the site of your presentation is this: **Visualize success in the room.** Picture yourself performing successfully in front of the audience. Imagine the positive response you will receive.

15.8.3 Rehearsal

We would like to emphasize just how important we think this is: REHEARSE! The time you spend going through the process **will** pay off: you will smooth out rough spots, you will improve your transitions between sections, and you will have a good estimate of how much time you will need to deliver your presentation. This is particularly important if you working with a group; everyone in the group needs to understand completely the pacing, the cues, the transitions, and each other's roles before the presentation begins.

When you rehearse, you will make your major errors in front of no one or in front of friendly faces. Finally, if you are unable to turn on the overhead projector, you will be able to figure it out without becoming flustered and risking disaster.

15.9 AUDIO AND VISUAL AIDS

The rationale behind visual aids is this: many people comprehend ideas better when they see them. Visual aids can enhance your presentation by allowing your audience to see your ideas as well as hear them. And, visual aids can help you by giving you something concrete to refer to during your presentation.

See Chapter 8, Including Graphics and Illustrations, for more information and examples on preparing your visual aids. We want to emphasize a few basic rules from that chapter; please refer to Fig. 8-2 in Section 8.2. Also see the shaded box for some specific advice on including visual aids in your presentations.

Keep visuals simple.
Use large type, large enough to be seen clearly in the back of the room.
Give your audience time to digest the data. Count to ten before you begin talking.
Work on making smooth transitions between your visual aids.
Talk to the audience, not to the screen where your visuals are projected.
Be prepared to refer back to a visual if a question arises from the audience.

15.9.1 Equipment

Modern technology allows us a wide variety of possibilities for our presentations. In this section we summarize some of the more popular media and some of the advantages and disadvantages of each.

TABLE 15-1 Guidelines for Selecting a Visual Medium

	Flipcharts	*Overheads*	*Slides*	*Computer Graphics*
Audience Size	under 20	10-75	10-250	10-250
Degree of Formality	informal	informal or formal	formal	formal
Design Difficulty	simple	simple	requires photographic skills	simple to very complex
Materials	easel and paper	projector and screen; shades	projector and screen; shades	computer; screen; hook-up to projector
Production Costs	inexpensive*	inexpensive*	relatively expensive	expensive start-up

* unless drawn by a professional

This is a modified version of a table in Marya Holcombe and Judith Stein's *Presentations for Decision Makers* (1983).

Overhead projectors: Still the most common medium employed in presentations, overhead projectors have one big advantage over other media: virtually every conference room has one. If you are traveling somewhere to give your presentation, you can be reasonably confident of finding an overhead projector. Transparencies are relatively easy and inexpensive to make, usually requiring only a photocopy and a thermofacsimile machine. (Note: most transparencies are smaller than a standard 8 1/2" by 11" sheet of paper.) Cardboard frames can be put around the transparencies for easy handling. Make sure you can locate the light switch since you will need a darkened room.

Slide projectors: A sequence of good-quality color slides is an excellent way of telling a story to your audience. You can organize your slides with a carousel tray, and advance the slides at your pace. Slides do require some expertise with a camera, and enough lead time to get the film developed; sometimes you have to re-shoot certain pictures.

Computer projections: The capacity to project the image from a small computer screen to a large audience is costly and requires extensive set-up. You need a large-screen projector and a conversion system. (The conversion may take the form of an interface cable or a signal splitter, among other options.) Portable computer projection panels work with overhead projectors to create a dynamic large-screen image.

Since you can do so many things with a computer, the possibilities for enhancing a presentation are as great as your imagination allows you. Make sure you test the system in the location of the presentation before the audience arrives.

Video cassette recorders: A television screen can be comfortably viewed by a small group of people. If your audience is above twenty, you will need a large-screen projector. A well-planned and well-executed recording can be a very effective way of capturing an audience's attention.

New technology: Within the next ten years, we will see a staggering number of new media. Keep an open mind to these new media and their possibilities for enhancing your presentations. (For example, an electronic "blackboard" is available that turns what you write on it into a handout to give to members of the audience.) You can enhance your presentations by integrating recorded speech and music, and by importing still and moving images into your computer from videodisc players, CD-ROM players, and image scanners. Hypermedia systems can control the integration.

While we advise you to explore the new technology, we don't want you to lose sight of the "old" media.

Posters: A well-designed poster is a most effective way of conveying your message to your audience. Lettering machines allow you to make crisp, large text that can be viewed from the back of a room.

Flipcharts: A very portable means of communication, flipcharts allow you to illustrate your ideas and write text in rooms with no media equipment. For example, if you are asked to present at a luncheon meeting in a restaurant, you should consider flipcharts as an option.

Microphones: For large presentations where more than 125 or so people will be in attendance, it may be necessary to use a microphone. We advise you to practice; be particularly aware of how close you need to be to the microphone.

Pointers: A tool so convenient and so helpful we wonder why they are not used more frequently by presenters. Pointers are inexpensive and portable, and they allow you to clarify your visual aids.

Handouts: By distributing key information to the audience, you allow your listeners to be readers, to follow along with the presentation, and to have some "hard copy" to take out of the room with them. Too much information can overwhelm the audience and become a distraction, so limit your handouts to essential information.

Chalkboards: An option in a small or medium-size room, chalkboards are rarely seen in the business world.

15.9.2 Some Advice on Using Equipment

Test the equipment. Make sure that it works and that it works in that particular location. Very few things are more embarrassing than being unable to turn on a machine in front of 150 people. Find the outlets in the room, and check whether you will need an adaptor to plug in your equipment.

Don't get carried away. Not everyone will be impressed by flashy visuals. Content is the crucial factor. A great-looking presentation enhances the message, but it doesn't take the place of your message.

Mixing media requires good timing. You can not afford delays while you switch from one medium to another. Work on your transitions. Try to practice them the way a relay team practices passing the baton. The smoother the transitions, the better your presentation will be.

Have options in case something does go wrong. Even if you prepare thoroughly, something could still go wrong. Don't panic. Have a back-up plan so that you can continue with your presentation. If the problem is out of your control, you will find that your audience will be very forgiving.

15.10 QUESTION AND ANSWER SESSIONS

It is important to have a positive perspective on questions from the audience: view them as an opportunity to make your case one final time. Handle the questions as a way of emphasizing the points you are trying to make. Do not become defensive.

Some presenters prefer the audience to break in whenever they have a question, believing that this practice makes for more spontaneous interactions with their audience. The drawback is that the presentation may be taken away from its intended course by a question that opens up a new area. The audience may follow this new path and make it difficult to bring the presentation back on track. If you as a presenter fear that this may happen, we recommend that you ask that all questions be held until the end and that you do this in your introduction. You want to make this request before someone asks a question; otherwise you may be seen as evading a question.

Whenever you receive a question, there are a few simple rules to follow:

- **Restate the question.** This allows the entire audience to hear the question and provides you with valuable time to frame your response.
- **Break down the question.** If a series of questions has been asked, answer them one at a time. If a single question is lengthy or involved, break the question down into its parts.
- **Don't argue with the audience.** Even if an individual is obviously wrong, let the audience decide who is right. You have nothing to gain

and the respect of the audience to lose by entering a heated exchange. If the questioner persists, suggest that he or she meet with you after the end of the presentation.

- **Try to predict questions from the audience.** This is a main aspect of preparing for a presentation.

As with the other parts of a presentation, you will become more skilled at responding to questions as you become more experienced. Even skilled presenters can antagonize people if they fail to maintain a positive attitude towards their audience.

15.11 A FINAL WORD

The concept of closure is important enough to mention again. You need to give your audience a sense that the presentation is coming to an end and that something has happened during the presentation.

You can refer back to your statement of purpose or you can return to the first sentences of your presentation. Either approach is a clear signal that you are coming to the end. (For example, you can say, "When I began my talk, I asked you to consider four factors in approaching this problem. . . .") Remember, frequently the last impression we as an audience have is the strongest one.

As part of our closure of this chapter, we have included a list of items to consider when making a presentation.

Prepare to be nervous.
Find out as much as you can about your audience.
Complete this sentence: The purpose of this presentation is to ________.
Get all of your audio-visual equipment ready well before you speak.
Rehearse. Rehearse all of your presentation.
Speak **to** your audience, not **at** them.
Stay within yourself. For example, don't try telling a joke if you are not comfortable telling jokes.
Keep to your time limit. An excellent presentation may be ruined by excessive chatter.
Prepare for questions from the audience.
End on a positive note.

Another way of gaining closure in a presentation is to give the audience a suggestion of things to do to follow up with the ideas you have presented. For example, you could suggest further reading to be done, such as reading Marya Holcombe and Judith Stein's *Presentations for Decision Makers:*

Strategies for Structuring and Delivering Your Ideas (Van Nostrand Reinhold Company, 1983). Or you could give the audience a task: If you could find a videotape of any of the presentations of Tom Peters, the author of *A Passion for Excellence* and numerous other works, you will be able to see an enthusiastic and informative speaker in action.

PRESENTATION EXERCISES

1. Locate a short videotape of a presentation. Use the Presentation Evaluation Form located in Section 15.3 to rate the presentation. After you have completed the evaluation, write a list of five things you have learned.
2. This exercise is called the press conference, and it is modeled after the standard press conference we see on television. The audience, role-playing members of the media, asks questions of a volunteer who stands in front of the room and tries to respond spontaneously. The volunteer should begin with a very brief introduction and then invite questions from the audience. The audience should attempt to elicit as much information as possible about the person in seven minutes. (Seven minutes may seem to be a short time until you try it.)

   At the end of the press conference, the audience should discuss *only* the positive aspects of the way the volunteer handled questions. In other words, you are to give positive feedback. Focus on the good things that the individual did.
3. Divide into groups of three or four individuals. Take fifteen minutes to discuss the following: What could be done for under $1500 to improve the facilities on campus? Focus on one improvement. Then discuss how you intend to present your proposal to the class. At the next class meeting, be prepared to make a presentation of your idea.
4. Sheila O'Brien comes to you with a request. She has a $15,000 budget to purchase computer equipment. Although you have some flexibility, she must stay within the budget. Ms. O'Brien needs 6 personal computers and 1 laser printer for her staff. She asks that you research the main differences between Apple Macintosh computers and IBM and IBM-compatible machines available in her price range. She would like you to make an oral recommendation to her staff. For more information on Sheila O'Brien, *consult Chapter 26, Communicating In An Organizational Environment.*
5. See Exercise 1 in Chapter 13, Communicating by Telephone. Make a ten-to-fifteen-minute formal presentation of your recommendations to your classmates.
6. See Exercise 4 in Chapter 13, Communicating by Telephone. Make a ten-minute formal presentation of your recommendations for a long-distance carrier for Gates and Associates.
7. See Exercise 1 in Chapter 26, Communicating In An Organizational Environment. Make a formal presentation of your recommendation of the one office site from part e. (You are not given a specific amount of time for your presentation. When you ask Sheila about this, she says, "Take as long as you need.")

8. See Exercise 2, part d in Chapter 26, Communicating In An Organizational Environment. Prepare a fifteen-minute formal presentation of your investigation and your recommendations.

9. Gates and Associates has contracted to investigate a serious accident with a tanker delivering oil to storage facilities in Chelsea, Massachusetts, and to provide an analytical report on ways to prevent future reoccurrence. (You may be asked to investigate an accident occurring in a location nearer where you attend school.)

Your team has been assigned to set up a temporary field headquarters in Chelsea. Land has been provided by the state government for your team to set up an office in a leased trailer. You will need to lease or rent equipment to furnish this office for a ninety-day period while your team investigates the accident and prepares the report.

You will need to rent or lease:

- trailer
- four 386 computers
- two fax machines
- telephones and lines for six persons
- plotter
- laser printer(s)
- office furniture

Additional equipment and resources can be provided on a needs basis. In other words, if your team can demonstrate the utility or necessity for equipment, the company will pleased to provide these.

Your team consists of five technical persons from Gates and Associates. In addition, you will have to hire a temporary person for full-time clerical assistance for the ninety days. This person will need to be familiar with standard business software and computer processing.

Mr. Gates wants to use this project to establish a standard set of operating procedures for setting up temporary field offices and he has arranged for your team to make a presentation to the whole company describing choices, costs, and problems in accomplishing this task quickly and efficiently.

Your team should prepare the presentation.

10. Develop a one-page questionnaire to be given to an audience to elicit feedback at the end of a presentation.

16

Producing Technical Descriptions

Note to Users: **This is a task chapter, intended to instruct you about the process of composing technical descriptions.**

16.1 INTRODUCTION

You should not be concerned about your feelings or value judgments when you write a technical description. A description communicates the physical, quantitative characteristics of objects, devices, mechanisms or processes. What a technical description calls for is objectivity, not your subjective emotional responses.

When you begin writing a technical description, your purpose is clear: you want to describe a technical object or process. You need to consider your audience. The style of your writing and the technical level of your writing will depend on your analysis of your audience. Are they a general audience? An informed audience? An expert audience? The words you choose and the depth or degree of difficulty of your coverage will depend on this analysis. If the audience is mixed, write for the least expert people in your audience.

In this chapter you will learn how to:

- Provide effective definitions.
- Include precise, quantified specifications.
- Use illustrations to detail component parts and stages.
- Put together the complete description.

16.2 DEFINITION

A technical description describes, precisely and quantitatively, a technical item or process. Such descriptions can stand alone as a separate document, or they can be incorporated into larger documents. Technical descriptions can be used in lab notebooks, user manuals and sets of directions, sales and promotional materials, reports, and proposals, among other documents.

Your technical description can take a number of forms, but you should include

- an exclusive title.
- a definition of the object or process to be described.
- a list of component parts.
- illustrations that point out the component parts.
- a set of specifications.
- detailed explanations of the function and the operating principle of the object or process.

Accordingly, this chapter will look at each of these features in turn.

16.3 THE TITLE

Your title should be as precise as you can make it. Distinguish the item or process you are describing with specific words, limiting the description to the one item or process being described. If you are describing a specific product your company manufactures, use the complete brand name of the product.

Example

Use *Starrett Satin Chrome Outside Micrometer Caliper 0-1 inch* rather than *A Vernier Caliper.*

16.4 THE IMPORTANCE OF DEFINING YOUR TERMS

In a sense, a technical description is a definition. To define is to set down the precise boundaries and limits of something, to determine the nature of something. You may want to view the entire description as an elaborate, extensive definition. Your task is to define the object or process precisely.

The first part of this task is to write a clear, accurate **sentence definition** of the item or process you want to describe. A sentence definition consists of the term to be defined followed by the particular class or group to which the term belongs followed by the distinguishing, identifying features of the term. What you are trying to do is focus on what makes this term unique, what distinguishes it from all other items in its class. Here are some examples:

Term	Class	Distinguishing features
karat	unit of measurement	indicates the proportion of solid gold in an alloy based on a total of 24 parts. 10-karat gold indicates a composition of 10 parts gold and 14 parts of other metals.
mean	calculation	obtained by finding the sum of a series of quantities and dividing it by the number of quantities.
pigtail	electrical wire	used between a stationary terminal and a terminal having a limited range of motion, as in relay armatures.

The sentence definition puts together the term, the class and the distinguishing features in a sentence: "A karat is a unit of measurement that indicates the proportion of solid gold in an alloy based on a total of 24 parts."

At other places in your technical description, you may want to include other sentence definitions to define terms that may be unfamiliar to your intended audience. This requires making decisions about the knowledge base of your audience. Be particularly careful with words that have a special meaning within your specialized field and with phrases, code-words, and slang expressions that have been developed by people you associate with.

You can also define your terms by offering **parenthetical definitions.** Simply place a word or brief phrase that accurately identifies the term in parentheses after the word.

Examples

This swage (die or stamp for shaping or marking metal) has 10 grooves.

The theodolite (surveying instrument for measuring vertical and horizontal angles) was invented by the English mathematician Leonard Digges.

You can also use parentheses to note an abbreviation that you will use later in your document.

Example

Vapor-phase soldering (VPS) entails transferring the latent heat of the vapor to the surface where it condenses. The advantages of VPS are. . .

As you become more of an expert in a particular field, you may find your conversations filled with jargon (the specialized words and phrases of those in the same line of work or way of life). You may not be aware of the extent of your use of jargon, but if you see puzzled looks on people's faces, be sure to clarify your terms. For example, if you have experience with computers, you may assume that the phrase "Boot the computer" is part of everyone's vocabulary. Imagine the confusion of someone unfamiliar with this jargon.

When you are writing, it is crucial that you define your terms. Your audience may not be able to reach you for an explanation, and if they fail to understand a key word, they may not be able to make sense of your entire document. So remember to include definitions for your readers.

> A Suggestion: If you are having trouble defining a term, describe your term **operationally.** That is, begin with a description of the function of the item. For example, an optical post clamp is used to join a pair of perpendicular posts supporting a laser, lens, reflector, or other optical component.

16.5 PARTS

We view items in terms of their parts in order to simplify. We can understand a complex item by breaking it down into its components; even a simple item can be better understood by looking at its parts.

When an item has more than fifteen parts, you should consider classifying the parts into categories or sections. This partitioning should be done on a logical basis, such as location or function, or sequence in a process. If you are describing a complex object, for example, you can group parts that are close together or which work together to perform a particular function. If you are describing a complex process, you can break the process down into sub-systems. You can describe the process chronologically (according to time) and group individual steps by the stages in the process.

Your parts list should be complete. It should include *all* parts. Any specialized terminology should be defined. The parts list should correspond to a diagram of the item that should be placed reasonably near the list.

Example

PARTS OF THE POST CLAMP WITH ATTACHED COLLAR

One (1) Post-Clamp Body
Two (2) E-Type Retaining Rings
Two (2) Post-Clamp Collars
Four (4) Quarter-Turn Screws

If you are describing a process, list all of the distinct stages.

16.6 ILLUSTRATIONS AND DIAGRAMS

A technical description *should* contain illustrations. The figure clarifies for the reader by allowing the reader to visualize what is being described. The more complex the item or process, the more helpful illustrations become. The illustration can take many forms:

photocopy With some practice and a good photocopier, you can insert illustrations in your description and place text around it so that most readers will never consider how it was done. Obviously, you need permission to photocopy someone else's illustration.

photograph Fast-developing cameras can take images clear enough for in-house illustrations. For higher quality, use a 35-millimeter camera; the camera can be mounted on a copystand to allow clear photographs of small objects.

drawing Depending on your audience, the drawing can range from a professional artist's rendition to your own rough sketch. Many companies have artists on staff to work with you on drawings and schematics.

computer graphic Computer-Aided-Design (CAD) software programs enable you to draw complex items and processes. Connected to a plotter, CAD systems allow you to produce very high quality illustrations. Wiring diagrams and diagrams of manufacturing, architectural and electronic systems, among others, are particularly well-suited for CAD software.

Which form you select depends upon decisions you make about your audience and about your purpose. Any document, such as a sales brochure, that will be viewed by individuals outside of your company, requires a high-quality illustration.

The illustration should be accurate and easily understood. A three-dimensional view is particularly helpful to a reader, as are views from the side or from a variety of perspectives. The illustration of an item should correspond to the list of parts. The name of each part should be typed or printed on the illustration, and arrows should direct the reader to the location of the part.

Note: Even an elementary word processor can draw an arrow by using dashes followed by a "greater than" sign

---------->

See the following illustration for an example of an easily comprehended three-dimensional view of an optical device. Note the arrows denoting the specific parts of the assembly. Also note the second part of the illustration which shows the completed assembly of the post clamp.

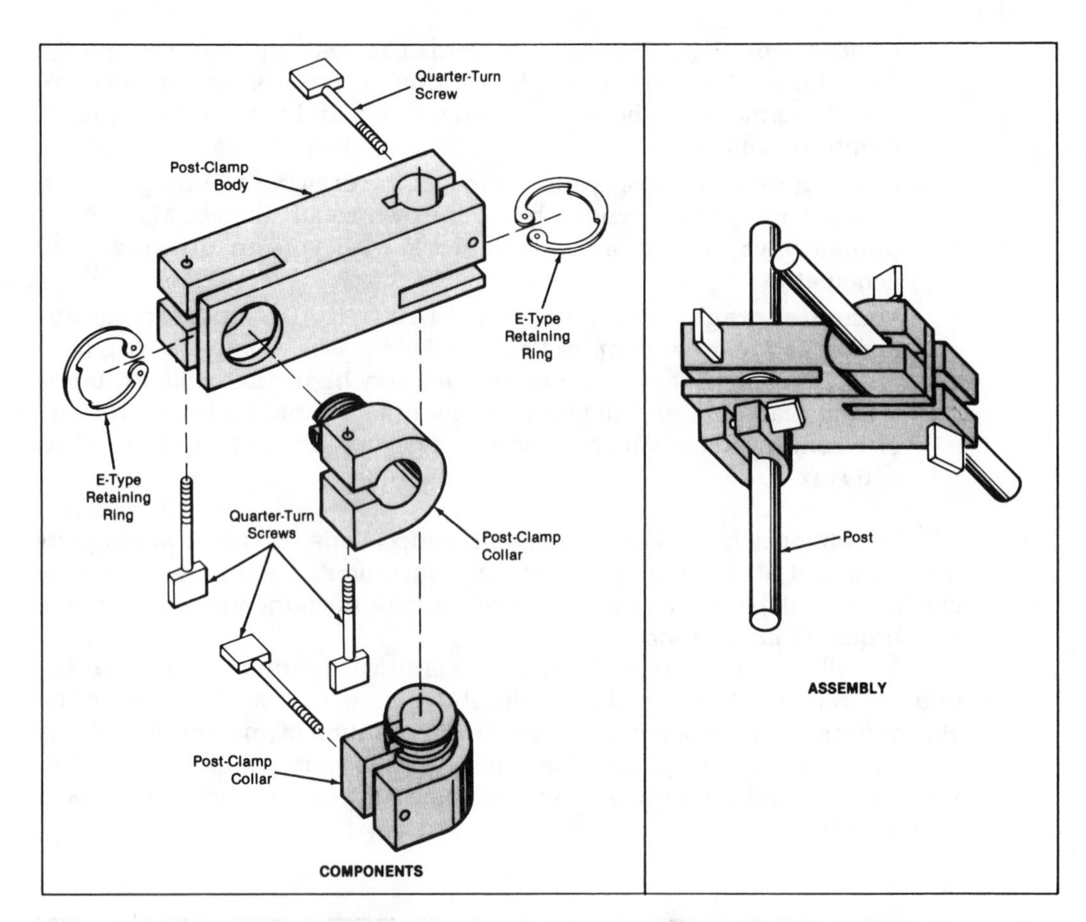

Figure 16-1 An illustration of the components and assembly of an optical post clamp.
Reprinted with permission of *NASA Tech Briefs.*

16.7 SPECIFICATIONS

A key element of all technical writing is precision. It should be present in technical descriptions. You need to include all the relevant details and you need to be accurate.

Specifications, or specs, are written for informed and expert audiences within a given field. Knowledgeable readers can tell a great deal about equipment and machinery from reading specs. Some experts can visualize the design and operation and, possibly, the flaws of a complex machine through reading the specs.

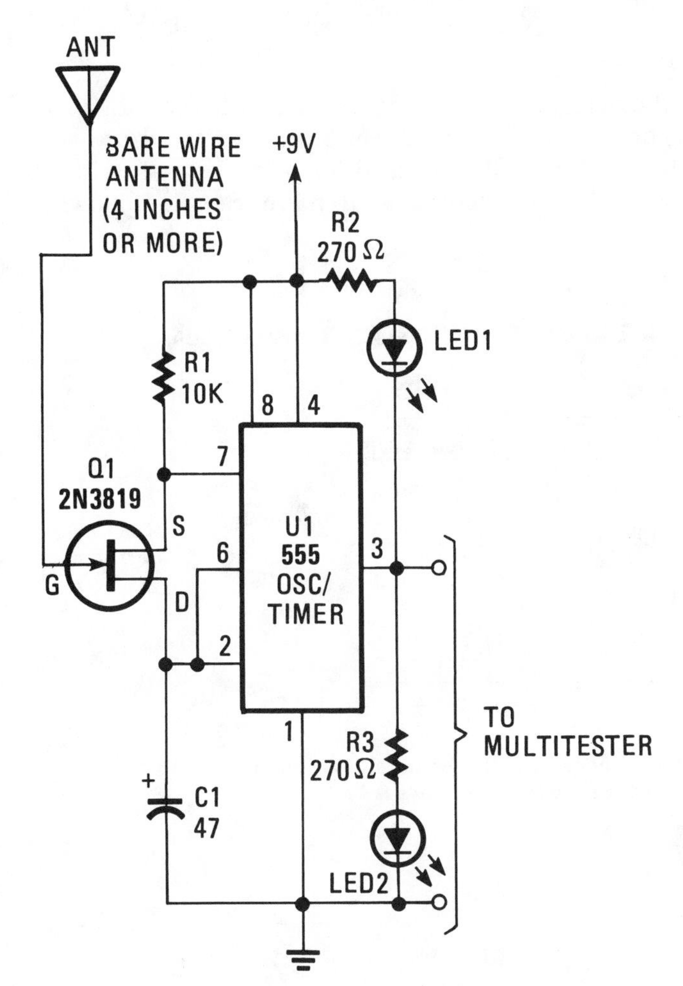

Figure 16-2 An electrical wiring diagram.
Reprinted with permission from *Popular Electronics*, January 1990 issue © Copyright Gernsback Publications, Inc., 1989.

16.7.1 What to Specify

What you specify depends to a very great extent on the object or process you are describing. Physical characteristics such as dimensions, weight, and shape are important in most specification lists, but such things as capacity, tolerance, temperature coefficient, tensile strength, clearance, electrical capacity, pressure ranges, bandwidths, and an endless list of other possibilities may be needed given the specific nature of the item or process you are describing. Price is crucial whenever you are selling a technical product. Look at the different items being specified in Figures 16-3, 16-4 and 16-5. Your specification lists may vary dramatically. As the expert, you must determine what information the readers need to know.

Double-check your items before you complete your document.

16.7.2 Ways to Specify

We recommend using a list or a chart. Try to avoid placing the relevant specifications with text: they are too easily confused or hidden within the words. Your specifications should be easily accessible, and they should be expressed in clearly recognizable units of measurement. See the following examples.

HEAVY DUTY STAINLESS STEEL THIMBLES (Type 302/304)

SPECIFICATIONS

Part No.	To Fit Cable Dia.	A1	A2	B	L1	L2	Thickness Comparison		Weight Each and Comparison	
							Heavy Duty (S)	AN Type (E)	Heavy Duty	AN Type
EY 18-3 *	3/32″	1/8	3/8	3/16	7/8	5/8	1/32	1/32	.41	.36
EY 18-4 *	1/8″	5/32	3/8	7/32	1	11/16	1/32	1/32	.65	.43
EY 18-5 *	5/32″	3/16	7/16	9/32	1 1/8	3/4	3/64	1/32	1.16	.57
EY 18-6 *	3/16″	7/32	1/2	5/16	1 1/4	7/8	3/64	1/32	1.45	.99
EY 18-8 *	1/4″	9/32	19/32	3/8	1 1/2	1 1/16	1/16	1/32	2.51	1.61
EY 18-9 *	9/32″	5/16	3/4	7/16	1 3/4	1 1/4	1/16	1/32	4.52	3.23
EY 18-10 *	5/16″	3/8	7/8	1/2	2	1 1/2	5/64	3/64	5.72	3.23
EY 18-12 *	3/8″	7/16	15/16	9/16	2 5/16	1 3/4	5/64	3/64	7.11	7.93
EY 18-16 *	1/2″	9/16	1 1/8	3/4	2 3/4	2 1/8	3/32	1/16	14.01	–

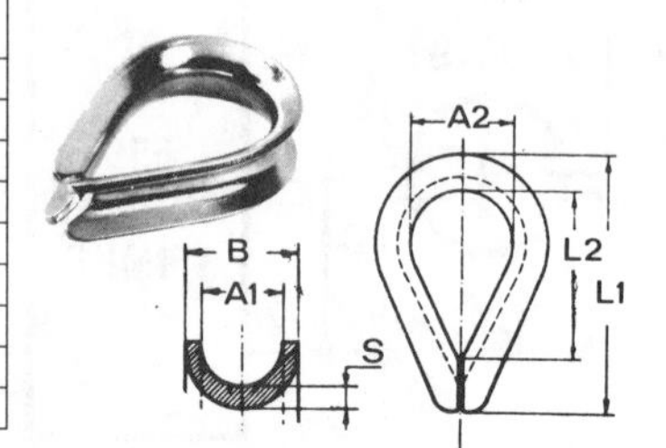

Figure 16-3 A specification list for stainless steel thimbles. **Reprinted with the permission of Loos & Co.**

Model	Nominal Voltage V	Nominal Capacity @ 20 hr. rate A.H.	Discharge Current @ 20 hr. rate mA	DIMENSIONS								Weight	
				Length		Width		Height		Ht. Over Terminal			
				in.	mm	in.	mm	in.	mm	in.	mm	lbs.	kg.
PS-445	4	4.5	225	1.93	49	2.09	53	3.70	94	3.86	98	1.5	0.7
PS-490	4	9.0	450	4.02	102	1.73	44	3.70	94	3.86	98	2.4	1.1
PS-605	6	0.5	25	2.24	57	0.55	14	1.97	50	1.97	50	.20	.09
PS-610	6	1.0	50	2.00	51	1.65	42	2.00	51	2.20	56	0.6	0.3
PS-612	6	1.2	60	3.82	97	0.94	24	2.00	51	2.13	54	0.6	0.3
PS-618	6	2.0	100	2.95	75	2.00	51	2.09	53	2.28	58	1.1	0.5
PS-626	6	3.0	150	5.28	134	1.34	34	2.35	60	2.56	65	1.5	0.7
PS-630	6	3.0	150	2.60	66	1.30	33	4.84	123	5.00	127	1.5	0.7
PS-640	6	4.0	200	2.76	70	1.89	48	4.02	102	4.02	102	1.8	0.8
PS-660	6	6.5	325	5.95	151	1.34	34	3.70	94	3.86	98	3.0	1.4
PS-665	6	6.5	325	3.86	98	2.20	56	4.05	103	4.05	103	3.0	1.4
PS-682	6	8.0	400	3.86	98	2.20	56	4.65	118	4.65	118	3.3	1.5
PS-6100	6	10.0	500	5.95	151	2.00	51	3.70	94	3.86	98	4.6	2.1
PS-6105	6	10.0	500	4.25	108	2.75	70	5.51	140	5.51	140	4.3	2.0
PS-6120	6	12.0	600	4.25	108	2.75	70	5.51	140	5.51	140	4.6	2.1
PS-6200	6	20.0	1000	6.18	157	3.27	83	4.92	125	4.92	125	8.2	3.7

Figure 16-4 A specification table for Lead-Acid Batteries. **Reprinted with the permission of Power-Sonic Corp.**

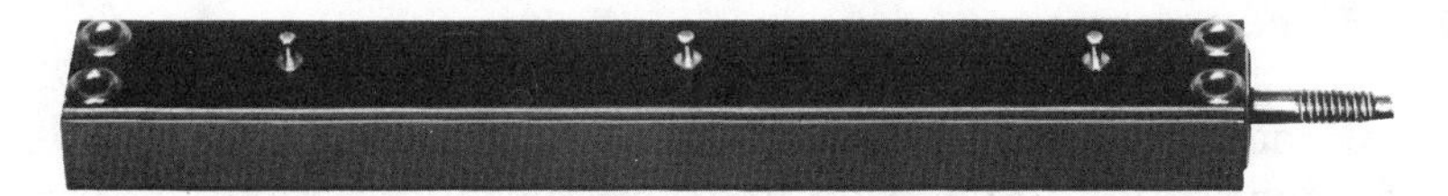

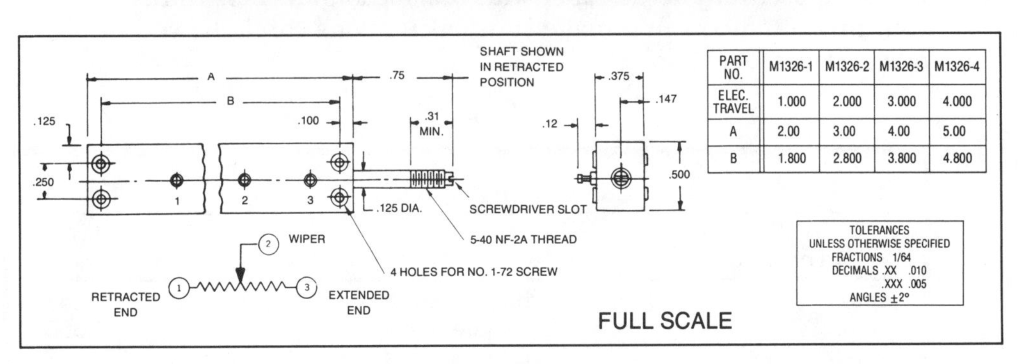

ELECTRICAL AND MECHANICAL CHARACTERISTICS

	M1326-1 STANDARD	M1326-2 STANDARD	M1326-3 STANDARD	M1326-4 STANDARD	RANGE OF SPECIAL DEVIATIONS
Total Resistance (Ohms)	25Ω to 25K	25Ω to 50K	50Ω to 75K	100Ω to 100K	10Ω to 250K
Total Resistance Tolerance	±5%	±5%	±5%	±5%	To ±1%
Theoretical Resolution (Inch)	.006 to .001	.010 to .001	.010 to .001	.010 to .001	To .001 Inch
Independent Linearity	±1.2% to ±.2%	±.6% to ±.1%	±.4% to ±.1%	±.3% to ±.1%	To ±.05%
Theoretical Electrical Travel (Inches)	1.000	2.000	3.000	4.000	To 12.000 Inches
Theoretical Electrical Travel Tolerance (Inches)	±.010	±.010	±.010	±.010	To ±.005 Inch
Total Mechanical Travel (Inches)	1.10	2.10	3.10	4.10	To 12.10 Inches
Temperature Coefficient (PPM/°C)	50	50	50	50	To 20 PPM/°C
Operating Temperature Range	−55° to +125°C	−55° to +125°C	−55° to +125°C	−55° to +125°C	−200° to 175°C
Power Rating (Watts at 40°C)	1	2	3	4	To 2 Watts Per Inch
Dielectric Strength (Volts, RMS to 60 Sec.)	750	750	750	750	To 2500 Volts, RMS
Terminals	3	3	3	3	To One Per .25 Inch
Taps	0	0	0	0	To One Per .25 Inch
Tap Tolerance (Inches)	±.010	±.010°	±.010	±.010	To ±.005 Inch
Life Expectancy (Cycles at 60 In./Min.)	500,000	500,000	500,000	500,000	To 5,000,000 Cycles
Shaft Frictional Force (Ounces)	4 Oz.	4 Oz.	4 Oz.	4 Oz.	
Weight (Ounces)	3 Oz.	4 Oz.	5 Oz.	6 Oz.	
Bearings	Nylon	Nylon	Nylon	Nylon	Bronze, Delrin

MAUREY INSTRUMENT CORP.
4555 West 60th Street • Chicago, Illinois 60629 • (312) 581-4555 • FAX (312) 581-2576

Figure 16-5 Electrical and mechanical characteristics of a wirewound potentiometer.
Reprinted with the permission of Maurey Instrument Corp.

Note how the specifications can be presented as tables. A large product line can be easily shown and compared, allowing the reader to select the best-suited item. Also note in Fig. 16-3 how an illustration accompanies the specifications to show where specific dimensions correspond to letters on the chart.

16.7.3 Measurement

Measurement requires exactness: precision measuring instruments allow us the opportunity to be exact with our measurements. You should take your measurements and make your calculations carefully; your audience relies on you for accuracy.

Conversion Factors

To change	To	Multiply by
acres	hectares	.4047
acres	square feet	43,560
acres	square miles	.001562
atmospheres	cms. of mercury	76
BTU	horsepower-hour	.0003931
BTU	kilowatt-hour	.0002928
BTU/hour	watts	.2931
bushels	cubic inches	2150.4
bushels (U.S.)	hectoliters	.3524
centimeters	inches	.3937
centimeters	feet	.03281
circumference	radians	6.283
cubic feet	cubic meters	.0283
cubic meters	cubic feet	35.3145
cubic meters	cubic yards	1.3079
cubic yards	cubic meters	.7646
degrees	radians	.01745
dynes	grams	.00102
fathoms	feet	6.0
feet	meters	.3048
feet	miles (nautical)	.0001645
feet	miles (statute)	.0001894
feet/second	miles/hour	.6818
furlongs	feet	660.0
furlongs	miles	.125
gallons (U.S.)	liters	3.7853
grains	grams	.0648
grams	grains	15.4324
grams	ounces avdp	.0353
grams	pounds	.002205
hectares	acres	2.4710
hectoliters	bushels (U.S.)	2.8378
horsepower	watts	745.7
hours	days	.04167
inches	millimeters	25.4000
inches	centimeters	2.5400

(continued)

kilograms	pounds avdp or t	2.2046
kilometers	miles	.6214
kilowatts	horsepower	1.341
knots	nautical miles/hour	1.0
knots	statute miles/hour	1.151
liters	gallons (U.S)	.2642
liters	pecks	.1135
liters	pints (dry)	1.8162
liters	pints (liquid)	2.1134
liters	quarts (dry)	.9081
liters	quarts (liquid)	1.0567
meters	feet	3.2808
meters	miles	.0006214
meters	yards	1.0936
metric tons	tons (long)	.9842
metric tons	tons (short)	1.1023
miles	kilometers	1.6093
miles	feet	5280
miles (nautical)	miles (statute)	1.1516
miles (statute)	miles (nautical)	.8684
miles/hour	feet/minute	88
millimeters	inches	.0394
ounces avdp.	grams	28.3495
ounces	pounds	.0625
ounces (troy)	ounces (avdp)	1.09714
pecks	liters	8.8096
pints (dry)	liters	.5506
pints (liquid)	liters	.4732
pounds ap or t	kilograms	.3782
pounds avdp	kilograms	.4536
pounds	ounces	16
quarts (dry)	liters	1.1012
quarts (liquid)	liters	.9463
radians	degrees	57.30
rods	meters	5.029
rods	feet	16.5
square feet	square meters	.0929
square kilometers	square miles	.3861
square meters	square feet	10.7639
square meters	square yards	1.1960
square miles	square kilometers	2.5900
square yards	square meters	.8361
tons (long)	metric tons	1.016
tons (short)	metric tons	.9072
tons (long)	pounds	2240
tons (short)	pounds	2000
watts	Btu/hour	3.4129
watts	horsepower	.001341
yards	meters	.9144
yards	miles	.0005682

Figure 16-6 A quick reference table of conversion factors. "Conversion Factors" table from *The 1987 Information Please Almanac* edited by Otto Johnson. Copyright © 1986 by Houghton Mifflin Company. Reprinted by permission of Houghton Mifflin Company.

16.8 FUNCTION

The function of an item refers to what it does. It may have one specific function, it may have a primary function and numerous secondary functions, or it may have a number of primary functions. A stapler, for example, has one specific function, whereas a hammer has primary and secondary functions and an integrated computer software package has many primary functions. See Fig. 16-7 for another example of the description of the functions of a specific micrometer caliper.

16.9 OPERATING PRINCIPLE

The operating principle refers to how an item works. Thus the function of an item is **what it does** while the operating principle is **how it works.** A pair of scissors is used to cut things; it works by placing something, such as a sheet of paper, between the blades, grasping the two handles and moving the handles towards each other. The blades will slice through the object. One blade rotates on a pin affixed to the other blade. See Fig. 16-7 for an example of a description of the operating principle of a specific micrometer caliper.

> Two excellent sources can make your task easier:
>
> *The Way Things Work* by David Macauley, published by Houghton Mifflin Company of Boston, MA.
>
> *How It Works: The Illustrated Science and Invention Encyclopedia,* in 24 volumes, published by H. S. Stuttman of Westport, Connecticut.
>
> If you use one of these sources, be sure to give credit.

The operating principle will become a very important part of your technical description when you are describing a process. You will describe the process from start to finish, and this will often entail how the process occurs. An illustration that shows the process step by step is particularly useful in this case.

16.10 PRODUCT DESCRIPTION

When the technical description is intended for the general public or a specific audience outside your company, you are concerned about the image of

STARRETT SATIN CHROME OUTSIDE MICROMETER CALIPER 0 - 1 inch

TECHNICAL DESCRIPTION OF
STARRETT SATIN CHROME OUTSIDE
MICROMETER CALIPER 0 - 1 inch

Submitted by

Scott E. Morris

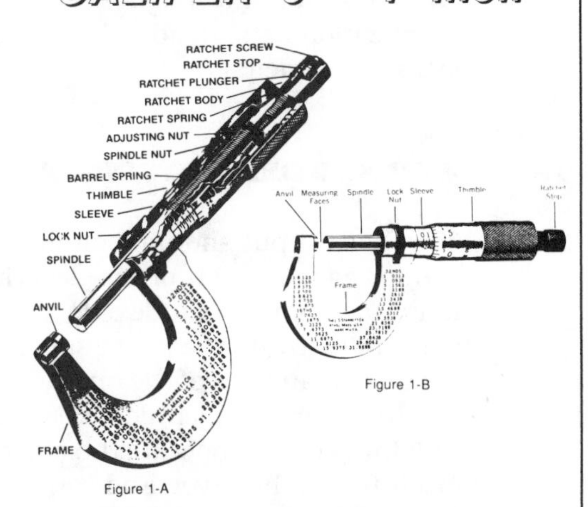

Figure 1-B

Figure 1-A

How To Read The Micrometer Caliper

Refer to figure 1-E

A. The "1" line on the sleeve is visible representing .100 in.

B. There are 3 additional lines visible, each representing .025 in. Thus, 3 x .025 = .075 in.

C. The "3" line on the thimble coincides with the *reading line* on the sleeve, each line on the thimble representing .001 in. Thus, 3 x .001 = .003 in.

D. The micrometer reading is:

.100 + .075 + .003 = .178

SLEEVE THIMBLE

READING .178"

Figure 1-E

An easy way to remember how to read the caliper is to think of the various units as if you were making change from a ten dollar bill. Count the figures on the sleeve as dollars, vertical lines on the sleeve as quarters, and divisions on the thimble as cents. Add up your change and put a decimal point instead of a dollar sign in front of the figures.

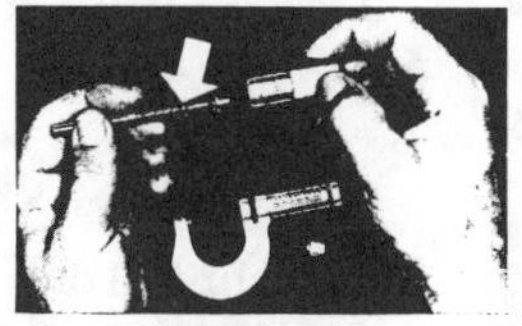

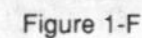

Figure 1-F

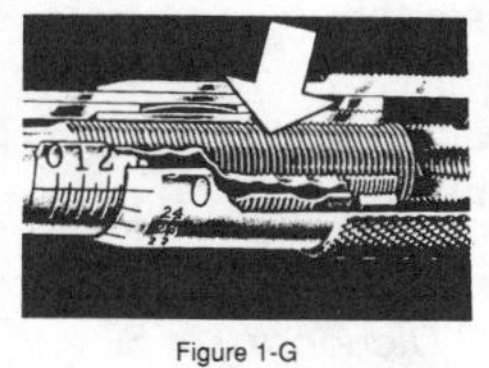

Figure 1-G

OPERATING PRINCIPLE

The micrometer caliper consists of a highly accurate ground screw or spindle (see figures 1-F & 1-G) which rotates in a fixed nut, thus opening or closing the distance between the two measuring faces on the ends of the anvil and spindle.

The pitch (distance between one point on one thread and the same point on the next thread) of the screw thread on the spindle is 1/40 in. or 40 threads per inch. This makes one complete revolution of the thimble advance the spindle face toward or away from the anvil face precisely 1/40 in. or .025 in.

STARRETT SATIN CHROME OUTSIDE MICROMETER CALIPER 0 - 1 inch

DEFINITION

A precision instrument used to measure distances from .001 to 1.000 inches through the amplification of screw threads and a vernier scale, commonly used in the metal working trades.

DESCRIPTION

Length: approximately 4 3/8 in.(5 3/8 open)
Width: approximately 2 1/4 in.
Weight: approximately 24 oz.
Material: hardened and precision ground tool steel
Texture: satin chrome finish
Min. Measurement Obtainable: 0.001 inches
Max. Measurement Obtainable: 1.000 inches

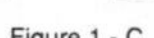

Figure 1 - C

Figure 1-D

FUNCTION

An outside micrometer caliper (commonly referred to as a "*mike*") is used to measure the distances between two parallel surfaces quickly and accurately. Some examples would be the thickness of a steel plate, outside diameter of tubing or round stock, width of a flange or step on a machined item (or *workpiece*), or external diameter of screw threads.

Measurements are obtained by placing the workpiece between the spindle and anvil faces, and slowly rotating the thimble counter-clockwise until the spindle creates "*positive contact*" (NOT CLAMPING PRESSURE) with the workpiece. See figures 1-C & 1-D. This positive contact is a "*feel*" that is acquired through experience and use of the mike. By turning the lock nut, the spindle becomes locked in place. Now the workpiece can be removed and the measurement can be read off the vernier scale located on the sleeve and thimble.

Figure 16-7 An example of a technical description
Reprinted with the permission of Scott Morris and the L. S. Starrett Co.

the company. The description can become part of the selling of a product. If this is so, the format of the description becomes much more important.

Given the importance of the document, it is highly likely that this will be handled by professional designers, either on staff or outside consultants. Rarely will you have to prepare a finished sales document. You may be consulted for guidance about technical matters, such as the accuracy of the specification lists, the detail of the illustrations, and the explanation of the operating principle.

16.11 A FINAL WORD: PUTTING IT ALL TOGETHER

Your final output should include a clear and specific title, consistent headings for each section of the description, and carefully positioned illustrations. You should include a definition of the object or process, a parts list, a labeled illustration, a set of specifications, an explanation of the function, and an explanation of the operating principle.

In general, a technical description will not include a set of directions explaining how to operate the object or complete the process. This is not an absolute rule, however, and you may be called upon to include a set of directions within your technical description. For more information on how to write directions, *see Chapter 17, Giving Technical Directions.*

Once you have completed all of the parts of the technical description, you can put it all together. An example of a technical description of a micrometer caliper is in Fig. 16-7.

DESCRIPTION EXERCISES

1. Write a sentence definition of the following items:

a) cryogenics	e) toggle switch
b) drill bushing	f) ohm
c) coining die	g) Mesozoic Era
d) peptone	h) operculum

Note: To write sentence definitions, you may need to consult a technical dictionary, such as the *McGraw-Hill Dictionary of Scientific and Technical Terms.*

2. Write a technical description of a paper clip. Include specifications, function, operating principle and an illustration of the paper clip.

3. Illustrate a stationary or static item. Make sure you label the parts of the item.

4. Illustrate a process. The process should have at least three steps. Make sure your illustration or diagram clearly shows the sequence of events in the process. Caution: Do not select an extremely complex process.

5. Select an object that has at least two moving parts and write a technical description of the object. You decide on the format for the assignment, but your description should include the following:

> An exclusive title
> A definition of the object
> A list of component parts
> An illustration of the object
> A physical description that includes
>   height, width, and depth
>   weight
>   texture
>   shape
>   other relevant physical details
>   function
>   operating principle
> A comparison with more familiar items, if helpful
> Definitions of key terms, when relevant

6. Find an example of a technical description. Photocopy the description. Analyze the strengths and weaknesses of the description. Make a list of at least three strengths and two weaknesses.

7. Choose one of the items on the following list. Investigate the item and write a technical description of the item.

a) electronic ferrule
b) dry particle separator
c) solenoid valve
d) image processor
e) pressure transducer
f) split box furnace
g) local area network

8. Find out how a multiplexer works, and prepare a comprehensive explanation of the operating principle of a specific multiplexer.

17

Giving Technical Directions

Note to Users: **This is a task chapter. It is designed to help you prepare instructions, procedures, and specifications which will guide and guard your audience.**

17.1 INTRODUCTION

Technical professionals, because of their specialized knowledge, are frequently required to explain how to do things, how to assemble something, complete a process, or repair broken machinery. The documents which result are variously called directions, procedures, instructions, and specifications.

All of these writing tasks have the same basic purpose, to describe a sequence of actions in which the audience will participate. Effective directions guide the reader through all the steps necessary to complete what they have started. People only read directions because they want to **do** something and do it correctly. Your job is to assist them.

This is a very important writing task. Faulty directions can cause losses of time and productivity and can put people in danger. You need to carefully balance the demands of the task and the needs of the users. When you write instructions, procedures, specifications, or directions, you are telling some person how to perform some activity or how to locate and attend to a very real problem.

In this chapter you will learn how to:

- Guide your readers and guard them.
- Design instructions, specifications, and procedures.
- Choose effective levels of detail.
- Use standard field specifications.
- Protect your audience with safety cautions.

17.2 DEFINITION

Technical directions are documents which provide a planned sequence of activities which the user expects to carry out. Whether your plans are called instructions, specifications, or procedures, you use the same writing technique: a detailed description of a process designed to guide your audience to complete a technical task or operation.

What distinguishes technical directions from other kinds of writing is the extent to which the audience cooperates. When you write directions, you are telling people what to do. Your style is direct and commanding. Your audience participates by following your orders—your directions. This gives you important responsibility for their safety and their success.

ETHICAL CONSIDERATIONS

Perhaps the strongest ethical responsibility is to provide reliable information. This means you have to consider the context of your technical communications. You need to know your audience, your message, and your strategy. You are legally obliged to provide accurate information, and you have an ethical obligation to provide that information in a useful fashion. You are not communicating to be appreciated, but to help the persons using your communication.

17.2.1 Instructions

Instructions are usually designed for beginners to do something for the first time. When considering your audience, it is logical to suppose that your reader is uninformed. In general, people familiar with a device or process do not use instructions. A new product or process, by definition, needs to be explained. The newcomer who needs a full explanation of what to do is the most common audience for your instructions.

There are two kinds of instructions: (1) explanations of how to complete a process or operation and (2) explanations of how to discover and correct a problem in the process. Whether you are providing operating or troubleshooting instructions, you need to provide step-by-step guidance to your novice readers.

Operating instructions tell readers how to assemble, install, maintain, and operate various tools, machinery, and equipment. These instructions are lists of simple steps, one action in each step, numbered for easy use, and carefully sequenced. In other words, you need to provide the steps in the order the readers will perform them. You can't add an ingredient after you've cooked the cake.

Troubleshooting instructions help the audience discover the cause of a problem, often equipment failure, and then to repair the problem. These instructions are more difficult to design because the order is less apparent. Unfortunately, equipment does not fail in chronological order and you have to direct your readers to consider all possibilities.

You need to imagine exactly what your readers will see, hear and feel as they analyze their problem. Remember this, it is someone's real problem, not just an abstract consideration in a comfortable office setting. Your audience needs to know what to do next.

What adjustments can they make to search for part failure, and what results should they expect? This is tricky because you are directing someone to perform a very complex task without any interaction from them. You need to anticipate what your readers might experience, then lay out their options. You are creating an if-then-else structure, where you repeatedly say: If these conditions exist, then do this; else, go on to the next condition. Consider using a table to present this information. *See Section 8.3 for guidelines on constructing tables.*

17.2.2 Specifications

Specifications are detailed and exact statements which define the materials, dimensions, quantities, and performance standards for something to be built, delivered, installed, or manufactured. Specifications are frequently found in contractual documents which set standards for goods and services, including design, testing, inspection, and delivery dates.

You can find information on military specifications in;
Specification, Types and Forms. Mil-S-83490.
Standardization Policies, Procedures and
Instructions. DOD-4120.3-M.

Both of the above are available from

Standardization Document Order Desk
700 Robbins Avenue
Building #4, Section D
Philadelphia, PA 19111-5094

17.2.3 Procedures

Procedures, as commonly defined in business and industry, are general descriptions of how trained people, already familiar with their individual tasks, cooperate together to complete an operation. The procedure writer outlines what each member of the audience does, but does not provide specific instructions for each task.

16.05 Launching

(1) The competent person participating in the launching of model rockets propelled by rocket engines shall make certain that these regulations are being adhered to by all present at the launch site and shall also be familiar with and conduct the launch in accordance with the instructions and NAR/HIA Safety Code supplied with the model rockets and the rocket engines.

(2) A device or mechanism shall be used during the launch which provides a suitable deflector to prevent the exhaust jet from impinging directly on the ground and shall restrict the horizontal motion of the model until sufficient flight stability shall have been attained for a reasonably safe, predictable flight.

(3) Launching or ignition shall be conducted by remote electrical means fully under the control of the person launching the model.

(4) A launching angle of more than sixty (60) degrees from the horizontal shall be used.

Source: "Procedures for Launching Model Rockets" Commonwealth of Massachusetts

17.3 TECHNICAL DIRECTION PARTS

Restrictive title. The more detailed and specific your title, the more it will restrict the use of your directions to persons who want to accomplish what you intend. Use your title to forecast. You want the title to convey the particular and precise purpose of your directions. This will help your audience to identify your document and discourage persons who want to do something else. Consider beginning with "How to . . ." and insert a precise description of the process to be followed.

Process overview. Directions will be carried out more effectively if the audience knows the reasons for the specific steps which will be taken. The best way to begin directions is with a cover section which explains, in general terms, the process described in the directions. Essentially, this section is there to provide necessary background to the reader. You are giving the reader a framework for the details which follow.

What are the directions meant to accomplish? How? You should provide the reader with a concise survey of what to expect including the level of difficulty and complexity of the task or process. Giving your readers a sense of where you are going makes it easier for them to follow your lead.

Materials list. Technical directions frequently contain lists of parts and materials which will be needed for the process which is described. Everything from bid solicitations to assembly instructions contain detailed inventories of necessary parts.

What parts and materials will your audience need to have on hand to finish the task or operation you are directing? A materials list needs to be specific and exact so that the people who use it cannot make mistakes. Quantities, sizes, amounts, measurements and descriptions must be correct and clearly stated with no possibility of confusion.

Equipment list. An equipment list is similar to a materials list; it lists items or tools needed for the process. The level of detail you need to use in your descriptions depends on what your audience knows, but you never want to leave your readers without a necessary piece of equipment. In other words, there is the same need for absolute accuracy.

You need to provide thorough and complete inventories of equipment and tools which will be used in the process you are directing. If you leave something out, most of your audience will not know. They will begin the operation and run into difficulties when the absent item becomes necessary.

Safety Precautions. Safety is your most important concern when you are designing technical directions. This is why imagination and care are so important to this part of the task. You need to think through your process step-by-step and insert warnings and other cautions wherever they can assist the users.

There are different levels of precautions which can be included in your technical directions. The most serious is a warning. Whenever there is any chance of a person being hurt or property damaged, you need to provide a warning. Your warning needs to be clear, visible, and understandable to the average person. Make it stand out from the rest of your document.

WARNING: DO NOT OPEN THE COMPUTER MONITOR CABINET UNDER ANY CIRCUMSTANCES. You may expose yourself to severe electric shock.

You should include cautions where there is a possibility that the reader will take the wrong step. A caution prevents harmful mistakes. These are signals to the users to take extra care at this point or to pay particular attention to the process because decisions will be required.

Caution: Do not place the computer monitor in direct sunlight. Excessive heat could damage the monitor.

Finally, you can provide notes for your audience: sidebars, parenthetical definitions, typographical signals, or useful commentary. These are obviously less important than warnings or cautions, but anything you can do to help your readers follow your directions smoothly will help your document design. Notes are commonly used to clarify and describe different options and alternatives.

Command steps. You need to design short, numbered, easy-to-follow, logical sequences. Each step should begin with a verb, telling the reader what to do. These steps should be organized in a way that is easy to follow.

Remember, you do not have your readers' undivided attention. They are trying to interpret your directions and perform the task at the same time. Begin each step with a direct command. Include only one task in each step. If your directions involve more than twelve steps, break them into stages.

The clearest, most effective directions are commands, written in the imperative.

Examples

Remove the four screws on the base.

Choose *Save a copy of this document* from the Files menu.

Illustrations. Clear and well-planned illustrations can help to clarify your technical directions and make them simpler to write.

You can use illustrations to avoid lengthy explanations and to clarify descriptions of parts. When you are writing directions for mechanical assemblies, include measurements in your graphic aids. As a general rule, include visuals wherever you want to help the reader picture what to do or locate. Place your visuals close to the steps which they illustrate.

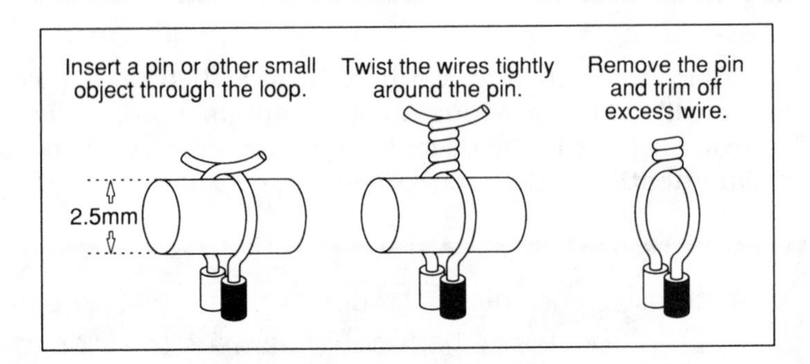

Figure 17-1 Illustrations can save text. This illustration shows how leads can be twisted to form a good thermocouple junction. **Reprinted with the permission of** *Circuits Manufacturing.*

17.4 HOW TO WRITE TECHNICAL DIRECTIONS

Begin at the end. Put together a physical specimen of the design of your document. What will your final product look like? What will people actually hold in their hands? What kind of document can you produce in the context of your schedule and resources? How will these directions be used?

Make a list of what your readers will want to get out of your directions. You need to get beyond your ease and closeness with what you are describing. You need to anticipate your readers' needs by imagining their situation. Pretend you have amnesia and make a list of everything you would need to know to use a product or complete a process.

Choose a familiar format so the structure of your document does not distract from your message. If your company has a preferred way of producing technical directions, follow it. If not, you should consider using milspec (military specification) standards or ASTM standards. See Section 17.9 for more details on the selection of format.

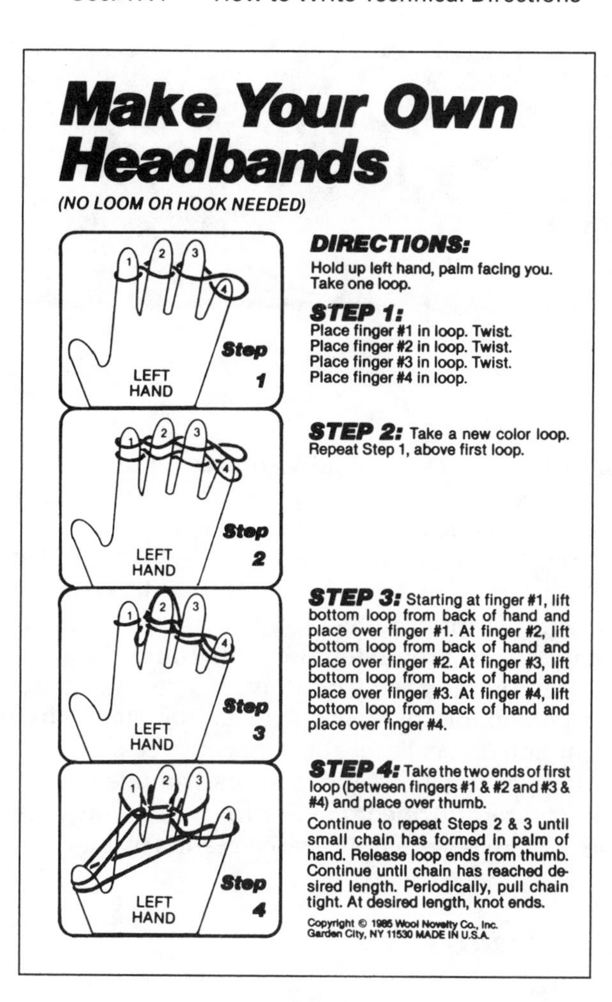

Make Your Own Headbands
(NO LOOM OR HOOK NEEDED)
DIRECTIONS:
Hold up left hand, palm facing you. Take one loop.
STEP 1:
Place finger #1 in loop. Twist.
Place finger #2 in loop. Twist.
Place finger #3 in loop. Twist.
Place finger #4 in loop.
STEP 2: Take a new color loop. Repeat Step 1, above first loop.
STEP 3: Starting at finger #1, lift bottom loop from back of hand and place over finger #1. At finger #2, lift bottom loop from back of hand and place over finger #2. At finger #3, lift bottom loop from back of hand and place over finger #3. At finger #4, lift bottom loop from back of hand and place over finger #4.
STEP 4: Take the two ends of first loop (between fingers #1 & #2 and #3 & #4) and place over thumb.
Continue to repeat Steps 2 & 3 until small chain has formed in palm of hand. Release loop ends from thumb. Continue until chain has reached desired length. Periodically, pull chain tight. At desired length, knot ends.
Copyright © 1985 Wool Novelty Co., Inc. Garden City, NY 11530 MADE IN U.S.A.
LEFT HAND
Step 1
LEFT HAND
Step 2
LEFT HAND
Step 3
LEFT HAND
Step 4

Figure 17-2 An example of well-placed illustrations.
Reprinted with the permission of Wool Novelty Co.

Revise your revisions! You need to keep on writing throughout the various stages of the document design. First draft, second draft—many technical directions go through more than a dozen drafts. After each revision, you need to thoroughly test your words and pictures.

Write with your reader in mind. Sometimes, because you are writing your directions so long before they will be used, it is difficult to imagine your audience. Focus on one individual, one user of your directions. Remember, the only purpose of technical directions is to help people accomplish something. No one will ever read them for pleasure or enjoyment. Directions are meant to be used. You can't forget the users.

Don't believe in rules you can't really remember. For example, some people insist that *shall* takes legal precedence over *will*. Shall, while generally accepted as an imperative, has no legal significance beyond legend. Will is perfectly acceptable. Trying to follow vague, half-remembered rules from a bygone grammar class keeps your mind off the job. Use common sense!

Begin at the end.
Make a list of your readers' needs.
Choose a familiar format.
Revise your revisions!
Write with a reader in mind.
Don't believe in rules you vaguely remember.

17.5 RESPONSIBILITY

You have a legal and professional responsibility to provide accurate and complete instructions, specifications, and procedures. Your directions need to be clear because your readers will act on them immediately. Few readers will complete reading your entire document before they begin the first step. More frequently, they will read your instructions as they work on the task.

When you are producing written directions you are responsible for maintaining the same standard of professional conduct and due care as your peers under similar circumstances. The question you need to ask yourself is what would a reasonable person do under the same circumstances, exercising prudent and due care for the consequences.

Legal liability is an issue here. Companies and individuals have been held responsible for inadequate and incomplete directions. You need to explain carefully because you are responsible for the results of your plan.

17.6 EDITING TECHNICAL DIRECTIONS

Editing your written technical directions, debugging them, is a two-stage process. Your first edit takes place right after you finish your first draft. You need to consider your directions from the viewpoint of a user. Do a physical walkthrough of the process, considering every possible variation of each step. Imagine what could go wrong. Inevitably you will find a number of errors in your document. As with computer programs, technical directions need to have all the bugs, ambiguities, and omissions removed before they will work correctly. Correct the mistakes you find and walk through the directions again.

Finally, when you think it's perfect, try your document on someone else. Then watch how many problems they encounter, problems you've overlooked. This is an important step. You should not revise technical directions by yourself. The assistance of a novice user, someone who relies completely on your written document for what to do, is a key part of your revising process. *See Chapter 23, Testing Your Output,* for a discussion of this procedure.

The importance of accuracy cannot be overstated here. Most persons take written directions as absolutely perfect. They follow what they believe they are being told to do, with little regard for consequence. You are the one who needs to anticipate the difficulties and the consequences and guard against them.

17.7 *THE BIGGEST PROBLEM: USERS*

When you design technical directions, your biggest problem will be your audience. People who use technical directions ordinarily do not use them very carefully. This means that you have to be exceptionally careful for them, reminding them what to do: step-by-step.

Your audience follows the directions which you have written to lead them to a desired outcome. If your design is too complicated, if you leave steps out, if your technical terms are obscure or undefined, the users will have very real problems following your plan. If you lose them, they will begin guessing. Then simple confusion turns into serious difficulty. You have to design a path which the audience can easily understand and follow.

You need clear writing and careful design. Layout, headings, indents, underlining, boldface, italics, fonts, and color—you should use whatever is available to make your plan obvious and clear. You don't want to overdo this so your directions resemble a ransom note. On the other hand, don't let your warnings fade into the text. Make your message stand out.

17.8 *PUTTING IT ALL TOGETHER*

Fig. 17-3 is an example of an effective set of assembly instructions from Bush Industries. Notice the format which includes:

- an overall diagram that shows the entire process.
- an illustration of the completed assembly.
- a parts list.
- the exact length and shape of screws and bolts.
- clear illustrations.
- a step-by-step breakdown of the assembly process.

17.9 *FORMAT*

Technical directions are complex documents which need careful design. They are structured instruction sets designed to help the user assemble a device, repair a machine, or participate in a complex activity. The best format for technical directions is the format which helps the users most effec-

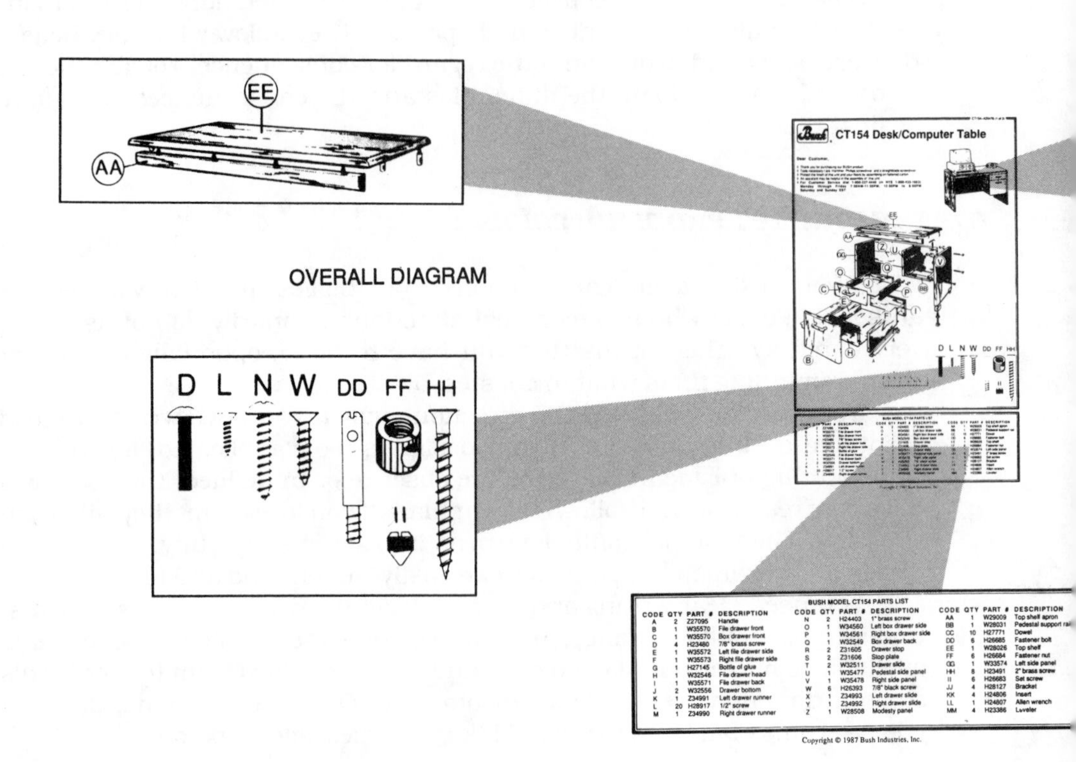

BUSH MODEL CT154 PARTS LIST

CODE	QTY	PART #	DESCRIPTION	CODE	QTY	PART #	DESCRIPTION	CODE	QTY	PART #	DESCRIPTION
A	2	Z27095	Handle	N	2	H24403	1" brass screw	AA	1	W29009	Top shelf apron
B	1	W35570	File drawer front	O	1	W34560	Left box drawer side	BB	1	W28031	Pedestal support rai
C	1	W35570	Box drawer front	P	1	W34561	Right box drawer side	CC	10	H27771	Dowel
D	4	H23480	7/8" brass screw	Q	1	W32549	Box drawer back	DD	6	H26685	Fastener bolt
E	1	W35572	Left file drawer side	R	2	Z31605	Drawer stop	EE	1	W28026	Top shelf
F	1	W35573	Right file drawer side	S	2	Z31606	Stop plate	FF	6	H26684	Fastener nut
G	1	H27145	Bottle of glue	T	2	W32511	Drawer slide	GG	1	W33574	Left side panel
H	1	W32546	File drawer head	U	1	W35477	Pedestal side panel	HH	8	H23491	2" brass screw
I	1	W35571	File drawer back	V	1	W35478	Right side panel	II	6	H26683	Set screw
J	2	W32556	Drawer bottom	W	6	H26393	7/8" black screw	JJ	4	H28127	Bracket
K	1	Z34991	Left drawer runner	X	1	Z34993	Left drawer slide	KK	4	H24806	Insert
L	20	H28917	1/2" screw	Y	1	Z34992	Right drawer slide	LL	1	H24807	Allen wrench
M	1	Z34990	Right drawer runner	Z	1	W28508	Modesty panel	MM	4	H23386	Leveler

EXACT
LENGTH
OF SCREWS
AND BOLTS

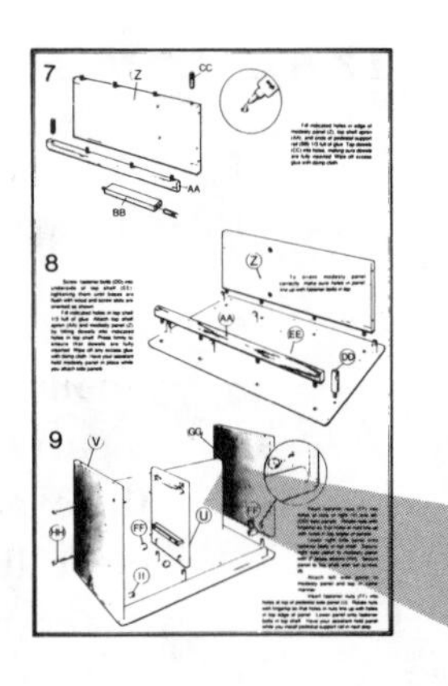

Figure 17-3 A complete set of technical directions. **Reprinted with the permission of Bush Industries.**

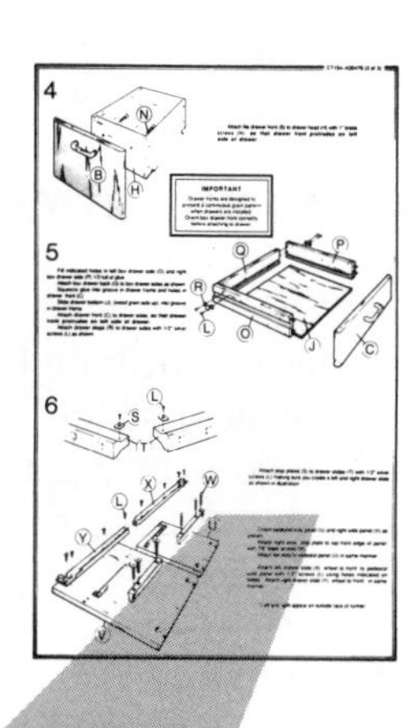

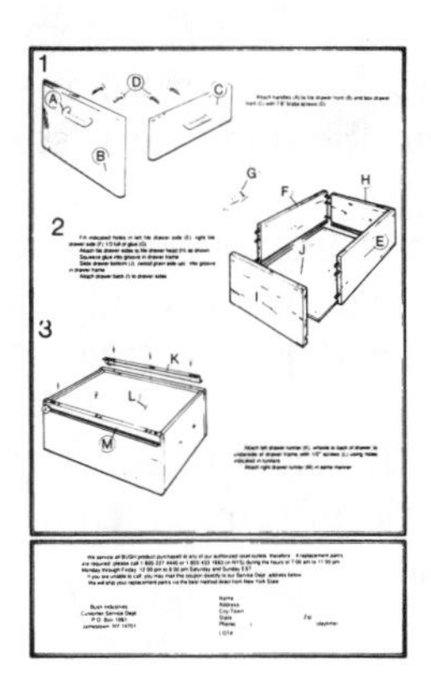

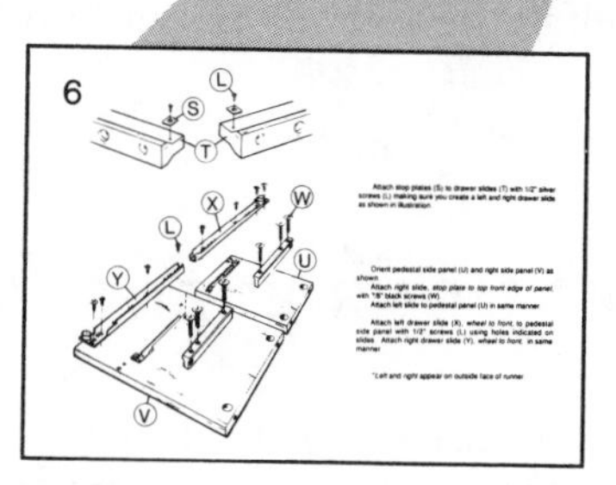

STEP-BY-STEP

COMPLETED ASSEMBLY

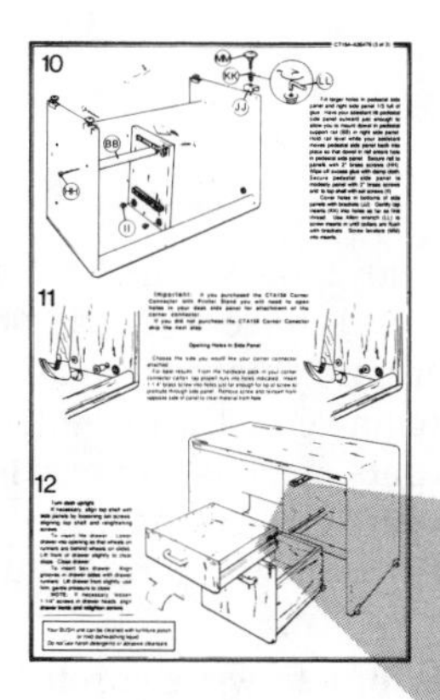

CLEAR ILLUSTRATIONS

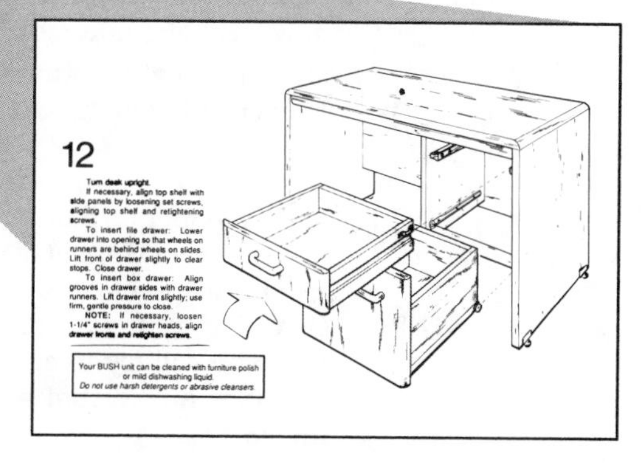

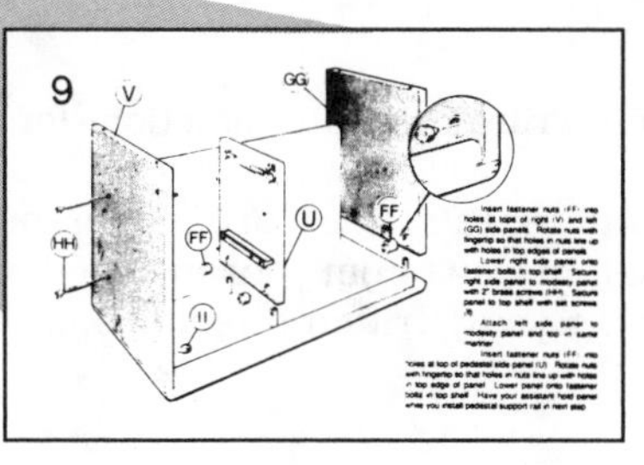

tively. All formats should be clearly labelled or titled, accessible with the minimal amount of technical detail, and clear to the most distressed and confused users.

17.10 A FINAL WORD

When you design written directions, you will often be producing a complex document long before it will be used. Your audience is distant from you in physical space as well as time. These circumstances can lead you to focus on your own immediate writing/designing/illustrating problems. You can begin to forget the people who will actually use the document sometime in the future.

Instead, we want you to remember that putting together a written set of directions is a complex activity which requires imagination, thought, and care. The people who will use what you produce depend on your ability to anticipate what they will need to do.

Your ability to imagine their difficulties is the most important skill that you need as their instructor. Technical directions require that you place yourself in the role of a patient and careful teacher.

DIRECTIONS EXERCISES

1. This exercise should be done in a room with moveable chairs and desks. Three participants are selected to create an impromptu maze with the objects in the room. When the maze is complete, three other participants are selected to work together to prepare a set of directions on how to navigate the maze. The directions must be specific. Common agreement on the units of measurement is needed; for example, the writers need to define what is meant by "five paces," "two strides," or "two feet." A volunteer is selected to navigate the maze, using the written directions as the only form of guidance.
2. One of the most practical laboratories for learning to write careful instructions can be performed by almost anyone. You begin with three pieces of 8 1/2″ × 11″ paper.

   STEP 1. Make a paper airplane.
   STEP 2. Use the second sheet of paper to write down your directions for building this airplane.
   STEP 3. Give your written directions and the third sheet of paper to someone else who will try to create your airplane from your instructions. Do not offer any verbal advice. Let your written directions speak for themselves.

3. The following series of illustrations uses no words to present instructions on how to modify a familiar outdoor item, the picnic table. Write a set of directions to accompany these illustrations. Remember to include a clear and limiting title, an overview, a list of tools and materials, and step-by-step instructions.

Reprinted with the permission of Roy Doty.

4. Select the operating or troubleshooting instructions for some mechanical or electronic device with which you are familiar. Evaluate these instructions using the standards presented in this chapter. Your evaluation should focus on how useful the instructions would be to a novice and uninformed user. Attach a photocopy of the instructions with your evaluation.
5. Write a set of format specifications for Gates and Associates. Specifically, describe the exact format which a business letter addressed from this firm must follow.
6. Choose a simple technical device with which you are familiar: a car door or a VCR, for example. Imagine that it will not work properly and that you are instructing someone how to figure out what the problem is, and how to fix it. Bring your troubleshooting instructions to another person in the class and ask them what you have left out.
7. Choose three sets of technical directions and compare their page designs. Which seems most helpful to the user? What helps users and what gets in the way? If the text were in a language you did not understand, which set of directions would be most useful?
8. Choose a familiar task and create a flow chart which illustrates the process step-by-step. Make sure that your graphic indicates every step where a decision is required. (Use a diamond to indicate decisions. See Fig. 8-8 for more flow chart symbols.)

18

Composing Short Reports

Note to Users: **This is a task chapter. It is designed to help you write short reports: organized, concise documents which circulate factual information through technical organizations.**

18.1 INTRODUCTION

Technical organizations rely on short written communications to keep projects coordinated, supplied, and on schedule. Important messages are communicated in writing so that there will be a record of them. Electronic mail, fax machines, and express mail have increased the pace of this activity. Today, even remote field-sites can stay in close written touch with headquarters.

Producing short reports involves collecting, organizing, and distributing written information. Entry-level professionals will be expected to provide technical and support information, specifications, activity reports, and other types of written documentation. Your ability to do this will have a direct impact on how your performance is evaluated.

In this chapter you will learn how to:

- Place your purpose up front.
- Use a problem → solution approach.
- Choose correct formats.
- Map the report with headers.
- Report recommendations persuasively.

18.2 DEFINITION

Short technical reports are written documents organized to supply the results of some activity; a trip report or progress report is an example of a short report. Technical reports can be defined by length, level of formality, or by specific function. Generally, the shorter the report, the less formal it will be. Short, informal reports are located somewhere between memos and long reports on the spectrum of written communications.

These documents vary from several paragraphs to several pages. Only the basic elements of a report are included: the introduction, the body, and conclusions and recommendations. This simple arrangement is useful for a variety of different reporting situations. Generally, short reports contain more information than a brief memo and the layout clearly reveals how the material is organized. They are usually arranged more carefully than memos and their style is more elaborate and less personal.

Short reports have a very practical purpose. People use them to make decisions. This goes on all the way up through the company hierarchy. The information in these reports is accumulated, refined, summarized, and rewritten as it moves through the communication channels of your organization. When your name is on a document which is circulated, the document represents you and stands in for you when you are not present. Sloppy, careless writing does not make a good impression. Everything you write should be correct and understandable.

One of the most important things that you should remember is that your short reports can circulate more widely than you expect. Reports sent to your immediate supervisors and co-workers can end up on the desks of persons throughout the hierarchy.

Writing a short report requires the kind of care and thought you would devote to writing an effective letter. Like other technical writing, these reports follow rules of style and format which are familiar to your audience. Writing good short reports is hard work, but this type of communication can be an important part of your job responsibilities.

18.3 TYPES OF SHORT REPORTS

Generally short reports are intended to inform or to persuade your audience. Apart from these two general functions, short reports are used for a variety of reporting purposes. There are as many types of short reports as there are uses for them. They provide a reliable and verifiable method of circulating information. Common types of short reports include:

Lab reports communicate information obtained through laboratory tests, a data-gathering process which is rigidly structured to provide objective results. Your description of the methodology provides credibility for the data that is discussed. You need to describe the equipment and procedures which were used in the investigation.

Lab reports begin by describing the purpose of the investigation. Follow your purpose with discussion of the specific methods of analysis, prob-

Sample Analysis Record page of

Gates and Associates

Client Name	Job #	Sample ID #

Sampling Location	Material Description

Lab Name	Method
Analyst Name	

Analysis	Date Reported	Date Billed
	Signature	

Figure 18-1 A lab report form.

lems experienced, and any conclusions or recommendations. Most technical organizations establish a particular format for their lab reports and you will be expected to follow this.

Progress reports, also called status reports, are used to keep supervisors and management informed about a project or activity. They want to know five things: the current status of the project, problems you have

Daily Progress Report ________ 199 __

Project Title ____________________

Institution ____________ Weather ____________

Contractor ____________ Temperature____________

Superintendent ________ A.M. Noon P.M.

1. What work was in progress today? Give location and a brief description of the work.
2. Provide the same for work of subcontractors.
3. What items of work were started today?
4. What completed today?
5. What is delaying progress?
6. Did any accidents occur? If so, describe what happened.
7. Were there any visitors to the site? Who? Why?
8. Were there any variations from the original plans?
9. Did anything else occur that should be reported?
10. What material and equipment was received? Did it meet specifications? Describe in detail.

Figure 18-2 An example of a progress report form.

encountered, and changes in the activity, the schedule, or the costs. The purpose of these reports is to make sure that managers will know whether they need to alter their budgets, provide new resources to meet deadlines, or make adjustments in the project plans. The information you provide will be crucial to their decisions.

A series of progress reports should stick with the same format. Each report should summarize the progress since the previous report, including any recommended changes in technical approach, schedule, or cost. Describe what you are doing to deal with problems, and provide a realistic estimate of what you expect to accomplish in the next reporting period.

Progress reports are similar to the **activity reports** which some organizations require from field-site employees. Generally you report what has been accomplished, what will be done next, problems you have encountered, requests for support and equipment, and a brief overview. Again, since these documents are part of a regular series, there is little need for introductory material.

As part of your job, you may be required to attend conferences and travel to various field sites. **Trip reports** provide your company with a written discussion of your activities. You want to provide specific information about your experiences and you also need to show the significance of your information to the company.

Short reports are frequently used to alert management to accidents, equipment failures, and property damages. **Trouble reports** are important because they may be used as evidence in insurance liability and compensation judgments. You need to include detailed information about what happened, when and where it happened, who witnessed the problem, and why the problem occurred. Your analysis of the cause of the trouble must be detailed and supported by evidence. Don't forget to include steps you have taken to avoid a similar problem or accident in the future. Keep an objective and factual tone and be as thorough and complete a reporter as you can be.

See Exercise 1 at the end of Chapter 22 for an example of a particularly inadequate accident report.

These are only some of the types of short reports that you might need to write. The types of short reports are as varied as the situations and circumstances which require them. All types share similarities of purpose and design. When you write a short report, you are trying to communicate factual information quickly and accurately. Your organization, style, and strategy reflect your ability to do this effectively.

GATES & ASSOCIATES

MEMORANDUM FORM

TO: Department Heads

FROM: Sheila O'Brien

DATE: February 25, 1993

SUBJECT: Trip Report: Strategies for Global Engineering

The purpose of this report is to describe my participation in a recent conference on ways in which American firms can become involved with international technology projects.

Last Thursday and Friday I was in Ypsilanti, Michigan where I represented Gates & Associates at the **12th Annual Conference on World Business and Technology**. Along with Wlodzimierz Brunansky, Hungarian Minister for Technical Development, and Patricia Becker from the Department of Commerce, I participated in a panel discussion on Doing Business in Central Europe.

The theme of the conference was "The 1990's: Strategies for Global Engineering." Frank Lazarovich, President and CEO of Maastricht Engineering, set the tone for the conference with his opening address. He maintained that engineering consultant firms must (1) understand the issues which shape the global economy, (2) view their companies in a global context, and (3) recognize that quality, service, and aggressive pricing are required for international success.

The panel session went well. I was surprised at the number of persons who attended, more than thirty companies were represented (see attachment). Mr. Brunansky used the Gates & Associates projects in Budapest to describe how joint ventures for technical development can be established within his country. Ms. Becker stated that the Commerce Department was pushing for increased export credits to support technical assistance projects in Central Europe. Representatives from several companies spoke to me after the session about possible participation in the Budapest project.

While at the conference I attended two other valuable sessions. Hilary Fratianni spoke about the role of developing countries in the strategy and management of high technology industries. Lawrence Hughes led a very interesting discussion on the effectiveness of international strategies for small technology businesses. I will circulate copies of these two papers.

My attendance at the conference convinced me that our attention to expanding our consulting to the world market for technology expertise makes excellent sense and that we should continue with this direction.

Figure 18-3 An example of a trip report.

18.4 PARTS OF THE SHORT REPORT

Your reports must first be accurate and correct. If they are to remain correct over time, while they are used by many different people, the reports must be clear and carefully designed. A well-constructed report consists of three fundamental parts:

(1) the Introduction, in which you tell the reader what is about to appear;
(2) the Body, in which you present a detailed report of the activity; and
(3) the Recommendations and Conclusions, in which you elaborate on what needs to be done next.

Your **introduction** should accomplish several things. It names the subject and tells the purpose of the report. Sometimes it gives the reader necessary background or context. Let the readers know with a glance what the report is about so they can focus on that message. State the objective. Summarize your findings, recommendations, or conclusions.

The top of your report helps the readers get their bearings and fit your report into a larger context. Help the reader to get started. Use a **title line** to telegraph key information. State the **purpose** of the report. **Map** the report by describing where to find specific facts. **Reference** your document to further sources of information. By the time your readers leave the introduction, they should have a good idea of how the report fits into their activities and where to go in the report for detail.

The **body** will contain the more detailed, complete information of the report. How much detail you will use depends on how much your audience knows and needs to know. Your report discussion needs to be complete, containing enough information for your readers to understand and judge your conclusions and recommendations. The text of your report needs to fully describe the technical details.

Effective short reports provide a clearly organized account of their message. A **problem** $\rightarrow$ **solution** structure works well. Begin by describing the problem and lead to your solution. Use headings so your readers cannot miss your organization or your main points. Support your text with illustrations, even in informal short reports. Frequent illustrations can make your meaning clear and shorten your explanations.

What goes into the **conclusions and recommendations** section depends on your purpose and what you were asked to do. If the purpose of your report is to provide information, then a simple summary will be sufficient. If your purpose is to persuade, however, to convince the readers that your solution is correct, then include your recommendations. Obviously, documents such as a feasibility report or short proposal require that you make your judgments clear. Be sure to make the evidence which supports your recommendations clear and convincing.

Remember, key conclusions and recommendations should also be included in your introduction. The end section of the short report provides this information in full detail. By stating your key points at the beginning and end, you improve your chances to get your message through.

18.5 FORMAT

The format for short reports is fairly flexible. Generally these reports will use a memo format. An informal and direct style works well for most internal communications. For external reporting or formal situations use a letter format and a professional style. Include a subject line if you report by letter and include headings for easy reference. All reports in a series should use the same format whether you choose memo or letter.

Short reports are supposed to provide easy access to information. The entire design—layout, headings, illustrations, and text—is intended to convey data to the readers as efficiently as possible. You can help your readers by making the report structure visible and easy to grasp. This means arranging your document into primary and secondary sections. Use distinctive headings and subheadings to label significant material and to indicate the relative importance of the report information.

Keep your reports short by using attachments. Your readers can decide for themselves if they want to read this supplemental information. Some people will be impressed by this data; others will ignore it. Be sure to separate your appendix material in an obvious way so you do not discourage readers who want the brief version. Package reference materials and detailed technical specifications differently.

18.6 CONSIDERING YOUR AUDIENCE

Few people would take the trouble to write or read short reports unless they were required to. This gives you a good starting place. Some person wants this report, plans to receive it, and probably has use for the information it contains. You are writing the report for this audience. Begin the document design process by discovering your purpose—WHY—from three whats:

WHAT do your readers expect to find out? If people are reading a report to extract information, they probably have a pretty good idea of what kind of information they are looking for. Progress, problems, scheduling, costs, and requests are just a few of the common categories of information transferred by short reports. You need to discover exactly what your readers require from the document.

WHAT do your readers already know? You have to include enough background information so that even an uninformed reader will follow what he is reading. If you assume that your readers are familiar with the report

Congress of the United States
OFFICE OF TECHNOLOGY ASSESSMENT
WASHINGTON, DC 20510–8025

JOHN H GIBBONS
DIRECTOR

For Immediate Release
September 26, 1990

OTA RELEASES REPORT ON INTEGRITY TESTING

Existing research does not clearly confirm or refute that honesty and integrity tests can accurately predict dishonest behavior in the workplace, concludes the congressional Office of Technology Assessment (OTA) in a report released at a hearing today by Rep. Matthew G. Martinez, chairman, Subcommittee on Employment Opportunities, Education and Labor Committee.

OTA also found that use of these tests raises important public policy issues. Public policy issues include the potential effects on individuals of errors in interpreting and using test results, uncertainties regarding possible discriminatory effects, and potential violations of privacy.

Several thousand U.S. firms use honesty and integrity tests to screen and select job candidates, primarily for non-managerial or less-skilled jobs such as convenience store employees or retail clerks. These paper and pencil tests are intended to help employers identify applicants with high tendencies to steal on the job or who behave in other "counterproductive" ways. Counterproductivity is often used to mean behavior such as tardiness, sick leave abuse, and absenteeism.

There are difficulties in defining the exact nature of these tests, says OTA, and accordingly there are ambiguities over the kinds of tests that fall into this category. Most experts believe that integrity tests are essentially personality tests, designed to identify individual traits and behaviors. There is disagreement, however, on several issues, such as whether or not honesty is a psychological trait, and if it is, how it should be defined in the workplace.

In addition to disagreement over the tests themselves, there is disagreement – and little solid data – on the actual extent and nature of employee dishonesty, and on the proportion of the population likely to engage in dishonest behavior. Also at issue is the extent to which an individual's attitudes and experience are shaped by the workplace situation.

Business has a clear interest in identifying the most honest and productive employees, and in having access to carefully developed methods for evaluation and selection. Integrity tests are marketed in response to this need.

Congress needs to weigh several factors: the potential gains to business of an effective tool to screen prospective employees and therefore gains to society; the potential harm to individuals, to business, and society of instruments that do not correctly identify individuals; and the disagreement within various research and stakeholder communities over the existing research data.

The OTA report responds to a request from the House Committee on Education and Labor.

Copies of the 96-page OTA report, *The Use of Integrity Tests for Pre-employment Screening*, for congressional use are available by calling 4-9241. Copies for noncongressional use can be ordered from the Superintendent of Documents, U.S. Government Printing Office (GPO), Washington, D.C., 20402-9325; phone (202) 783-3238. The GPO stock number is 052-003-01216-2; the price is $4.00.

OTA is a nonpartisan analytical agency that serves the U.S. Congress. Its purpose is to aid Congress in dealing with the complex and often highly technical issues that increasingly confront our society.

Figure 18-4 This short report is a summary of a long report. Readers interested in more specific information can request the 96-page report. **Reprinted with the permission of the Office of Technology Assessment.**

topic, then you should make this assumption explicit. Write it out! Let the readers know what you assume they bring to the report.

WHAT do you want your readers to do? This question does not always apply. If you are writing to inform, you do not particularly care what readers do with the information. If you are writing to persuade, however, then you need to be clear about what you want the readers to do. This means stating what you want to happen directly.

18.7 FORMS

When the reporting schedule becomes routine it may be a good idea to develop a preprinted form to collect and assemble data. Many organizations use forms for informal reporting. Using preprinted forms allows you to control the information that is gathered. This allows for easy reporting and quick tabulation. A simple instance is the telephone message form.

Date ______________ Time ______________

To ______________

WHILE YOU WERE OUT

M ______________

of ______________

Phone ______________
area code number

TELEPHONED		WILL CALL AGAIN	
RETURNED YOUR CALL		WANTS TO SEE YOU	
PLEASE CALL		*URGENT*	

Message ______________

Signed ______________

Effective forms make it easy to supply and retrieve information. This means they must be carefully planned. You need to give a good deal of thought to the design of the form, the kinds of information you request, and the ways it will be supplied. A well-designed form can provide a consistent, uniform method to report routine data which needs to be collected in an unvaried fashion. There are several software packages which allow you to create custom forms for almost every purpose.

If you are filling out forms which are important, grant applications for instance, we recommend that you create photocopies and generate several rough drafts before completing the actual document.

18.8 A FINAL WORD

Technical professionals who progress in their companies usually write more every year. Even entry-level employees are expected to communicate effectively through a variety of short reports. Eventually you will have to supervise the writing of your subordinates. Their reports will often go out under your signature and will reflect on the quality of work done under your supervision. Clear reporting is a valuable skill for any career.

What you want to develop is the capacity to write fluently. Good functional writing is clear, concise, and direct. When you are asked to write something, you want to respond quickly and flexibly. Learning to write this way will save you significant amounts of time, establish your credentials as a communicator, and let you help the persons you supervise.

SHORT REPORT EXERCISES

1. Mr. Gates thinks you might make a good field-site specialist. Write a short report describing how to set up field-site offices for project teams who have temporary assignments (one-to-three months) in places where engineering failures have occurred. See Chapter 26, Communicating in an Organizational Environment, for background information.
2. To test you out for the job described in Exercise 1, Gates would like you to design a progress report format which field-site employees could use to keep headquarters informed. You should include ways of reporting time periods, of requesting support and equipment, and of documenting work completed.
3. Design a preprinted form which collects information from your fellow students or co-workers. Decide what information you will request, and write a survey form that contains at least five questions. See Fig. 4-1 in Chapter 4, Conducting Research, for examples of different types of survey questions. Distribute the form to at least ten persons, and present the information you collect in a table or graph. Write a short report explaining what the information means.
4. Write a progress report using Fig. 18-2 as a model. Describe the present status of some ongoing activity you are involved in. If you have difficulty choosing a topic, you might describe your progress in becoming an effective writer and presenter.

5. At some point in your life you have been involved in an accident where there was some sort of physical injury. Write an accident report about the event, making sure to provide specific details. If you are unable to remember specific times and places, you can use your imagination. Include a brief section of ways to avoid similar accidents.
6. Make a list of the things that are wrong with the accident report in Exercise 1 in Chapter 22. Discuss the list in groups of four people, and then try to design a form which would force people to provide full and detailed information about accidents.
7. Interview three persons who work as technical professionals and find out what types of reports these persons have to write. Then use this information to write a one-page report describing your findings.
8. Write a short report to your class on a new or emerging technology in your field. Include at least one table or graph to convey key information. Where will you place the illustration for maximum effectiveness? How much explanatory text will you include?
9. Choose a country in South America or Africa and write a short investigative report in which you describe the language, climate, population, and current political situation. Be sure that your information is accurate and concise, and acknowledge your sources in the report.
10. Write a feasibility report to Mr. Gates on the possibility of installing an exercise room at Gates and Associates. The room would be a place where personnel could go to work off stress. You want to include several exercise machines and perhaps a punching bag for bad days, and any other equipment you consider appropriate. Be sure to provide a simulated budget.

19

Producing Long Reports

Note to Users: **This is a task chapter. It is designed to help you organize long reports which your readers will use as the basis for complex and involved decisions.**

19.1 INTRODUCTION

Producing long reports involves collecting, organizing, and distributing written information. Your audience will use your report to decide what to do or how to do it.

Technical organizations regularly use long reports as decision-support tools. Complex decisions require a careful analysis of all available research and factual data. A formal written report provides a database for this sort of activity. Readers will expect to use your long report as an information source which will provide them sufficient background material to make involved decisions.

Skillful writers of long reports recognize that productive reports must be useful. The people that use the reports have to find the information they need quickly and efficiently. At the same time, they need to be confident of the facts upon which they are basing their decisions. Your reports need to be both accurate and credible.

In this chapter you will learn how to:

- Use outlines as document blueprints.
- Design for multiple audiences.
- Choose effective organizing strategies.
- Consider realistic production schedules.
- Include appropriate report parts.

19.2 DEFINITION

A long report is a formal document with three basic parts: the front materials, the body (containing information and analysis), and the end materials.

Long reports are more widely distributed than short reports. They are reviewed and used by more people and they stay in circulation for extended periods of time. Long reports are often kept and referred to for future decisions and action.

Long reports serve a variety of functions within technical organizations, and they are sent outside the organization when a formal written document is appropriate. Some of the functions of long reports include:

- presenting research
- forecasting future plans
- reviewing projects
- supporting recommendations

In other words, long reports are used when the decision is important and all available information needs to be considered, often by many different people.

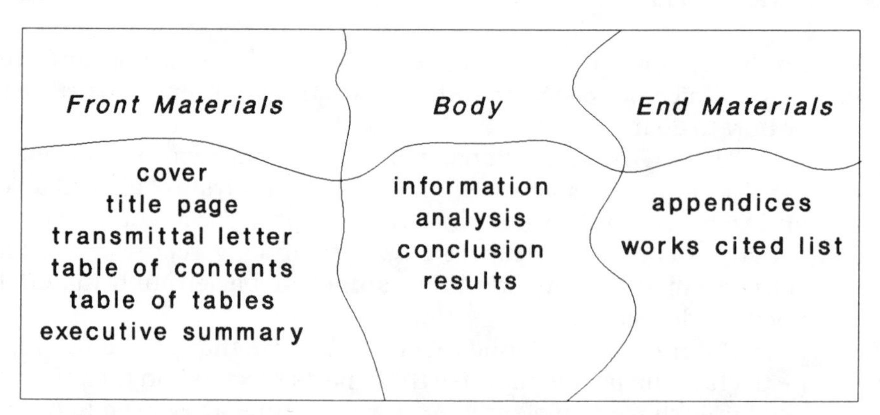

Figure 19-1 Putting together the pieces of the three major parts of the long report.

Thinking clearly about your audience is particularly important when working on long reports. You need to remember that a long report will have many different readers and each will bring specific interests to your document. Busy managers, for example, may not have time to go through the whole report. They will focus on the summary and recommendations. Other individuals, especially those who will have to carry out the recommendations, may have a strong interest in the exact details of the report.

When you are thinking about the purpose of the report, think about:

- What were you asked to do?
- Who will use the report? How? Why?

Since long reports may have several different purposes and since your readers will bring their own purposes and interests to the document—not all readers will want to read your entire report—your task is to help them use the document efficiently.

Formal in appearance and style, long reports attempt to provide readers with enough background, data, and information to support informed decision-making. Readers expect to use these reports to determine a course of action. Your report should be helpful to them in making this choice.

19.3 SOME TYPES OF LONG REPORTS

Feasibility reports are used to consider whether a course of action is practical. These reports assemble information which is used to decide whether an idea is workable and suitable, and whether to go ahead with it.

Research reports describe the results of investigations, studies, and experiments. Empirical information resulting from actual test data is an important product for technical organizations in an information economy.

Field reports describe the findings of on-site investigations where information is collected directly from its source, categorized, and sent to the organization for further analysis.

Failure Analysis reports are used to explain why something broke down or didn't work as expected. These are very important to technical organizations because technology learns from its failures and mistakes. Every plane crash, for example, is investigated and a cause determined. This information is presented in formal reports and used for future technical decisions.

Progress reports are used to describe how a complex project is going. Generally these will be short reports but the more complex the project, the more complex the reporting requirements.

Periodic reports provide information at regular intervals on the activities and condition of an organization or projects. Annual reports, cost/

MASSACHUSETTS DEPARTMENT OF PUBLIC WORKS
REQUEST FOR PROPOSAL
SUPPORTIVE SERVICES TO DISADVANTAGED BUSINESS ENTERPRISES

D. Reports

1. Three copies of twelve monthly reports and one final report will be due the fifth working day of the following month to the MDPW EEO Coordinator's office.

   The monthly reports will present the Contractor's activities which were designed to fulfill requirements of the supportive service contract and include an analysis of these efforts. Recommendations affecting the Scope of Work should be presented in this report. Receipt of the monthly report shall be a condition of payment.

Figure 19-2 Information on requirements for reports.

schedule status reports, accounting statements, and other types of calendar-scheduled documents qualify as periodic reports.

19.4 PARTS OF THE LONG REPORT

Long reports are generally designed in three sections, each with a number of different parts. Some of these parts are optional and others necessary. You should be aware of all of these parts so that you don't overlook something which would improve your own report's credibility and usefulness.

We are going to describe these parts from front-to-back, the order they would usually occur in a long report. We do not mean to suggest, however, that this is the order in which they should be written, assembled, or prepared. The front material will often be written last. It would make no sense, for example, to write a transmittal letter before you write your report. The purpose of the letter is to direct the attention of the readers to the report itself, to focus their interest, and guide them efficiently through the document. The report needs to be written first.

On the other hand, some material can be prepared or assembled before long reports are actually written. Some of the appendix information can be collected with the idea that it will be used to provide detailed data in future reports.

19.4.1 Front Materials

Your report begins at the cover. You should use the **cover** and **binding** of your document to create interest and a positive impression. Base your selection on how the document will be used. Considerations such as your audience, length and number of copies, and the need for users to photocopy sections are some of the factors which should influence your decision.

Expensive covers are not always the best idea. The message they convey can be interpreted as extravagance. An attractive cover invites the reader to continue, but it can be reasonably priced. When you are choosing bindings, think about whether people will need to remove or insert new pages, and especially consider whether they will need to photocopy sections.

Your **title page** should draw readers into the report. Professional reports are not required to have dull titles, so you should attempt to be informative. Subtitles can add interest by providing detail or motivation for the reader.

Coping With An

Oiled Sea

Background Paper

An Analysis of Oil Spill Response Technologies

Congress of the United States Office of Technology Assessment

Foreword

In the aftermath of the *Exxon Valdez* oil spill in Alaska in March, 1989, a myriad of investigations were initiated to evaluate the causes of that accident and to propose remedies. The Office of Technology Assessment was asked to study the Nation's oil spill clean-up capabilities and to assess the technologies for responding to such catastrophic spills in the future. The request for this study came from Senator Ted Stevens, a member of the Technology Assessment Board, and from Congressman Billy Tauzin, Chairman of the Subcommittee on Coast Guard and Navigation of the House Committee on Merchant Marine and Fisheries. This background paper presents the results of OTA's analysis. It discusses the current technologies and capabilities in the United States and abroad and evaluates the prospects for future improvements.

Cleaning up a discharge of millions of gallons of oil at sea under even moderate environmental conditions is an extraordinary problem. Current national capabilities to respond effectively to such an accident are marginal at best. OTA's analysis shows that improvements could be made, and that those offering the greatest benefits would not require technological breakthroughs – just good engineering design and testing, skilled maintenance and training, timely access to and availability of the most appropriate and substantial systems, and the means to make rapid, informed decisions. One must understand, however, that even the best national response system will have inherent practical limitations that will hinder spill response efforts for catastrophic events – sometimes to a major extent. For that reason it is important to pay at least equal attention to preventive measures as to response systems. In this area, the proverbial ounce of prevention is worth many, many pounds of cure.

JOHN H. GIBBONS
Director

Figure 19-3 An example of a title page.
Reprinted with the permission of the Office of Technology Assessment.

Figure 19-4 An example of a transmittal letter; in this case, since the report is being delivered to the general public, the letter is called a Foreword.
Reprinted with the permission of the Office of Technology Assessment.

The **transmittal letter**, sometimes called a cover letter, presents the report to a reader. (Sometimes, the transmittal letter is placed before the report, in front of the cover.) Any report which you do not deliver personally should have an explanatory memo or letter attached. Within the organization, a memo is usually sufficient.

Begin with a paragraph explaining what is being sent and to whom. Specify the title and purpose of the report. Depending upon your communication strategy you may want to direct the reader to a recommendation, conclusion, or other information you think is important to the reader. You should not repeat the summary or abstract.

A carefully prepared **table of contents** is a map which provides your readers with access routes to the specific information they are looking for. The more detailed you are, the easier it will be for the persons using your report to find the facts they need. A table of contents is necessary for any report longer than ten pages. This table should list all of the headings used in the report along with their location. Attached materials should also be listed and located for immediate access to readers. A **table of tables** or list of exhibits is sometimes useful in a major report. You want to tell the readers where to find figures and tables so these can be referred to quickly and easily. This is particularly important if your report has lots of illustrations.

Contents

Chapter 1. Overview 1
Chapter 2. Findings 3
Chapter 3. Oil Spill Response Technologies 11
    Introduction 11
    Mechanical Spill Response Technologies 12
    Dispersants 19
    Burning, Bioremediation, and Other Techniques 22
    Quantity and Distribution of Response Equipment 24
Chapter 4. U.S. Oil Spill Response Issues 27
    Introduction 27
    Technology: Which Cleanup Method? 27
    Decisionmaking: Democratic and Authoritarian Approaches 28
    Responsibility: The Polluter or the Government? 28
    Logistics: Essential to Quick Response 29
    Beach Cleanup: How Much Emphasis? 30
    Federal Regulations and Emergency Situations 31
    Equipment Testing: How Realistic? How Effective? 32
    Personnel, Training, and Drills 33
    Research and Development 34
Chapter 5. Selected European Oil Spill Policies 39
    Introduction 39
    French Oil Spill Policy 39
    Dutch Oil Spill Policy 41
    United Kingdom Oil Spill Policy 43
    Norwegian Oil Spill Policy 44
    General Comments 46
Appendix A. Mechanical Containment and Cleanup Technologies 51
Appendix B. Quantity and Distribution of U.S. Equipment 58
Appendix C. Acknowledgments 68
Appendix D. Acronyms 70

Figure 19-5 A Table of Contents.

Contents

Figures

Figure		Page
3-1.	Spread of an *Exxon Valdez*-Sized Oil Spill	11
3-2.	Vessel-of-Opportunity Skimming System	17
3-3.	West German THOR	18
3-4.	Schematic Drawing of the Navy's Class V Oil Skimmer	18
A-1.	Components of a Boom	51
A-2.	Weir Skimmer	52
A-3.	Floating Suction Skimmer	53
A-4.	Schematic Drawing of a Weir Skimmer	53
A-5.	Vortex Skimmer	53
A-6.	Disc Skimmer	54
A-7.	Rope Mop Skimmer	55
A-8.	Paddle Belt Skimmer	56
A-9.	Sorbent Belt Skimmer	56
A-10.	Sorbent Lifting Belt Skimmer	56
A-11.	Submersion Belt Skimmer	57

Tables

Table		Page
2-1.	Large Oil Spills: A List of 66 Spills Greater Than 2 Million Gallons, 1967 to Present	4
3-1.	Equipment Required To Deploy Response Elements	20
4-1.	Suggestions Received by the Coast Guard	36
4-2.	Selection of Planned R&D Efforts by Federal Agencies and Industry	38
5-1.	Summary of Oil Spill Response Arrangements	47
A-1.	Boom Classification According to Freeboard and Draft	52
B-1.	Major Types, Quantities, and Location of U.S. Navy Spill Response Equipment	59
B-2.	Major National Strike Force Equipment of the U.S. Coast Guard	59
B-3.	Major Sources of Skimmers in the United States	60
B-4.	Major Sources of Containment Boom in the United States	61
B-5.	Major Sources of Offloading Pumps in the United States	62
B-6.	Major Sources of Dispersant Delivery Systems in the United States	63
B-7.	Alyeska Response Equipment	64
B-8.	Equipment Based at the Oil Spill Response Center, Southampton, U.K.	64
B-9.	Petroleum Industry Response Organization Regional Center Capital Equipment Cost	66
B-10.	U.S. Oil Spill Co-Ops	67

Figure 19-6 A Table of Figures and a Table of Tables.
Reprinted with the permission of the Office of Technology Assessment.

Perhaps the most important thing that you write will be the **executive summary.** This will be the most frequently read section of your report. Many readers will form their judgment of the whole report from this one page. You need to convince potential users that the report will be worth their time. The executive summary, used to compress the main points of the report, is very important to busy technical professionals who need to use their reading time productively. The summary uses the same organization as the report itself, stating results and recommendations in the same sequence. Provide only enough information for the user to determine whether to read the report itself.

19.4.2 Body

The **body** of the report needs to be carefully planned. Your report discussion needs to be complete, containing enough information for your readers to understand and judge your conclusions and recommendations. The text of your report needs to fully describe the background, purpose, and technical detail.

Begin by telling the user how the information in the report was collected. Describe the sources of the data and the technical processes used to

Chapter 1

Overview

The March 24, 1989 *Exxon Valdez* oil spill in Prince William Sound, Alaska dramatically illuminated the gap between the assumed and actual capability of industry and government to respond to catastrophic[1] oil spills. There are many reasons why this gap wasn't better appreciated before March 24: elaborate oil spill contingency plans had been prepared and approved; oil spill equipment had been developed and stocked; major damaging spills had occurred infrequently, and almost never in the United States; and a nebulous faith had existed that technology and American corporate management and know-how could prevent and/or significantly mitigate the worst disasters.

The *Exxon Valdez* accident shattered this complacency. In the aftermath of the spill a small army of people has been put to work around the country studying how the United States can do a better job preventing spills and how it can be better prepared to fight one that does occur. In this backrgound paper, OTA examines the state-of-the-art of oil spill technologies and response capabilities. On an encouraging note, it appears that improvements can be made in oil spill cleanup technology and, perhaps even more, in the way we organize ourselves to apply the most appropriate technologies to fight oil spills. Such improvements should result in a reduced risk of significant damage from a major spill in the future.

However, the unfortunate reality is that, short of eliminating oil transportation at sea entirely, there is no perfect solution to offshore oil spills. **It is certain that oil spills will occur again.** If improvements in prevention technology are made, the frequency of major spills may decrease, but improvements are unlikely to eliminate oil spills entirely, and a very large spill under adverse conditions could still overwhelm our capacity to respond effectively. Even using the best technology available and assuming a timely and coordinated response effort, it is not realistic to expect that a significant amount of oil from a major offshore spill could be recovered, except under the most ideal conditions. **Historically, it has been unusual for more than 10 to 15 percent of oil to be recovered from a large spill**, where attempts have been made to recover it. With improvements in technology and response capability, it should become feasible to do much better, but it is unlikely that technical improvements will result in recovery of even half the oil from a typical *large* spill.

It is not feasible to be prepared for all contingencies: each oil spill is unique in terms of location, weather, oceanographic conditions, time of occurrence, characteristics of the oil, equipment available, and experience of response personnel. Accidents are unpredictable. They may be caused by "acts of God" or human error, both of which are impossible to fully anticipate or control. The ideal conditions in which cleanup technology would be most effective rarely occur in the real world.

The U.S. industry has concentrated its efforts in developing technology to fight the numerous small spills in harbors and protected waters. On the one hand, industry has oversold its ability to fight major spills, and the government has largely relied on private capabilities; on the other, the public's expectation about what can be accomplished once a major spill has occurred has been too high. *Prevention* of major spills, although beyond the scope of this study, must be a high priority.[2]

[1] For the purposes of this report the terms "catastrophic," "major," and "large offshore" spills refer to discharges in excess of 1 million gallons of oil that occur in open waters subject to rough seas, high currents, or other adverse environmental factors.

[2] For a detailed discussion of prevention measures, see U.S. Congress, Office of Technology Assessment, *Oil Transportation by Tankers: An Analysis of Marine Pollution and Safety Measures* (Washington, DC: U.S. Government Printing Office, July 1975).

1

2 • Coping With an Oiled Sea: An Analysis of Oil Spill Response Technologies

It is important to put the environmental impacts of a major oil spill into perspective. Such a spill is indeed a catastrophe, but oil spills are not the worst type of pollution with which Federal and State authorities have to deal. In terms of threats to human health and persistence in the environment, spills of hazardous chemicals or radioactive waste can be far larger problems, and accidents involving dangerous materials can cause significant loss of life. Nevertheless, it is a serious problem when a large quantity of oil is spilled in a coastal or near-coastal area. The public is particularly concerned about large spills in sensitive areas because the effects on the local ecosystem are acute, often initially devastating both to biota and economic activities. Oil can be toxic to organisms that come into contact with it and can cause major problems with recreational or other uses of coastal regions, such as commercial fishing in Alaska. If large amounts of oil reach the shore, the oil may persist for long periods, even though natural degradation mechanisms do assist recovery.[3]

As bad as the *Exxon Valdez* accident was, it could have been far worse: **only about one-fifth of the crude oil the tanker was carrying was released**. Fortunately, the rest was offloaded. The *Amoco Cadiz* did spill its entire cargo off the coast of France in 1978, a cargo roughly the same size as that carried by the *Exxon Valdez*.[4] Significantly, neither the oil industry, the Federal Government, nor the State of Alaska were prepared to deal with a spill the size of the *Exxon Valdez* spill. It was fortunate, in a sense, that the spiller in this incident was a major international oil company capable of marshalling significant resources, rather than a small tanker company.

[3] In 1974 the supertanker *Metula* spilled some 16.2 million gallons of oil after grounding in the Strait of Magellan. Essentially, no cleanup occurred and at least half of the oil lost washed onto about 50 miles of shoreline. A study by the National Oceanic and Atmospheric Administration about 6 years after the spill concluded that much of the oil remained in sediments, along beaches, and in marshes. In heavily oiled, sheltered areas, it seems likely that the oil will persist for more than 100 years.

[4] The *Amoco Cadiz* spill released 68.4 million gallons of light crude oil off the Brittany coast in France. Prevailing winds kept the slick near the coast for 1 month, eventually oiling about 200 miles of coastline. In a 1985 report, *Oil in the Sea*, the National Research Council estimated that it would take decades before the environment recovered.

Figure 19-7 An example of an executive summary or overview. **Reprinted with the permission of the Office of Technology Assessment.**

obtain the information. Your reader needs to know where the information comes from. This step is particularly important if the report uses survey or laboratory data.

The organization of the discussion section should reflect the strategy you have chosen to communicate your information. For example, if you are asked to produce a recommendation report, where your task is to choose among alternatives, you need to use a comparative strategy. Use graphics to compare and contrast your selection criteria and your conclusions.

Make the report interesting with lots of visual support. Diagrams and drawings can help to make your meaning clear. Remember to use well-writ-

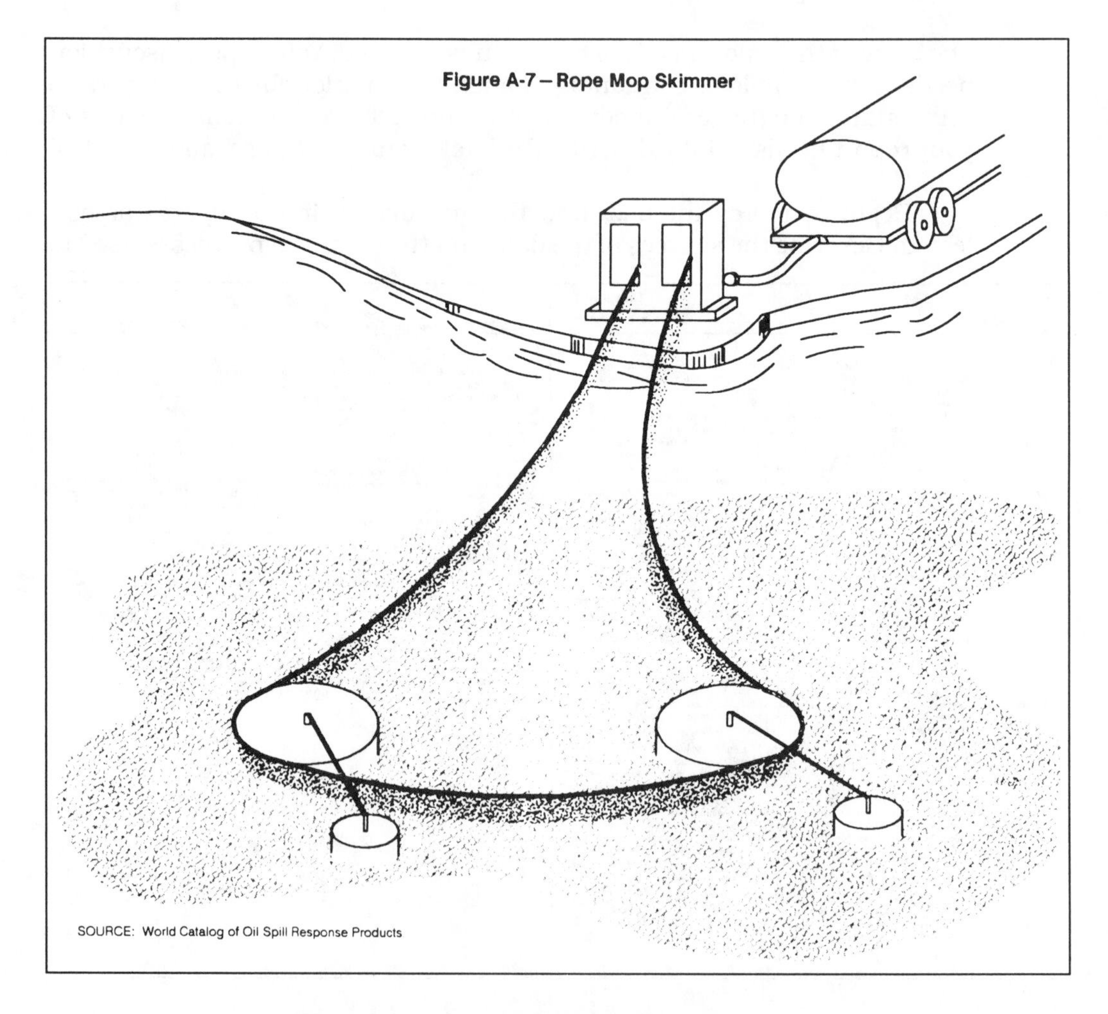

Figure 19-8 An example of an illustration.
Reprinted with the permission of the Office of Technology Assessment.

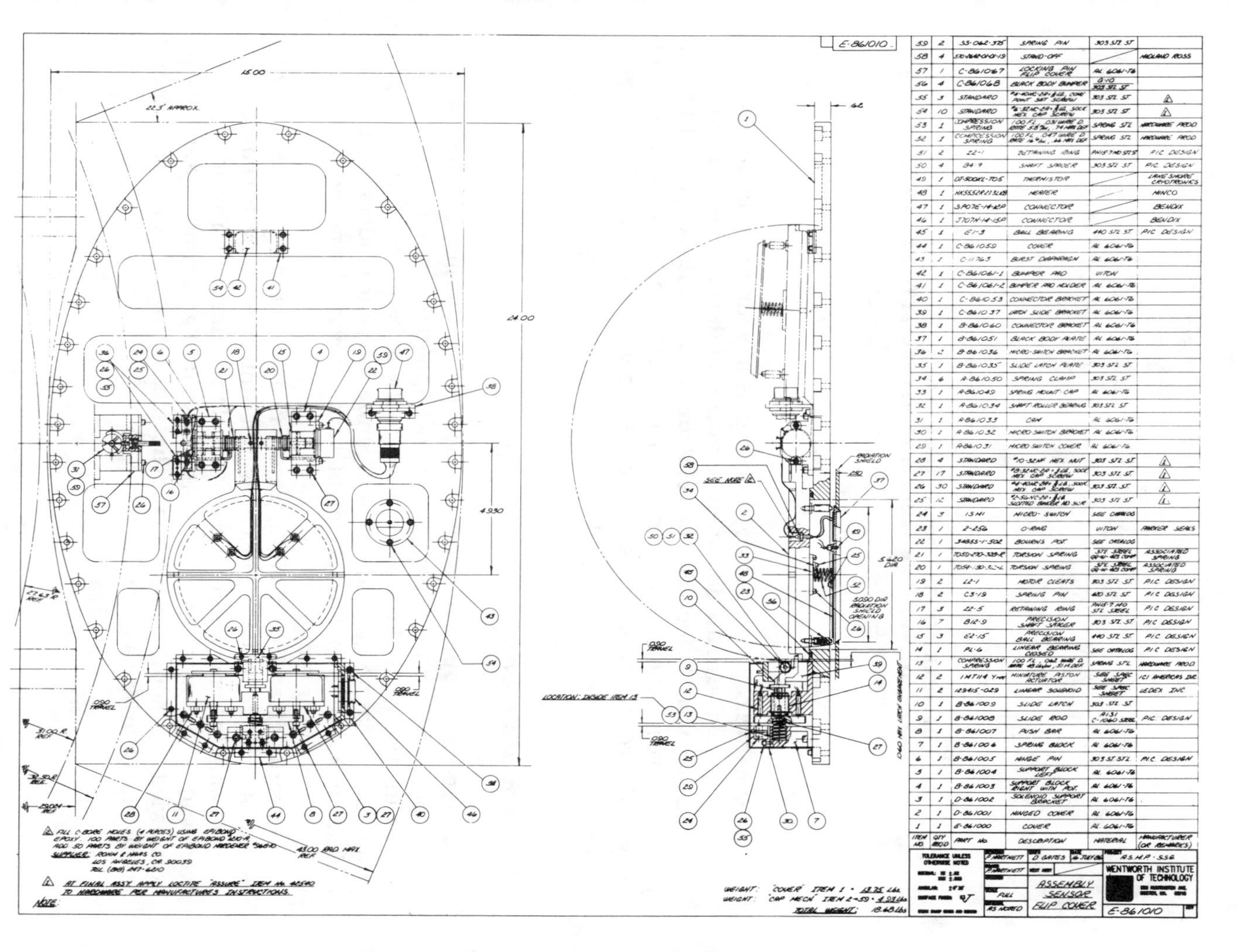

Figure 19-9 A more complex illustration.
Reprinted with the permission of Wentworth Labs.

Appendix D
Acronyms

API	American Petroleum Institute
CEC	Commission of European Communities
CEDRE	Center for Documentation, Research, and Experimentation on Accidental Pollution (France)
DOT	Department of Transportation
DOI	Department of the Interior
EPA	Environmental Protection Agency
MPCU	Marine Pollution Control Unit (United Kingdom)
MMS	Minerals Management Service
NOFO	Norwegian Clean Seas Association for Operating Companies (Norway)
OHMSETT	Oil and Hazardous Materials Simulated Environmental Test Tank
OSR	Oil Spill Response Ltd. (Southampton, U.K.)
OSC	On-Scene Coordinator
PIRO	Petroleum Industry Response Organization
USCG	United States Coast Guard
SPCA	State Pollution Control Authority (Norway)
VOSS	Vessel-of-Opportunity Skimming Systems

☆ U.S. GOVERNMENT PRINTING OFFICE : 1990 0 - 273-068

Figure 19-10 An example of an appendix.
Reprinted with the permission of the Office of Technology Assessment.

ten, informative captions for the illustrations in your report because many persons will just scan the document, glancing at the pictures and captions.

You may want to include a **conclusions and results** section. Although you may have mentioned your outcomes in the summary, introduction, and body of the report, this allows you to restate your recommendations and provide general reasons for them.

19.4.3 End Material

The **appendix** or **appendices** contain all types of information which would interrupt the flow of the report. Letters of support, computations and data sheets, interviews, and questionnaires ought to be provided when appropriate; use separate appendices to organize for your readers. You should include information that people might want to refer to immediately in order to verify a claim or computation.

A **works cited list** or **bibliography** is necessary to give credibility to verbal and secondary data. Most of us judge information by its source. Your readers want to know where your information comes from. You do not, however, want to interrupt your text with constant digressions. If you want to acknowledge a source use a **parenthetical citing.** Mention the source briefly and clearly in a parenthesis:

> The United States is a member of the ATA Carnet system which allows business travelers to temporarily bring commercial samples and scientific/technical equipment into other countries without paying duty. (U.S. Council of International Business, 1212 Avenue of the Americas, New York, NY 10036)

Your bibliography or works cited section needs to follow standard formats. The persons who use it are looking for some information and they do not want to figure out your system. Let them look where they expect to find things.

19.5 ORGANIZATION

The format of a long report is the logical arrangement you use to assemble the various parts of the document. Your format will depend on which strategy you choose to present your information. In turn, your strategy will depend on two things: your assignment and your audience.

If you have been asked, for instance, to provide recommendations, then you need to do so. If you believe your audience will not be pleased with these recommendations, however, you may choose to lead up to them rather than stating them at the beginning. Before you can begin to outline the report or start to write the different sections, you need to decide how your document will be structured.

Two basic patterns are **particular** → **general** (induction) and **general** → **particular** (deduction).

Use induction when you need to build a case for a general rule or guideline for action; start by establishing a number of specific cases until you have constructed a general rule. For example, if you want to explain why company profits exceeded expectations in the last quarter, you could look at each individual account, and build towards a general rule.

Use deduction when you have an established rule; apply a specific case to the general rule and draw your conclusion. If, for example, you had already established that all of your computer software clients need an upgrade after 14 months, you could apply a specific case, Company X, and conclude that you need to supply them with an upgrade in November.

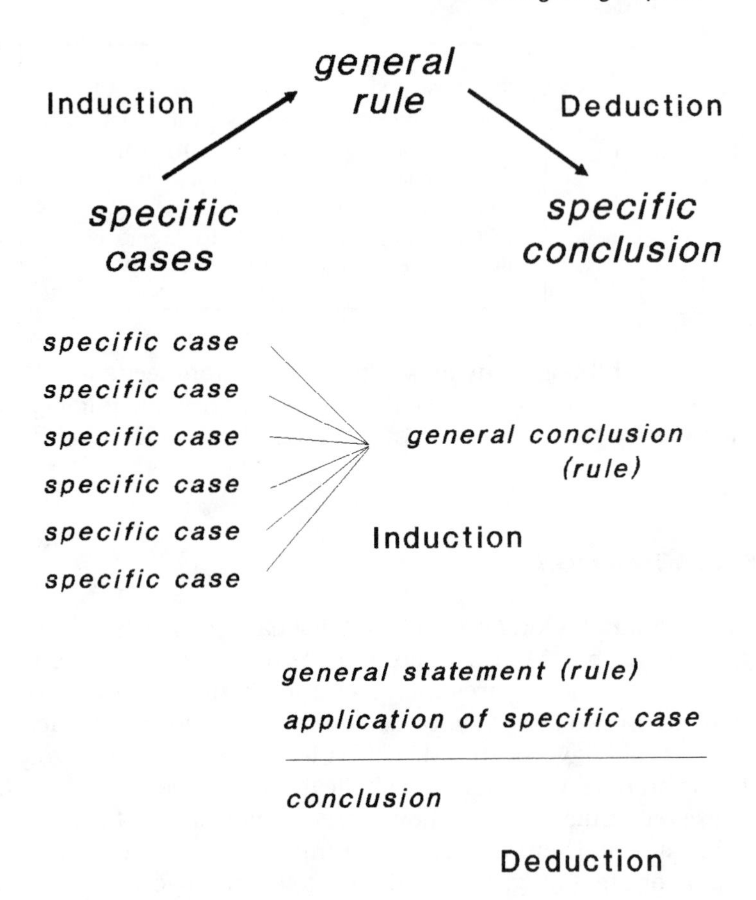

The most common structure for recommendation reports is **problem → solution.** You begin with the problem, stating it as exactly as you can. Then you describe your method of investigating the problem and propose a variety of solutions, demonstrating that you have given thorough consideration to the issue. Finally you discuss the solution you recommend and explain your reasons.

Many companies provide formats for long reports. In this case, it is essential that you follow the company format closely. If you are designing the format, however, you must make sure that it is complete and useful for its various readers. The logical arrangement of your material should reinforce what you say in the report by making it easy for all of your readers to find what they need.

19.6 USEFUL TECHNIQUES

What we have discovered is that most effective report writers begin by considering the preparation of a long report as a carefully planned and scheduled activity. In most instances long reports aren't simply written; they are deliberately and closely managed documents. Many people are involved and their activities must be coordinated.

Collaboration. Long reports are commonly created by teams of technical professionals. This type of writing is a collaborative activity where many people work together to create reports that represent their combined information, viewpoints, and decisions. An engineering project team, for example, might have an architect, a civil engineer, an environmental engineer, and technicians with a variety of specialties, all working on the same activity. Who will write the report? Every one of them will be involved in the crucial activity of reporting.

Performing this task means working in a group environment where you will be required to brainstorm, share assignments, search out and evaluate alternatives, and package your information in an attractive and useful format. If you are going to do a good job, you need to work well with others.

Outlining. Careful document design is essential to the process of creating effective long reports. This means outlining. You need to keep the whole report in mind at each step, through research and data collection, writing and editing, and the preparation of front and end materials. When you are writing a long report you are assembling a complex structure with many parts. As with other types of building, you need to consult your blueprint at each stage of construction.

You don't want to leave out important information. At the same time, you don't want to clutter your report with irrelevant details. Detailed information which will interest only a few persons can be included in the end material as an appendix. Material that is unnecessary should be left out. This is why you need to keep your audience firmly in mind and create your document for their use.

An outline is a tool which lets you work with the structure of your document before you begin the actual writing. A **topic outline** uses short phrases to indicate the order and importance of different sections. **Sentence outlines** use complete topic sentences to summarize each section. These sentences are explained and elaborated to provide the paragraphs for the rough draft.

Complex writing projects such as long reports often involve the creation of two outlines. First, a topic outline is brainstormed; then the sentence outline is developed from this beginning. At the outline stage, you can easily change the pattern or structure of your document. As you add detail and revise the outline, you can experiment with the order and placement of your information. You can review your document blueprint for balance and organization.

Scheduling. A complex formal report will often require that the efforts of many different persons be coordinated. Schedule these efforts

TOPIC OUTLINE
PRELIMINARY REPORT
CHELSEA OIL SPILL

1. Introduction
2. Background Information
3. The Problem
4. Location and description of spill
5. Available resources
6. Necessary equipment and/or expertise
7. Conclusion
8. Recommendations

PARTIAL SENTENCE OUTLINE
PRELIMINARY REPORT
TANK FAILURE ANALYSIS

I. The quick introduction of the Gates & Associates response team to the oil tank failure site in Bellville, MI, will allow us to determine the comparative effectiveness of two tank inspection procedures before data is removed from site.

A. The purpose of this investigation is to serve as a field site analysis of the magnetic field and x-ray techniques for tank inspection.

1. The oil tank failure which occurred on January 14 was the result of metal fatigue caused by the stress of contractions in extreme cold.

2. The heavy viscosity of the petroleum provided an unexpected resistance to the normal structural contractions because of three main factors.

a. There was too much oil in the tank.

b. The tank was not properly maintained.

c. Relief valve was sealed shut by rust.

B. We will conduct a series of three test procedures on the sheared structure of the failed plates using the traditional and new procedures.

Figure 19-11 An illustration of a topic outline and a sentence outline.

carefully. Create a writing plan which lays out all of the tasks and assigns responsibilities and deadlines. Use the schedule to maintain these deadlines.

The final version of the agreed-upon schedule ought to be expressed graphically so that participants can see what they are agreeing to provide and when. Flow charts, activity charts, and work breakdown schedules (WBS) all provide ways to understand visually what will be expected and when.

Storyboarding. Storyboarding is an effective method for coordinating multiple authors in designing, planning, and editing successful reports. This is a graphics-driven approach to document design, an elaborated version of the cut-and-paste technique used by many writers.

A storyboard is a visual blueprint where you create a model of your entire document before the majority of writing is done. You begin by breaking your report topic into information passages of about two hundred words each. Each passage is treated as a module or series of specifications which can be reviewed before the final text and graphics are completed.

Each module has four elements. The **headline** is a phrase that describes the contents of the spread. The **summary** provides a brief abstract of the module content, and enough information for the reviewers to verify the accuracy and importance of the module. You should leave a place for **notes** by the review team where they can leave comments, questions, and suggestions. Finally you have **exhibits** to supplement your text.

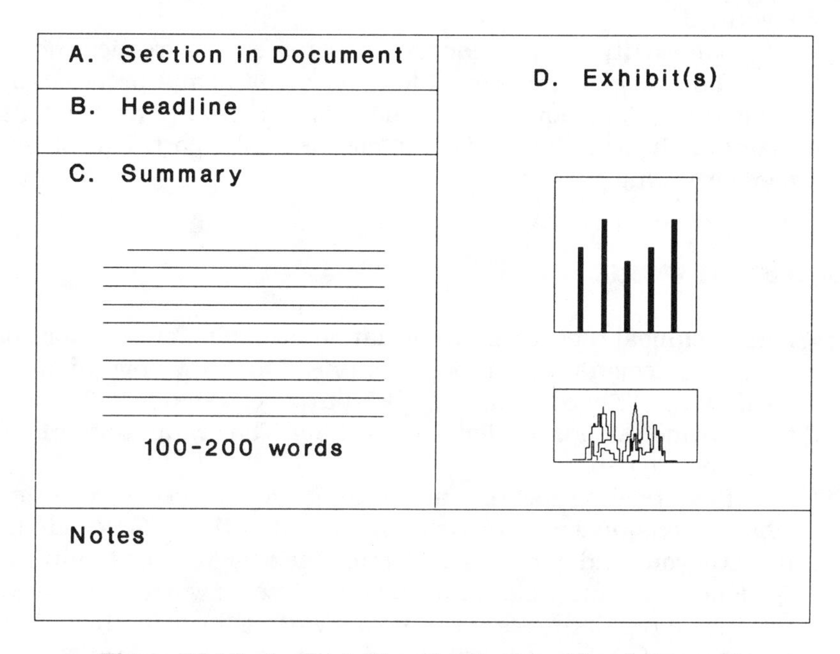

Figure 19-12 An illustration of a storyboard for a long report.

These can include tables, drawings, bullet lists, photographs, or any other type of graphics. You can simply describe your exhibits or provide a rough sketch.

At this point, with all the modules drafted, you should schedule a *walkthrough* or review session. All the members of the team need to participate. The modules should be taped to a wall and the team should carefully consider each one. You should look at the order of the modules and the amount of information included in each one. You can determine if the graphics support the text. Is your information accurate and complete? The purpose of this session is to make changes before the draft report is written.

Group revision. Everyone makes mistakes. When many persons are working on the same writing project, there is an even greater chance for mistakes to slip into a document. The whole team needs to be involved in revision. Mistakes cost credibility. Even simple mechanical errors can create the impression that your report is careless and flawed. Be sure that you carefully proofread every section of the report. Review captions, table of contents, index, numbering, and appendices. Every reader will look at something different and every section needs to be error-free.

Software tools can make revision and editing simpler. Some word processors have a helpful feature called *redlining* mode where comments and new text are highlighted to distinguish them from the original. Other software programs allow dated and initialed comments to be inserted throughout a document. In the end, however, you are responsible for the report your team provides. The only way to be sure that your work is perfect is to check it yourself.

Some writing teams choose a designated proofreader and editor, a person whose responsibilities include assembling and proofreading each circulated revision. Some sort of head writer or editor will be necessary to maintain the schedule and complete the final report. This person will manage the writing project.

19.7 A FINAL WORD

Long technical reports are information products, created for specific audiences. To design useful reports you need to know how the product will be used, who will use it, and for what purposes. No one will read your report for enjoyment. Your audience is working. They want a product that is easy and quick to use.

Easy reading means hard writing. Report writers need to anticipate what various readers already know and what they will need to find out. The needs of your audience will determine the design of the report. Some of your audience will simply glance at the report, some will study it, and others will use it on a regular basis. This means writing for *all* of them. Your ability to communicate effectively and your technical competence will be judged by your technical reports. This means writing, editing, testing, and rewriting. Technical reports must be designed to explain technical information to a wide variety of audiences. Don't confuse your interest in the technical task with your need to produce a report that addresses these multiple users.

LONG REPORT EXERCISES

1. Design two different title pages for a long report. The report describes the abilities of Gates and Associates to respond to oil spill disasters. Review the article in Chapter 12, Reading Technical Material for general information. If you have not already done so, review Chapter 26 for information on Gates and Associates.

   Make up any details you need. Discuss the different impressions the titles cause with your work group or classmates.
2. Create a table of contents for a book about your life. Try to organize this with some consistent patterns or categories, such as time, work roles, or personal interests.
3. Prepare the following parts of a long report:

   Title Page
   Table of Contents
   Table of Illustrations
   Executive Summary
   Transmittal Letter

   For the purpose of this report, assume that some organization of your own choosing has asked Gates and Associates to solve a problem. Your report should contain an analysis of the problem, its causes and its effects; a discussion of possible solutions, and their advantages and disadvantages; and the supporting evidence for your recommended solution.

   You do not have to write the entire report, but you will need to think through the problem before you begin. Your table of contents must be complete, with at least two levels of headings. If a topic is not listed in your contents, it can be assumed that you did not intend to introduce that topic in the report.
4. Conduct a survey of use of fax technology within your company (or class). Write an outline for the body of a long report that is based on the results of your survey: Discussion, Results, Recommendations, and Conclusions should all be written up in a sentence outline.
5. Write a memo to the persons in your office or classroom describing and explaining some ways which appendices can be designed so people will be more likely to read them.
6. Prepare a letter of transmittal, a table of tables or figures, and a glossary which defines five related terms with which you are familiar. What do these documents tell you about the imaginary report upon which they are based? How do these support elements influence the way that readers will view the report itself?
7. You are part of a project team which will be preparing a long report recommending the use of expert systems to identify the secondary effects of toxic chemicals

rapidly and accurately. Sheila O'Brien asks you to come up with a project schedule. You should read *Chapter 26, Communicating In An Organizational Environment*, for more information on Sheila O'Brien, your role in Gates and Associates, and some of the problems toxic chemicals can create.

Develop a ten-week production schedule for this report.

8. Using the advice in this chapter, write a long technical report of at least ten pages on some topic that you find interesting and are currently involved with at work or in another course. Be sure to include all of the parts of the long report; do not overlook graphics and illustrations.

9. Working with a group of at least three persons, create a topic outline for a long report. Then turn the topic outline into a sentence outline. Discuss the difficulties you encountered, both with the outline formats and in collaborating with other writers.

10. Examine the figures within this chapter which illustrate the various parts of the report "Coping with an Oiled Sea: An Analysis of Oil Spill Response Technologies." With the other members of your group, discuss your impressions of the title page (Fig. 19-3), the foreward (Fig. 19-4), the table of contents and table of figures (Fig. 19-5 and Fig. 19-6), the overview (Fig. 19-7), the illustration (Fig. 19-8), and the appendix (Fig. 19-10). What do you think about the format and design of the document from the parts reproduced in this chapter?

20

Designing User Manuals

Note to Users: **This is a task chapter. You should use this chapter if you are asked to participate in the planning, writing, editing, or distribution of user manuals.**

20.1 INTRODUCTION

In today's competitive markets, technical products and complex systems need detailed and elaborate explanation. Carefully designed user manuals can increase customer satisfaction, improve training, and reduce requests for user assistance.

Written support material for a technical product or process requires a lot of effort, but it is an important part of the product and system performance. All types of manuals—reference, office, safety, and procedure manuals—are used by persons who need to find some information or to do something. These manuals need to be designed, written, edited, tested, maintained, distributed, and used.

A well-designed manual is an information device, an important part of a technical system. User manuals should be engineered for use. This means the design considerations must emphasize the viewpoint of the persons who will be working with the manuals. Usable manuals are often essential to the safe and successful application of thousands of products and services. Your ability to contribute will be noticed and appreciated.

In this chapter you will learn how to:

- Develop document standards.
- Distinguish between reference and instruction.
- Design for easy revision.
- Use a systems approach and structured design.

20.2 DEFINITION

User manuals are task-oriented documents which provide detailed specifications and operating instructions for complex technical systems or products. These information devices, often called *user documentation*, fill two important functions: instruction and reference. In either case, the purpose of user manuals is to provide rules and information for people who want to use technical products and systems.

In this chapter we illustrate the text with sections from a user manual for a technical product which ought to be in every house and apartment. You want to remember, however, that there are many different types of user manuals: quality assurance, safety procedure, service, and training manuals, to name a few. All have in common the need to help the persons who will be using them. No one will consult a manual unless they need to do so. When they do consult the manual they expect you will have anticipated their problems.

Operations manuals tell people how to do things. These documents are used to instruct users. This means demonstrating how to operate products and how to follow the procedures in complex systems. Operations manuals are organized according to the tasks which the reader will carry out. Complex assembly instructions and installation plans, for example, require very elaborate procedures (sets of activities in sequence) which must be followed exactly.

Reference materials provide users with information they might want to have but do not need in order to use the system or product. Typically these documents will contain detailed technical information. Inventories, parts lists, and catalogs are examples of reference manuals which require complete, current, and accurate information.

Whatever the function, user manuals exist for users. Their styles and strategies should be transparent, so that readers don't even notice the document. Users are interested in meaning more than style. Manuals should come equipped with search tools. The persons who use your manuals want to know about something or how to do something, and they need to find this information quickly.

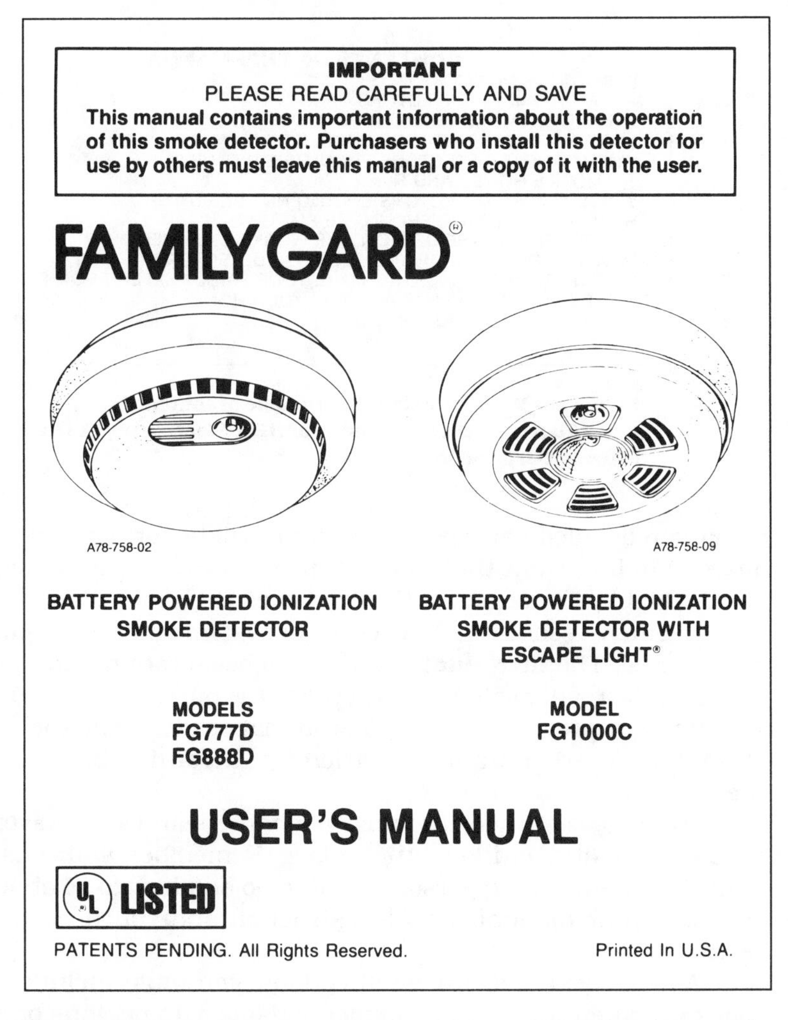

IMPORTANT
PLEASE READ CAREFULLY AND SAVE
This manual contains important information about the operation of this smoke detector. Purchasers who install this detector for use by others must leave this manual or a copy of it with the user.
FAMILY GARD
A78-758-02
A78-758-09
BATTERY POWERED IONIZATION SMOKE DETECTOR
BATTERY POWERED IONIZATION SMOKE DETECTOR WITH ESCAPE LIGHT®
MODELS FG777D FG888D
MODEL FG1000C
USER'S MANUAL
UL LISTED
PATENTS PENDING. All Rights Reserved.
Printed In U.S.A.

Figure 20-1 An example of a user manual cover. Note that the manual applies to three different models.
Reprinted with the permission of BRK Electronics, A Division of Pittway Corporation.

20.3 PARTS OF THE USER MANUAL

Your **cover** and **title page** should draw readers into the manual. Indicate the purpose of the manual in the title. This means that the title must be informative in a few short words. Practical considerations such as printing and binding costs will affect your choice of cover, but it should be attractive and appropriate. Durability may also be an important design factor.

A carefully prepared **table of contents** is a map which provides your readers with access routes to the specific information they are looking for.

CONTENTS OF THIS MANUAL

Figure 20-2 An illustration of a table of contents
Reprinted with the permission of BRK Electronics, A Division of Pittway Corporation.

The more detailed you are, the easier it will be for the persons using your manual to find what they need. It is also a good idea for you to provide a detailed table of contents at the front of each chapter.

Begin by describing the product or system and then explain how to use the manual. The **introduction** provides background and special information that the reader will need to perform the operations and use the manual effectively. It provides an overview so that readers will know how to locate what they need. Establish a friendly, helpful tone with clear, direct statements.

Overviews at the beginnings of chapters and **summaries** at the ends help readers find and keep their place. Remember, your chapters are lessons. You are teaching persons what to do and how to go about it. They will be using your manual for self-instruction and you need to respect their learning.

As with sets of technical directions, you must include **warnings** and **cautions** when there is any danger of damage to persons or property. This is a legal as well as an ethical obligation. Persons who use the product carefully will rely on the user manual to protect them from possible harm. Include **notes** when you want to provide the reader with some useful information that does not involve any danger.

Provide **lists** of simple steps, one action in each step, numbered for easy use, and carefully sequenced. In other words, you need to present the steps in the order the readers will perform them. Organize your material according to what the users will want to do. Then tell them what to do. Each step needs to be a command.

Reference sections will be more useful if you organize your material in **table** format. Rows and columns make information easy to locate. Again, your task is to help the reader. Use standard formats for bibliographic information and other reference materials.

A **glossary** will provide definitions for users who do not want to search

WHAT THIS SMOKE DETECTOR CAN DO
This smoke detector is battery powered. It is designed to sense smoke that comes into its sensing chamber. It does not sense gas, heat, or flame.

This smoke detector is designed to give early warning of developing fires at a reasonable cost. This detector monitors the air. When it senses smoke, it sounds its built-in alarm horn. It can provide precious time for you and your family to escape before a fire spreads. Such **early warning is only possible,** however, **if the detector is located, installed, and maintained as described in this User's Manual.**

Figure 20-3 An example of an introduction.
Reprinted with the permission of BRK Electronics, A Division of Pittway Corporation.

through the index and manual. If your manual will be used by people who are not familiar with many of your technical terms, then prepare a glossary. Select a list of terms that need to be defined and explained. Don't assume, however, that all readers will use your glossary to teach themselves definitions and terms. This is a reference which many persons will never consult, so also include definitions of key terms within your text.

You may want to include an **index.** This is an alphabetical listing of all the major topics discussed in the manual, along with their location. Some users always begin at the back with the index. Others will judge your book

⚠WARNING **This smoke detector is designed for use in a single residential living unit only.** In other words, it should be used **inside** a single-family home or apartment. It is not meant to be used in lobbies, hallways, basements, or another apartment in multi-family buildings, **unless there are already working detectors in each family unit. Smoke detectors placed in common areas outside of the individual living unit (such as on porches or in hallways) may not provide early warning to residents. In multi-family buildings, each family living unit should have its own detectors.**

Detectors designed to be linked together should be interconnected within one family living unit only. If detectors are interconnected between living units, nuisance alarms will occur in other units when detectors are tested. (Models FG777D, FG888D, and FG1000C cannot be interconnected.)

⚠CAUTION (As required by the California State Fire Marshall)

"Early warning fire detection is best achieved by the installation of fire detection equipment in all rooms and areas of the household as follows: A smoke detector installed in each separate sleeping area (in the vicinity, but outside of the bedrooms), and heat or smoke detectors in the living rooms, dining rooms, bedrooms, kitchens, hallways, attics, furnace rooms, closets, utility and storage rooms, basements and attached garages."

Figure 20-4 An example of a warning and a caution.
Reprinted with the permission of BRK Electronics, A Division of Pittway Corporation.

HOW THIS DETECTOR SHOULD BE PUT UP

- This smoke detector is made to be mounted on the ceiling, or on the wall if necessary.
- This model is a single-station detector that cannot be linked to other detectors.
- Before installing this detector, read "Where Smoke Detectors Should Be Put Up" and "Where Smoke Detectors Should **Not** Be Put Up." Then decide where to put up this detector.

To put up your detector, follow these steps:

1. Hold base firmly, and pull up on cover tab or arrow marked "OPEN HERE." This will open the hinged cover. See Figure 7.

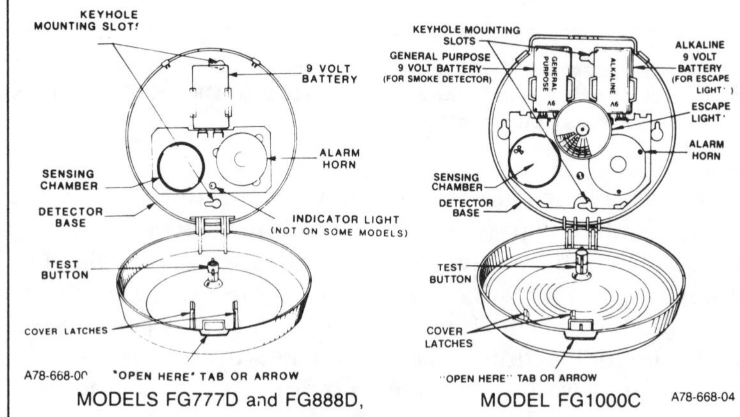

Figure 7: SMOKE DETECTORS WITH COVERS OPEN, SHOWING COMPONENTS

NOTE: Hinged cover may unsnap from base if opened too far. This will not damage the detector. Simply snap cover hinge back onto base if this happens.

2. Place detector base against the ceiling (or wall) where you want to mount it. Use a pencil to trace around the inside of two of the key-hole shaped slots used for mounting. See Figure 7.

CAUTION: This smoke detector comes with cover latches that will prevent the smoke detector cover from closing if a battery is not installed. This is to warn you that the smoke detector will not work until a new battery is installed. If you mount the smoke detector on a wall, be sure the test button is at the bottom or this feature will not work properly.

3. Put detector where it won't get plaster dust on it when you drill holes for mounting.
4. Using a 3/16-inch (5mm) drill bit, **drill holes in the centers of the small ends of the two key-hole slots** outlined on the ceiling (or wall).
5. Push plastic screw anchors (in the package with the screws) into the holes so that tops are flat against the ceiling (or wall).
6. Tighten screws all the way into the two screw anchors. Then unscrew each of the screws two full turns.
7. Put detector base back on the ceiling (or wall) so the large ends of the key-hole slots slip over the screw heads. Carefully slide base so the small ends of the key-hole slots are around the heads of the screws.
8. Tighten screws.

Figure 20-5 Note how each step begins with a command verb: hold, place, put, drill, push, tighten.
Reprinted with the permission of BRK Electronics, A Division of Pittway Corporation.

by the entries in the index. Try to include all of your major topics in the manual and include common synonyms so that the reader does not have to use your term to find a topic.

20.4 WRITING USER MANUALS

Long documents such as user manuals require a systems approach and a structured design. A systems approach views the entire process (design,

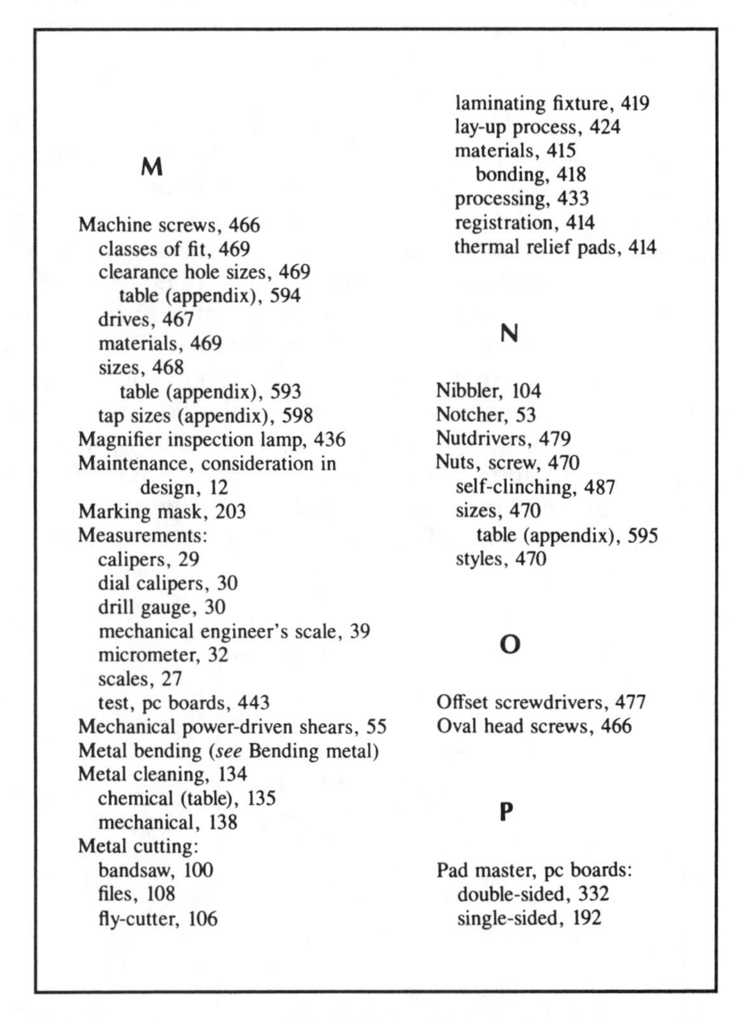

Figure 20-6 An example of a segment of an index from *Electronic Techniques: Shop Practices and Construction* by Robert S. Villanucci, Alexander W. Avtgis, and William F. Megow, 4e, © 1991.
Reprinted with the permission of Prentice Hall, Englewood Cliffs, NJ.

writing, assembly, and production) as a single complex operation which you need to manage carefully. The efforts of many people need to be coordinated and combined into a coherent manual.

Structured design leads to structured documents. You want to place your message in an accessible and reliable sequence so your readers can quickly extract the information they need. The organization of your manual must be clear and easy-to-follow. Multiple writers can use the structured design to connect their individual contributions to an organized plan.

Essentially, user manuals are long reports designed for a very specific audience, persons who need to know how to locate, do, or learn something. If you are involved in working with a manual you should *read Chapter 19, Producing Long Reports*, for advice on scheduling, collaboration, and other issues of a complex document with multiple authors, audiences, and editors.

Use outlines to build models of the document before you create the document itself. The outline is a blueprint which can be reviewed and revised. It is easier to rework models than change the completed manual. Your outlined model provides a complete, precise description of the manual. This is used as a basis for writing and assembling the finished document.

If a structured outline is going to be effective, it must be substantive and detailed. The outline must provide enough information for the editing/testing process where team members and users examine the model to see if it is what they want. Remember, it is important to include users in the review of the model because their needs should determine the final design.

Begin this complex writing project by defining your audience. Who are they and what will they need to know? What tasks will they want to perform? Design your manual by considering their needs first. Try to develop an overall plan which will unfold its details on the readers' schedule.

4.0 Using your phone

- 4.1 Dial tone
  - 4.1.1 Tone definitions
  - 4.1.2 Busy signals
- 4.2 Outgoing calls
- 4.3 Incoming calls
- 4.4 Phone numbers
  - 4.4.1 Long distance
  - 4.4.2 Area Codes
- 4.5 Phone directories
  - 4.5.1 Phone books
  - 4.5.2 Operator
  - 4.5.3 Yellow pages

Figure 20-7 A segment of an outline for a telephone system

The Five Stages for User Manual Design

1. Needs Analysis
2. Design
3. Assembly
4. Editing
5. Maintenance

Organize by what users need to do. If you use your product or system as a framework, users will have to jump all around the document, looping and branching to get what they need. If you pattern your manual on the logical sequence of user activities—installation, frequent applications, and custom features—your audience will recognize that your design has them in mind.

Manual writing is a five-stage process. You begin with a **needs analysis,** defining and deciding what documents or publications will be needed. You are providing your audience with information devices and it is important to determine exactly what they might need, what you can afford, and what you can produce in time for product or service delivery.

Your second stage is **design,** where you develop a structured outline in complete detail, including headings and sub-headings. You need to develop and polish your outline or storyboard so that it can be presented for technical verification and approval to proceed.

The third stage is **assembly** where a whole variety of elements, including tables, graphs, charts, text, and illustrations, have to be pulled together and connected in a draft document. This means coordination of a planned schedule where many persons work together, where difficulties are anticipated and resources provided. Keeping many people working together on the same writing project is like directing the symphony—a whole lot of hard work.

Editing is a group process when a document is created by numbers of people, but even if you are the only editor, you need to consider the views and needs of different persons who will be using the document. Accuracy and clarity are the most important things you can contribute to this process.

The final stage of user manual design is **maintenance.** Out-dated, unreliable information will cause problems for users and every new edition multiplies the versions in circulation. Maintaining the manual needs to be a part of the original design considerations. This is important so we have a full section to press this point. A plan for keeping the user documents current and accurate is an essential part of product service.

20.5 *THE THREE G's*

Most user manuals want to assist the user to:

Get Going
Get Better
Get Help

Get going . . . Users will be impatient to start. Your manual should begin by helping the user start to perform some activity. This means providing lots of information at once. You need to introduce the technology, explain how it works, and warn your users of potential problems and dangers, all at the same time. Don't confuse your own interest in the technology with what the user needs to get started. Simplicity is the key.

Focus on the persons who will be using the manual and organize it according to what they need to do. Don't describe the product part-by-part or the system step-by-step. Begin with what the reader is motivated to do.

Get better . . . Here is where you tell the users ways to improve their performance. Give them advice, tips, and shortcuts which will make the product or system more productive. Use this part of the manual to describe advanced features, special ways to customize and adapt the product, and potential new applications.

Get help . . . Many people avoid manuals. For these persons, the most important function will be to *help* when problems occur. You need to identify common difficulties and their solutions. You also need to tell your readers when they need expert assistance. Troubleshooting tables can provide quick access to needed information.

Troubleshooting Matrix

Trouble	*Probable Cause*	*Possible Solutions*
Ratchet won't turn.	Reverse lever is stuck in mid-position.	Turn reverse lever to one side or the other.
Socket won't fit.	Wrong size socket.	Use a 3/8" socket or use an adaptor allowing use of 1/4" and 1/2" sockets.
Socket won't release.	Release button is stuck.	Lubricate release button and try again.

Figure 20-8 A troubleshooting matrix.

20.6 CONSISTENCY

Your readers should not have to deal with sudden and unexplained changes in the manual, which is frequently their only source of information. Careless inconsistencies can confuse your readers, frustrate them, and cause them to improvise. New words should not appear while they are puzzling out how to do something. You don't want your readers confused by inconsistent terms. Users should not have to branch, loop, switch volumes, or move around the document unless it is unavoidable. Use familiar formats so readers can find information quickly.

User manuals, whether their purpose is reference or instruction, need to be consistent throughout. Everything has to match. This includes vocabulary, cross-references, structure, and design. When one of these elements changes, the readers expect that the change means something. Continual changes will confuse and discourage your readers. Stick to the same format.

Repetition works very well in real life. Look at advertising—or engineering! Important messages are repeated several times to verify that they have been communicated correctly. Redundancy is an accepted engineering principle. Double your chances by saying it twice.

Frequently, there are different ways to describe the same process. Inputting, keyboarding, and entering are all words that describe the same computer function. Every effective manual establishes a consistent and simple style for itself. Don't use synonyms. Choose a term and stick with it.

Keep your manual consistent with its own design and structure: Don't wander from one design to another. Don't vary the way you organize the chapters or sections unless there is a very good reason. Develop document standards and keep to them.

The most important consistency is with the product or process itself. You should actively field-test the manual by trying it out with novice users. No matter how elegantly your manual is written, it is a device which needs to communicate complete, accurate information to beginners. It is better to discover problems before the manual is produced or distributed.

20.7 READABILITY

Many organizations use readability scales to measure the level of reading difficulty of their customer-oriented documents, especially user manuals. The U.S. military, for instance, uses a revised version of the Flesch READ scale. These indexes, which attempt to provide some objective measure for text analysis, use various formulas to determine how much effort will be required to read a particular document.

At the least, try to improve the readability of your manuals with wide margins, varied type sizes, repetition of key points, frequent illustrations, and an overall professional appearance.

A List of Useful Ideas

- **Include lots of illustrations.** Probably, every section of your manual could use some sort of illustration. Often your graphics will duplicate the information that is in your text. This kind of repetition ensures that your message is received.

- **Put detectors as close to the center of the ceiling as possible.** If this is not practical, put the detector on the ceiling, no closer than 4 inches (10 cm) from any wall or corner. See Figure 4.
- If ceiling mounting is not possible and wall mounting is permitted by your local and state codes, **put wall-mounted detectors between 4 and 6 inches** (10 and 15 cm) **from the ceiling** See Figure 4.
- **Some rooms have sloped, peaked, or gabled ceilings.** If yours do, **mount detectors 3 feet** (0.9 meter) - **measured horizontally - from the highest point of the ceiling.** See Figure 5.

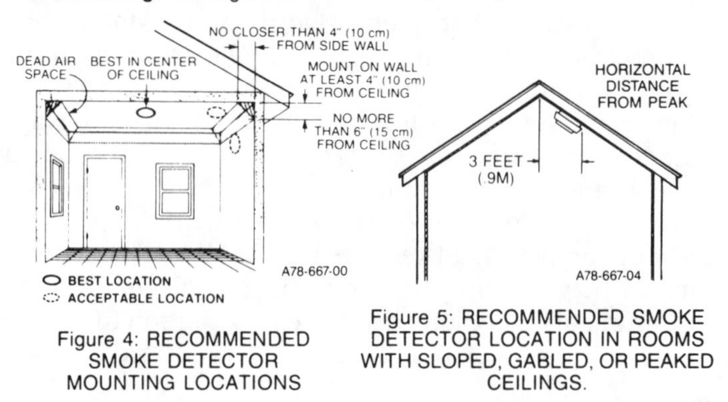

Figure 4: RECOMMENDED SMOKE DETECTOR MOUNTING LOCATIONS

Figure 5: RECOMMENDED SMOKE DETECTOR LOCATION IN ROOMS WITH SLOPED, GABLED, OR PEAKED CEILINGS.

Note: All figures in this list are reprinted with the permission of BRK Electronics, a Division of Pittway Corporation.

- **Add definitions.** Don't assume your readers know what you mean or intend by what you say. Define your terms. Let your readers know exactly what you are talking about. Explain by giving thoughtful definitions to your words.

(continued)

WARNING
GENERAL LIMITATIONS OF SMOKE DETECTORS:
WHAT SMOKE DETECTORS CANNOT DO

Smoke detectors have played a key role in reducing home fire deaths in the United States. However, according to the Federal Emergency Management Agency (an agency of the U.S. Government), they may not go off or give early enough warning in as many as 35% of all fires. What are some reasons smoke detectors may not work?

Smoke detectors will not work without power. Battery operated smoke detectors will not work without batteries, if the batteries are dead, if the wrong kind of batteries are used, or if the batteries are put in wrong. AC powered smoke detectors will not work if the power supply is cut off for any reason. Some examples are a power failure at the power station, a failure along a power line, a failure of electrical switching devices in the home, an open fuse or circuit breaker, an electrical fire, or any other kind of fire that reaches the electrical system and burns the wires. If you are concerned about the limitations of either batteries or AC power for your smoke detectors, **install both types or AC/DC detectors for more security.**

- **Prepare summaries.** Include plenty of summaries, short, condensed versions of important and key material. Frequent summaries allow readers to review and consolidate what they have learned.

BASIC INFORMATION
ABOUT YOUR SMOKE DETECTOR

- **Put detectors outside of every bedroom area and on every floor of your home.**
- **Put the detector as close to the center of the ceiling as possible.**
- The detector will beep when you put the battery in it. If the indicator light on the detector **flashes once a minute, the detector is receiving power from the battery.**
- **Test the detector weekly by holding the test switch button in for about 10 seconds until the alarm sounds. The alarm may not sound immediately** when you press the button.
- Model FG1000C has the Escape Light' that switches on when the alarm sounds to help light your way to safety.
- **Replace battery(ies) if the detector beeps once a minute.**

- **Tell your audience where to go for further information.** Direct your audience to sources of further information. This is very important if they are likely to need details which you do not provide.

⚠WARNING This detector is **not** meant to be used in non-residential buildings. Warehouses, industrial or commercial buildings, and special purpose non-residential buildings require special fire detection and alarm systems.

This detector alone is not a suitable substitute for complete fire-detection systems in places which house many people, like hotels or motels. The same is true of dormitories, hospitals, nursing homes, or group homes of any kind, even if they were once single family homes. Please see NFPA 101, **The Life Safety Code,** NFPA 71, 72A, 72B, 72C, 72D, and 72E for smoke detector requirements for fire protection in buildings not defined as "households".

(continued)

- **Include examples.** One of the best ways to communicate difficult technical concepts is to include examples where you model applications and problems which the typical user might encounter.

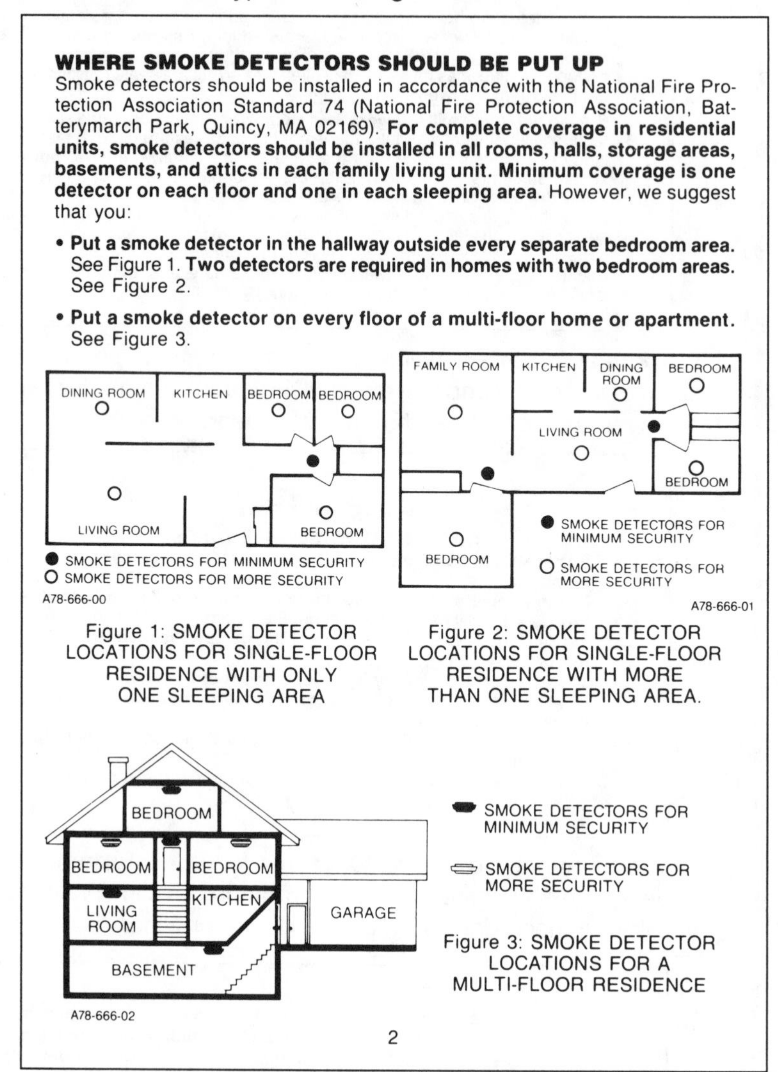

WHERE SMOKE DETECTORS SHOULD BE PUT UP

Smoke detectors should be installed in accordance with the National Fire Protection Association Standard 74 (National Fire Protection Association, Batterymarch Park, Quincy, MA 02169). **For complete coverage in residential units, smoke detectors should be installed in all rooms, halls, storage areas, basements, and attics in each family living unit. Minimum coverage is one detector on each floor and one in each sleeping area.** However, we suggest that you:

- **Put a smoke detector in the hallway outside every separate bedroom area.** See Figure 1. **Two detectors are required in homes with two bedroom areas.** See Figure 2.
- **Put a smoke detector on every floor of a multi-floor home or apartment.** See Figure 3.

Figure 1: SMOKE DETECTOR LOCATIONS FOR SINGLE-FLOOR RESIDENCE WITH ONLY ONE SLEEPING AREA

Figure 2: SMOKE DETECTOR LOCATIONS FOR SINGLE-FLOOR RESIDENCE WITH MORE THAN ONE SLEEPING AREA.

Figure 3: SMOKE DETECTOR LOCATIONS FOR A MULTI-FLOOR RESIDENCE

2

20.8 EDITING AND TESTING USER MANUALS

Reliable manuals are the result of reliable testing. What many involved authors forget to remember is that the purpose of the testing is to cause failures which then can be corrected. We'll say it again. **The purpose of**

testing is to cause failures! The more failures you discover and correct, the better you are doing your job.

The first failures you should look for are technical. At every step of the design and writing process, you should check for factual accuracy. This is essential. If your facts are wrong, the manual is useless and possibly harmful. Every factual statement should be verified.

Search your draft document for places where the users could lose their way. Watch out for detours. You don't want them to become confused and set out on their own. During your first review of the document, revise language, clarify content, and identify inconsistencies. You are trying to discover and remove misleading information. Watch out for the ripple effect. Changing one section of the manual can have surprising and unanticipated problems.

Edit for mechanical errors—grammar, spelling, and punctuation. If you want your manual to have credibility with the readers, find your mistakes and fix them. It may be an unfair perception, but readers will equate errors with incompetence.

20.9 MAINTAINING USER MANUALS

How easily can your manual be corrected, updated, or improved? This is an important question. What many people forget is that user manuals need to be maintained. They need to be constantly kept up-to-date and current or they will not be useful. A well-crafted manual can be easily supplemented and modified as the product or system changes.

Designing a user manual should always include planning for future modifications. If your document will need frequent revisions, for example, consider using a loose-leaf binder. Try to include information which you know will change in the appendices. Date your documents and supplements clearly to avoid confusion.

Usually some person is assigned the task of maintaining the manual over time, making sure that it changes as the product or system does. This means changing the document when the product is modified or new features added. Changes in policy and system procedures also need to be communicated as they occur. Often the upkeep costs of user manuals will be more than their development costs. This means that you should consider how the document will be maintained and modified during the design process. It is too late to think about this when the manual is complete.

20.10 INFORMATION PROBLEMS

There are three kinds of information problems which turn up in manuals and cause trouble for users. The first is **missing information.** Your audience has a right to expect the manual which they rely on will be thorough and complete. Novice users will not know the information is missing until they are required to carry out a task or procedure. Unnecessary telephone support, service and warranty complaints, and customer aggravation will be the result.

Incorrect information can lead to even more damage. You have an important professional responsibility and a legal obligation to provide accurate and correct information, as well as complete information. People and property can be injured by careless misinformation. Simple proofreading errors can multiply into endless hours of technical explanation and apology. Make sure that it is correct the first time.

Ambiguous information, where the presentation confuses the audience and produces unintended options and choices, is not that uncommon. Clarity, almost as important as accuracy in technical messages, requires that you make your meaning as clear as is reasonably possible. User manuals impart a special obligation for direct, understandable, and crafted language. Your audience is dependent on your ability to anticipate problems and explain solutions in ways which cannot be misunderstood. This is a design task of the first order.

20.11 A FINAL WORD

Remember that poorly designed products or systems will not be compensated for by the wonderful manual your team creates. A poor procedure will be difficult to follow, no matter how carefully you write it down. A bad product, clearly explained, is still a bad product. Well-designed systems and products need less documentation, and the writing problems you experience may be a reflection of what is in your mirror: the product or system your manual supports.

Well-written user manuals return more than their cost to the companies which make the effort to create and maintain them. Good documentation can reduce expensive customer services, including field service, phone support, and training. Good writing can also add value to a product or service by making things easier for users. Your manual ought to be a friendly user-oriented publication designed for the convenience of your audience.

It is difficult work trying to explain complex technical ideas to people who know a lot less than you do. You need to think of your readers as confused, error-prone, and dependent. You have to adapt to the vocabulary and technical skill-level of your readers. Users are looking for clarity and reliability. Good manuals provide both.

USER MANUALS EXERCISES

1. Take a simple product with which you are familiar—a product which needs to be assembled. The product will be shipped to another country where English is **not** spoken. Someone else will write the manual in that country's language, working from an outline which you will produce. Supply as much detail as you can.
2. Design a structured outline for a manual which describes how to assemble and use a touch tone phone. Choose one section of the outline and create a storyboard for that section, using the format below.

A. Title of Manual	Volume Number Date
B. Section in Manual	F. Illustration(s)
C. Headline	
D. Summary	
E. Text	

3. Have every person in your group bring a user manual of some sort to a meeting where you will rank order the different manuals in terms of their usefulness and appropriateness of their design.
4. Choose a user manual for a computer software product to which you have access. Rate the manual for each of the Three G's. Does the manual help the user to get going? Get better? And get help? How would you re-design the manual to do these things more effectively?

5. Assume you are a field-site specialist on assignment for Gates and Associates. Create a short report in which you develop three reporting procedures for field personnel to keep the home office informed and up-to-date. See Chapter 26 for more information on Gates and Associates.
6. Imagine that you have been assigned to create the information devices which will accompany a portable FAX machine. Draw up a list of the documents you would include, and a list of the parts which would make up these documents.
7. Design a troubleshooting section for a user manual for a technical process or device. Ask a member of your group to role-play an inexperienced user and conduct a *step-through* (trial run) of the user manual. The user must follow your troubleshooting guide exactly to solve problems.
8. Find a user manual you like. Establish criteria, such as page layout, visual aids, style, and strategy, to defend your selection of this user manual. Now write a short report to recommend this format to your co-workers. Attach the user manual to your report.
9. Write two descriptions of a procedure, one for a general audience and one for an expert audience. Review you document with members of your group to determine if you have met the needs of your audience.
10. Write a short report in which you specify the maintenance requirements for a projected reference manual. Assume that the manual is for a software product which is revised on a six-month schedule. In other words, you need to design a set of procedures for keeping the reference manual up-to-date every time there is a scheduled change in the product. (Hint: determine what will need to be updated, then determine the order of the updates.)

21

Preparing Technical Proposals

Note to Users: **This is a task chapter. It is intended to show you ways to design and write technical proposals which will hold attention and win approval.**

21.1 INTRODUCTION

The ability to write successful proposals is probably the most valuable technical writing skill you can develop. Successful proposals mean sales. Every proposal sells something: a product, a service, or an agreement. A proposal is a bid for business.

Technical, business, cost, and grant proposals, all determine who will be the contractor and the terms of the contract. Thorough, persuasive, and competitive proposals bring business and contracts.

An organization's ability to respond to *Requests for Proposals (RFP)* and *Invitations for Bid (IFB)* may become the key factor in its profit base. Your ability to contribute to this important process will be noticed and appreciated.

In this chapter you will learn how to:

- Target your proposals.
- Choose a strategy that focuses on decision-makers.
- Use the cumulative effect to improve your chances.
- Define objectives which your evaluators can measure.
- Develop effective support materials.

21.2 DEFINITION

A proposal is a report written for the special purpose of selling something: an idea, a product, or a service. Its chief function is to persuade a very specific audience to make a decision. Successful proposals are designed to help these readers arrive at a positive decision.

Proposals are sometimes defined in terms of their audience. **Internal proposals** are made within an organization and are usually designed to solve some immediate problem. Typical in-house proposals might recommend the location of a new facility or the purchase of new equipment, for example. They are directed to individuals higher up in the company for decisions and funding.

External proposals are sent outside of the organization to potential clients. Both **government** and **sales** proposals represent an offer to provide goods or services. Technical proposals are strategic documents which define an organization's offer to deliver these items.

These documents need to be comprehensive organizational, technical, and business plans which will persuade your audience that yours is the best offer. Your chief writing task is to make your offer look good to the people who will make the decisions. This means that you must stress benefits to the readers.

21.3 PROPOSAL FORMATS

There are different formats which can be used for written proposals but in general you are designing a persuasive report. Some funding sources provide no specific guidelines for proposal content. Many, however, have very specific instructions for formatting and organizing your content. Follow these directions exactly.

Some companies or agencies put out a **Request for Proposals** (RFP) or **Invitation for Bid** (IFB). These documents describe in detail what is to be

3.4 Phase I Proposal Format

All pages shall be consecutively numbered and the ORIGINAL of each proposal must contain a completed red copy of Appendix A and Appendix B.

a. **Cover Sheet.** Complete RED COPY of Appendix A, photocopy the completed form and use a copy as Page 1 of each additional copy of your proposal.

b. **Project Summary.** Complete RED COPY of Appendix B, photocopy the completed form and use a copy as Page 2 of each additional copy of your proposal. The technical abstract should include a brief description of the project objectives, and description of the effort.

c. **Identification and Significance of the Problem or Opportunity.** Define the specific technical problem or opportunity addressed and its importance. Begin on Page 3 of your proposal...

Figure 21-1 An example of specific guidelines.

bid and they often contain very precise directions for responding. Solicited proposals will not be successful if they do not follow these directions exactly. The proposal team's first job is to know what the audience is looking for.

Generally, a small organization will have fewer specifications than a large one. Government (local, state and federal) has numerous requirements on bidding for awards, contracts, and grants. These regulations are complex and time-consuming but they are designed to keep the public's business in public view.

21.4 THE PROPOSAL WRITING PROCESS

Both sales and government proposals are generally created by teams of technical professionals. Performing this task means working in a group environment where you will be required to brainstorm, delegate tasks, search out and evaluate alternatives, and package your research into a persuasive presentation.

21.4.1 The Team

Managers, scientists, engineers and technicians are often assigned to proposal writing teams. Preparing winning proposals is a collaborative activity where many people work together to create documents that represent their combined information, viewpoints, and decisions.

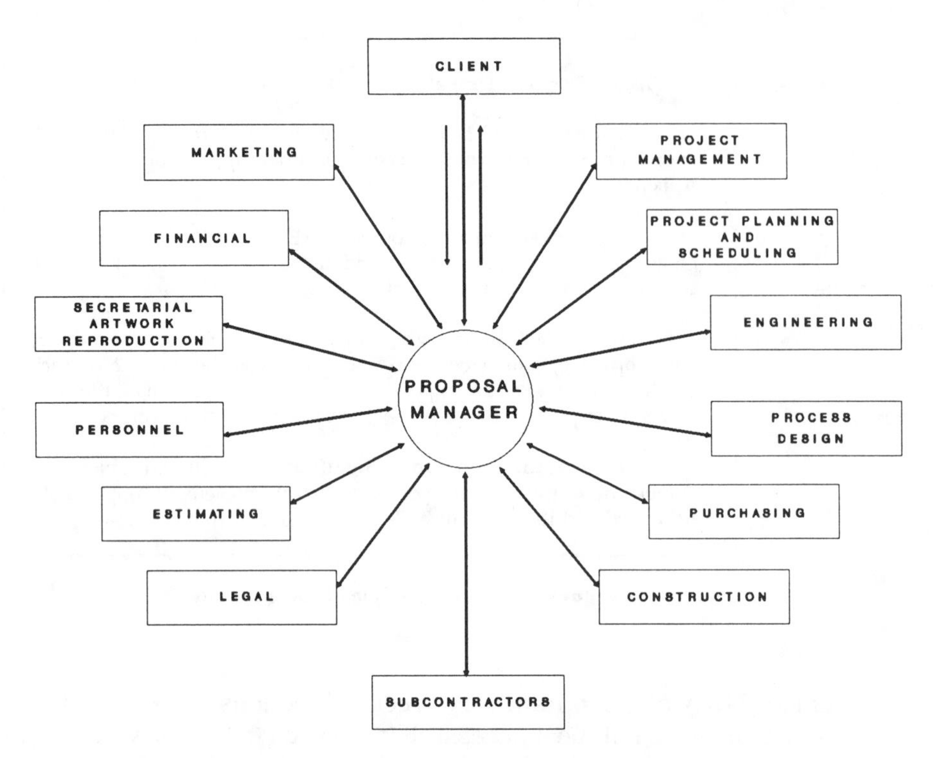

Figure 21-2 The role of the Proposal Manager
Reprinted with the permission of Van Nostrand Reinhold. Source: *Project Management: A Systems Approach to Planning, Scheduling and Controlling* by Harold Kerzner

The team, no matter how many members it contains, necessarily fills a number of different roles—proposal manager, contract officer, text editor, graphics editor, writers, and production editor. Sometimes these roles may be filled by only one or two persons, but all of these jobs need to be done.

21.4.2 The Plan

A systems approach with specific stages and steps is necessary to move the proposal toward completion. Most proposals are complex group efforts which need to be managed very carefully. Research, writing, graphics, and production responsibilities need to be assigned to competent people. RFP deadlines are inflexible and a schedule needs to be created and maintained.

Good proposals require good planning. This means that serious program planning is an essential element in the preparation process. Competitive proposals need to contain a careful and thoughtful inventory of your organization's goals and objectives, methods and evaluation process, personnel, and abilities. Assembling this material, editing it through multiple drafts, and seeing it through production: all require organizational commitment and a detailed writing plan.

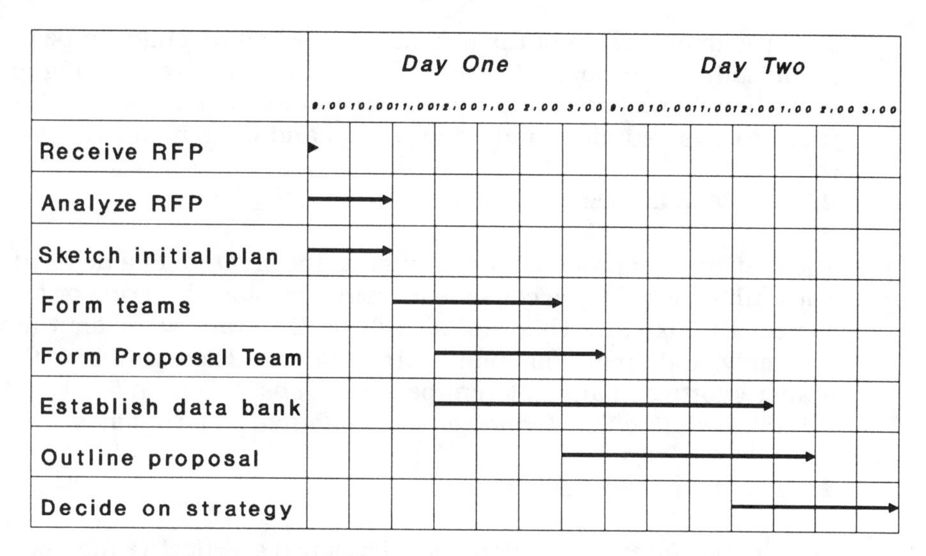

Figure 21-3 An example of a writing plan for the preliminary stages of a proposal preparation.

A writing plan lays out all of the tasks, assigns responsibilities and deadlines, specifies the number of edits, and, in general, describes the procedures for preparing and producing the proposal.

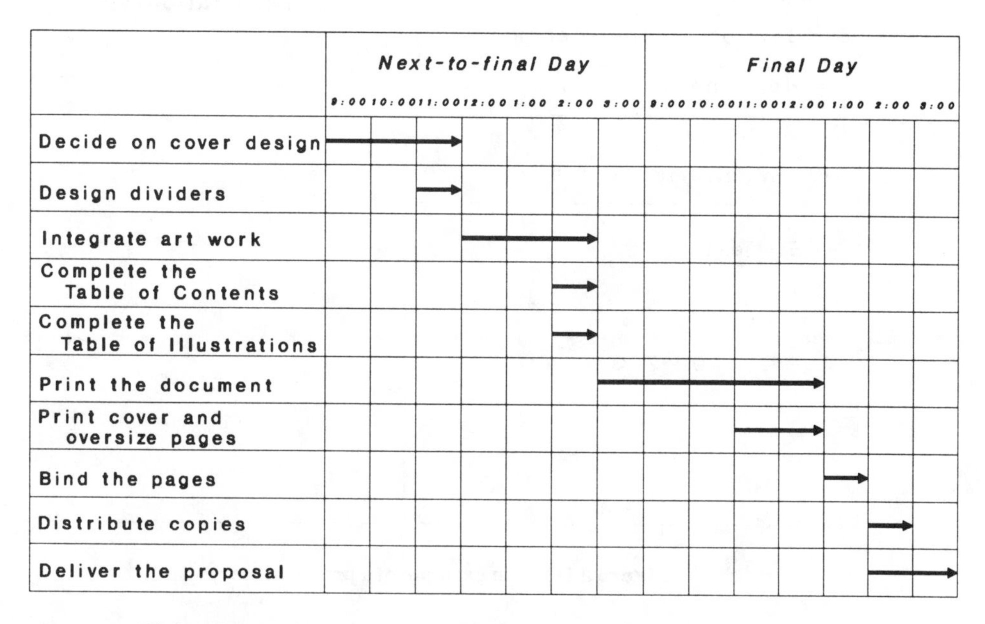

Figure 21-4 An example of a writing plan for the final stage of a proposal preparation.

The final version of the agreed-upon schedule ought to be expressed graphically so that participants can see what they are agreeing to provide and when. Flowcharts, activity charts, and work breakdown schedules all provide ways to understand visually what and when items will be expected.

21.4.3 The Outline

The first stage of proposal preparation is the creation of a detailed outline. The outline needs to be consistent, using the same lettering or numbering system throughout. This outline will be the framework for the writing, assembly, and production phases and will provide continuity to the collaborative efforts of many different persons. *Consult Section 5.4.1 in the chapter on Scheduling Your Writing for more information on outlines.*

21.4.4 The Storyboard

The Hughes Aircraft Company has developed an effective method for coordinating multiple authors in designing, planning, and editing successful proposals. Their method, called "storyboarding," is a graphics-driven approach to document design, an elaborated version of the cut-and-paste technique used by many writers.

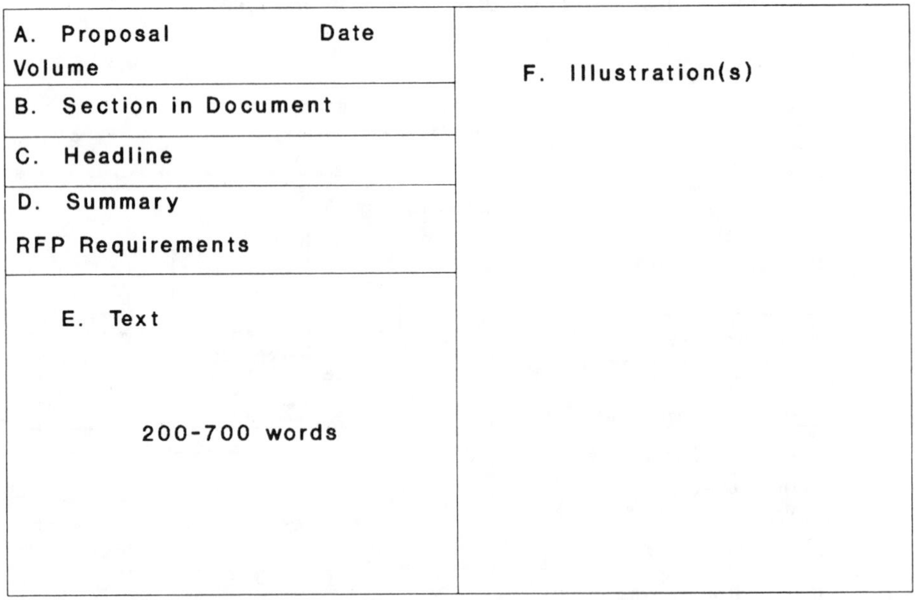

<table>
<tr><td>A. Proposal Date
Volume</td><td rowspan="5">F. Illustration(s)</td></tr>
<tr><td>B. Section in Document</td></tr>
<tr><td>C. Headline</td></tr>
<tr><td>D. Summary
RFP Requirements</td></tr>
<tr><td>E. Text

200-700 words</td></tr>
</table>

Figure 21-5 An example of a proposal storyboard.

21.5 PARTS OF A TECHNICAL PROPOSAL

Technical proposals, whatever the scope of work or size of the project, basically contain the same functional parts:

- Introduction and summary of conclusions.
- Statement of the problem you propose to solve.
- Statement of the work you will perform.
- Your management approach and performance schedule.
- Your capabilities and qualifications for the job.

The **front matter** of a proposal generally includes a title page, a table of contents, a list of illustrations, and a list of tables. This front section should be used as a map to help the readers quickly discover and cross reference the information they are seeking. Very often a notice of "proprietary information" states that the proposal information should not be given to other organizations.

Transmittal letters often accompany government and sales proposals. Make sure that your letter references the RFP you are responding to. Point out that the proposal is in compliance with all procurement requirements. Do not expect the letter to accompany the proposal through its evaluation process.

Your **table of contents** needs to support the entire proposal. It is there to help your audience. You must make sure that it is sufficiently detailed, with meaningful headings and subheadings. This is one of the access points to your proposal and it should provide both substance and direction to the readers.

The **summary** needs to be concise and responsive. Often this will be the only section of your proposal which is reviewed. Your summary must communicate the essential message by identifying key points and focusing the readers' attention on results, conclusions, and recommendations. Busy readers want to get enough information from your summary to know what is significant and what decisions will need to be made. Try to understand the information needs of your audience and then arrange your summary to be clear, focused, and helpful.

Your **introduction** provides a description of your qualifications and ability to deliver what you offer. Use this section to relay your understanding of the problem and your responsiveness to the request. Substantiate any claims you make for technical expertise, key personnel, or unique facilities. In other words, support your technical proposal with as much evidence as you can.

A **compliance matrix** (a table directing the reader to specific pages) helps your audience complete their evaluation. *See Fig. 21-6.* Essentially this is a cross-reference page which visually displays your compliance with the request. What you are trying to do here is make it easy for your readers to see that you have met all of the conditions they have required. Make sure that you list all of the deliverables and number the objectives for discussion purposes.

The next section is frequently a problem for proposal designers—the **technical discussion.** You don't want to lose your audience at this point. Remember to include enough background so the evaluators can analyze the scope of work and your anticipated difficulties, alternatives, and solutions.

Trade-off standards must be explicitly stated and reliability considerations have to be clear. Other issues including safety and quality control have to be addressed. Testing requirements, evaluation procedures, and other measurements may need specification. Service conditions, subcontractor and supplier roles, as well as milestones and decision points all need to be arranged.

Successful proposals usually contain a graphic illustration of the management organization, frequently a chart which displays team access to decision-level managers. You need to illustrate a management control plan. (PERT and Gannt charts allow you to visualize what must be done to accomplish a desired objective on time. See Section 8.4 in Including Graphics and Illustrations for more detailed definitions of PERT and Gannt charts.) Outline your reporting procedures, and describe your **management experience.** Very often your organization's qualifying past performance serves as a key to your selection.

All of your customers will be interested in **schedules.** When will the project be completed? How will progress be monitored? Work breakdown schedules, frequent reporting requirements, quantitative scrutiny of the completion steps, all provide assurance to the contractor.

Accounting is another area of client concern. The **costs** section provides the reviewers with evidence that your level of expertise, supervision, and accounting methods will insure completion of the contract on the delivery schedule. Travel, options, and alternative cost implications must be discussed. Costs are usually kept separate from the main body of the proposal because they are considered incidental to the plan of work.

You need to show your readers what you mean, to let them see what you are saying. Appropriate **illustrations** break up the monotony of plain text, provide support for your words, and help the readers see your main points. Strong graphic support is essential to the successful proposal.

Tables, charts, spreadsheets, work breakdown schedules, even equations—all of these can be displayed visually. Many proposal teams try to achieve a ratio of 60/40 between text and graphics. Whenever you can support your text with an illustration, it is a good idea to do so. *See Chapter 8, Including Graphics and Illustrations,* for suggestions.

For long proposals an **index** can provide assistance to your readers. Cross-references lead them from topic to related topic. The index should be a useful entry point to your proposal, guiding the reader who is looking for a particular section.

The main entries repeat your headings and sub-headings and emphasize the points which you have already made. You need to ask yourself where you would look for a term if you were unfamiliar with the subject. Be careful not to scatter references under various synonyms.

21.6 PROPOSAL BASICS

Make It Readable

Your proposal should be designed for use. The few people who read it need to make a decision. Anything you can do to help your readers arrive at a positive decision is good design. This means your offer must be neat, attractive, and readable. Careful and thoughtful document design is an essential part of your planning process.

This includes providing your readers with illustrations, glossaries, charts, tables of contents, figures, indexes, appendices, cover letters, letters of support, checklists, summaries, and detailed headings. All these devices are used to make reports more readable, interesting, and persuasive.

Use Headings to Label Your Ideas

Every page in your proposal must be visually attractive. Headings can break up solid chunks of text, provide white space on the page, and give your proposal a professional appearance. Most readers develop an attitude toward your document from the way it looks, even before they begin reading. Appearance can assist clear writing in appealing to the persons who will evaluate your proposal.

Headings and sub-headings make information easier to find. Proposal readers are often looking for specific information, and it is up to you to help them with plenty of labels. Readers can skim and find the sections they are particularly interested in. Headings make your proposal more accessible and useful to busy readers.

Include a Compliance Matrix

We recommend that you prepare a compliance matrix for your proposal. This should list each subsection of the RFP, and next to it the part of your proposal that responds to that section. This guarantees that your document is fully responsive to the RFP.

Compliance Matrix for RFP Requirements

	Section	*Remarks*
Subline Item 0001 AA Three TIGER Payloads		
Subline Item 0001 AB Four SNAKE Payloads		
Subline Item 0001 AC Four MLLES Payloads		
Subline Item 0001 AD Six TOPPER Payloads		
Subline Item 0001 AE Three MIROR Payloads		
Line Item 0002 Data		
SPECIFIC CONTENT		
(a) Statement of available plant, equipment and test facilities proposed for use on this contract.		
(b) Statement of additional plant, equipment and test facilities required for this contract.		
(c) Does the offeror demonstrate a thorough knowledge of engineering analysis and modeling using 2–dimensional theory/software, applied to structural, model and thermal (steady state and transient) problems on aerospace frames?		

Figure 21-6 An illustration of a part of compliance matrix

Including this graphic display of compliance in your proposal makes it easier for the evaluators to see that your offer conforms to all requests. Proposals which do not comply completely are automatically disqualified.

Take a Positive Approach

Hard luck stories will impress your readers that you have needs, not competence for the task. You need to present a positive message which demonstrates your capacity to deliver what you are offering.

Remember, people are looking to choose a successful applicant, not a source who cannot complete the task or provide the product. An organization's reputation for success is based on the ability to choose applicants who will achieve what they set out to accomplish.

21.7 EDITING THE TECHNICAL PROPOSAL

Editing a technical proposal is a group effort, the same as is the writing. A series of proposal reviews should be scheduled at various stages of the preparation process. All of the parties to the proposal should be involved in the editing. You don't want to jeopardize the proposal's credibility because of poor writing or technical mistakes.

You need to edit for compliance as well as grammar, but remember, both can cost you the contract. The source selection board or evaluators may mistake your mechanical errors for larger incompetence. Simple errors cannot explain themselves because you are not present while the proposal is being reviewed. Every detail needs to be checked and re-checked.

ITEM	TYPE	COMMERCIAL OFFER			QUALIFICATIONS		
	FORM	LETTER	BASIC	DETAILED	LETTER	BASIC	DETAILED
TABLE OF CONTENTS			●	●		●	●
INTRODUCTION OR SUMMARY			■	■		■	■
SCOPE OF SERVICES		●	●	●			
TERMS OF COMPENSATION		●	●	●			
GUARANTEES		▲	▲	●			
INVESTMENT COSTS		▲	▲	■			▲
ELEMENTS OF PRODUCTION COSTS		▲	▲	▲			▲
COMPANY DIVISION ORGANIZATION AND BACKGROUND EXPERIENCE			■	■	▲	■	■
PROJECT ORGANIZATION AND EXECUTION		▲	●	●			▲
RESUMES			■	■			▲
MASTER PLANNING			▲	●			▲
DETAILED ESTIMATE			■	●			
PROCESS DESCRIPTION AND DESIGN CRITERIA		▲	●	●			▲
FLOW DIAGRAMS AND PLOT PLANS			■	●			▲
EQUIPMENT LIST		▲	■	●			▲
DETAILED EQUIPMENT LIST				●			▲
PROJECT STAFFING			■	■		▲	▲
PROJECT SCHEDULE			■	■			
CONTRACT COMMENTS		■	■	●			
DRAFT AGREEMENT		▲	■	●			

● MANDATORY
▲ OPTIONAL
■ RECOMMENDED

Reprinted with the permission of Van Nostrand Reinhold. Source: *Project Management: A Systems Approach to Planning, Scheduling, and Controlling* by Harold Kerzner.

We recommend a formal proposal review process where all parties to the process are brought together to review the finished document, page by page. Only about one in five proposals is accepted, and winning government proposals generally score above 95 on the one-hundred-point scale which is used to determine awards. The closer you are to zero-defect, the greater your chances of success.

21.8 THE FOUR C's

Generally a C performance is only adequate and something to be avoided, but when you are writing a proposal four c's can be a perfect score. Trying to achieve four c's can be a successful strategy if you can match your client's need for clear understanding, conformity, competence, and above all, credibility.

Comprehension Your clients, your audience, need to know that you understand exactly what their problem or need is. There is only one way to do this: explain it back to them while weaving your solution into their story.

Compliance These same clients want to make sure that your offer exactly matches their specifications, procedures, and expectations. They want this exactly, no matter how carelessly they have defined what they want.

Capability Occasionally, and too frequently, organizations propose to carry out activities for which they do not have the capability: defined, simply, as the facilities, personnel, equipment, detailed plans, and appropriate people to carry out the work. Be realistic! Don't take on jobs you can't do as a way to build an organization. It doesn't work that way.

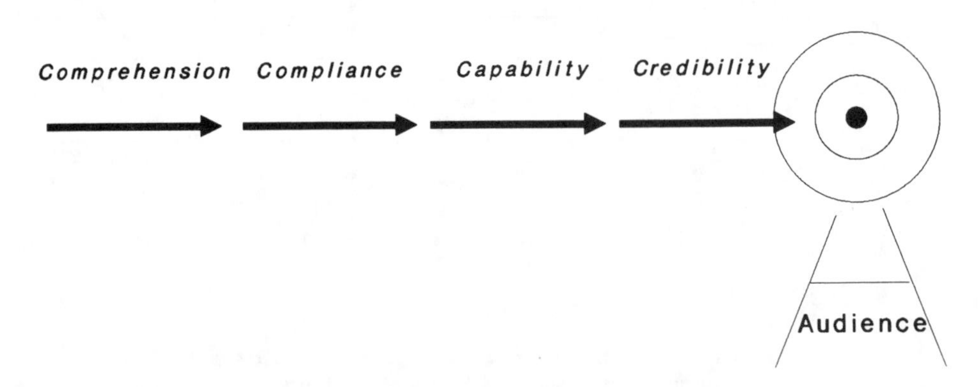

Figure 21-7 The 4 C's

Credibility Lastly, and most importantly, your offer needs to be believable. When you are doing business with strangers, the last thing they want to do is to make a mistake. Try being very clear: Use words which can be measured.

21.9 THE CUMULATIVE EFFECT

Proposal writing is a costly way to market your services and products. It is a time-consuming and labor-intensive process which requires a great deal of effort, particularly the first time through. However, there is a cumulative effect which makes subsequent proposals easier to prepare.

Even an unsuccessful proposal can improve the quality of an organization's goals, result in better planning and financial management, improve program credibility and program management structures. Preparing a written proposal can help your organization to assess its operating strategies. This effort will contribute to the next proposal you prepare because much of this process will not need to be repeated.

Learn from your mistakes. Many companies and agencies will discuss your unsuccessful proposal with you. You can ask why you did not win the award or contract, and frequently you will get an answer. Use this information to improve your next effort.

Another useful outcome from the cumulative effect is *electronic boilerplate,* reusable text which can be stored on computer for future proposal efforts. This is text which describes your organization's management structure, for instance, or your related experience.

There are some parts of all proposals which can be prepared ahead of time and simply plugged in to each new proposal. These include personnel abilities, company accounting policies, and other details which change very infrequently. As you develop each proposal, you should be accumulating a library of reusable text, illustrations, and supporting materials which can be used again and again.

21.10 THE GRANT PROPOSAL

Grant proposals are very similar to other types of technical proposals. The major difference is that you are seeking an award, rather than looking for business. It often comes down to the same thing. Grant proposals require the same good planning and careful preparation. Even though you are an applicant, requesting funding from a grant-making agency, you still need to persuade the funding source that your organization can best provide the services they are looking for. Figure 21-8 is an example of a grant proposal form; you may apply for this grant.

If possible, you want to show that your proposal is potentially beneficial to persons beyond your own community. You want to provide measur-

THE CHARLES A. LINDBERGH FUND, INC.

1992 LINDBERGH GRANT APPLICATION

INSTRUCTIONS: Send 7 copies of the typewritten application, following the format provided on pages 1-4 (using the actual form is not required), and 2 stamped, self-addressed envelopes (does not apply for applicants outside the U.S. unless U.S. postage stamps are available) to: Marlene K. White, The Charles A. Lindbergh Fund, 708 South 3rd Street, Suite 110, Minneapolis, MN 55415. All applications and Endorser's Reports are to be postmarked by Tuesday, June 18, 1991. It is the responsibility of the applicant to submit a complete proposal and separate Endorser's Reports.

1. CATEGORY OF GRANT APPLIED FOR (see page 4): ____________________

2. PROJECT TITLE: "____________________

____________________"

(State in terms of the concept of "balance")

3. PRINCIPAL INVESTIGATOR:

Name & Title (last name first): ____________________

Affiliation: ____________________

Office Address: ____________________

____

Home Address: ____________________

____

Telephone (office): (______)__________ (home): (______)__________

FAX: (______)__________

4. INSTITUTION OF AFFILIATION DURING PROJECT: ____________________

5. AMOUNT OF FUNDING REQUESTED: __________ (maximum of $10,580)

6. DATES OF PROJECT (funding from 6/92): ____/____/____ to ____/____/____

7. PROVIDE A CONCISE STATEMENT OF HOW YOUR PROPOSED WORK WILL PROMOTE A BETTER BALANCE BETWEEN TECHNOLOGICAL GROWTH AND OUR HUMAN/NATURAL ENVIRONMENT:

8. NAME, TITLE, ADDRESS AND TELEPHONE OF TWO ENDORSERS:

____________________ ____________________

____________________ ____________________

____________________ ____________________

____________________ ____________________

9. SIGNATURE OF PRINCIPAL INVESTIGATOR: OFFICIAL OF AFFILIATED INSTITUTION (if applicable):

____________________ ____________________

Date __________ Date __________

10. HOW YOU FIRST LEARNED ABOUT THE LINDBERGH GRANTS PROGRAM: ____________________

____

Figure 21-8 The Lindbergh Grant form.
Reprinted with the permission of the Charles A. Lindbergh Fund.

-2-

11. PROJECT SUMMARY: These statements must be in non-technical terms, yet sufficient to accurately describe your proposed work. This section is important in the final review process by the Board of Directors. You are encouraged to limit your responses to the space allotted.

(A) Statement of the problem you seek to resolve, how you plan to proceed, anticipated results, and practical applications of your project:

(B) Statement of how the project will achieve a better balance between technological growth and preservation of our human/natural environment (No application will be considered further which does not contain this statement.):

Figure 21-8 *(continued)*

SECTIONS 12-18: You are encouraged to observe the suggested number of pages for each section wherever possible.

12. REVIEW OF LITERATURE (one-three pages): Provide a review of the literature, with citations which may be on a separate page(s), relating to the most recent advances made in your area of work and how your proposed work will contribute. Document originality of the proposed project.

13. BUDGET (one page): Include a line itemization of anticipated costs of your proposed project and all other sources of funding (received and anticipated). Also specify if the research for which you are applying is dependent upon additional funding you are still seeking. If the full grant is required, a statement of justification should appear on the same page. The Fund does not support overhead costs of institutions. Unexpended funds remaining upon completion of your project must be returned to the Fund. Grants will be considered for salaries (including employee benefits), research assistantships (however, we cannot fund tuition relating to graduate research), computer time, supplies and equipment, field work, reasonable secretarial and technical support, travel, and other items necessary to the successful completion of the proposed work.

14. METHODOLOGY (one page): State how the project will be conducted, including time frame (up to 12 months, beginning as early as June 1, 1992).

15. EVALUATION (one page): State how results will be analyzed and overall project evaluated. Outline plans for utilizing the results.

16. PERSONNEL (one cover page, plus attachments): To include a listing of all personnel working on the proposed project and the amount of time each person will be spending on the project. Attach an abbreviated curriculum vitae and a list of relevant publications for the Principal Investigator and all professional personnel. Please note that the Lindbergh Fund prefers that the individual actually heading the project work team be identified as the Principal Investigator. (The Fund has no educational requirements for who may be a p.i.)

    For non-U.S. citizens: Citizenship ____________________
    Type visa if in U.S. ____________________ Place of birth ____________________

17. ENDORSER'S REPORTS (one page each): ENDORSER'S REPORT FORM TO BE PREPARED BY APPLICANT. Send one copy of your application to each of the two endorsers you have indicated on page 1 of your application, along with a one-page Endorser's Report form, which is to include the following items at the top of the page:

    (a) ENDORSER'S REPORT

    (b) Principal Investigator's Name
    (c) Project Title
    (d) Instructions: The Lindbergh Fund requests that the endorser evaluate the attached proposal using this form provided. Your evaluation should be based upon the proposed project's ability to solve the problem and the solution's ultimate practical application. Please state your opinion as to the applicant's ability to carry out the proposed work successfully. The Fund's concept of balance between technological growth and preservation of our human and natural environment is key in the selection process. It is based on the conviction of Charles A. Lindbergh: "The Human Future depends on our ability to combine the knowledge of science with the wisdom of wildness." The endorser is to send 7 copies of the completed endorser's report only directly to: Marlene K. White, The Charles A. Lindbergh Fund, 708 South 3rd Street, Suite 110, Minneapolis, MN 55415, (612)338-1703 (phone) (612)338-6826 (FAX). They must be postmarked by June 18, 1991. If one copy of the report is sent by FAX, 7 hard copies are required via mail or delivery. All reports remain confidential information to the Fund.
    At the bottom, include spaces for:
    (e) Signature
    (f) Date
    (g) Name
    (h) Address
    (i) Telephone Number

    If applicant or endorser desires delivery verification, arrangements must be made directly with delivery service of your choice.

18. APPENDIX (Required for the Arts and Humanities category only): Include 6-12 color photographs representative of your work. Seven copies of each photograph are required for our review process.

SELECTION PROCEDURE: (A) Applications/requests for funding are screened administratively for completeness and appropriateness. (B) They are then sent for two "Balance Reviews" by members of the Fund's Board of Directors and former grant recipients and are reviewed for the project's ability to address the balance between technological growth and man's human/natural environment. (C) Next, the applications are sent to an independent Technical Review Panel for evaluation of the project's ability to solve the stated problem, the originality of the approach, and its practical application. All reviewers' comments are confidential information to the Fund. (D) Applications receiving the most favorable balance and technical reviews are sent to the board's Grants Selection Committee. It is this committee that recommends the most outstanding projects for Lindbergh Grants and Certificates of Merit. (E) The full board acts on the recommendations at its spring meeting.

NOTIFICATION: For notification purposes, all applicants are required to enclose two stamped (only if U.S. postage is available), self-addressed business size envelopes with the application package. Receipt of your application will be confirmed in writing no later than July 15, 1991, and you will be informed of the final selections by April 1, 1992.

Figure 21-8 *(continued)*

able objectives, clear deadlines, and responsive evaluation steps. The more specific you can be in your funding request, the more confidence you will create in the evaluators.

Begin with a **summary,** a clear and short statement of the entire proposal. Then you need an **introduction** which describes your qualifications and ability to provide what your proposal offers. Then a **needs assessment** or problem statement will specify the needs which will be met or problems solved by your proposal.

You must define both **objectives** and **methods** in ways which can be measured. Possible outcomes must be predicted and activities which will lead to these results carefully described. You also need an **evaluation** section, a plan to determine which objectives have been met and which methods were particularly successful.

Lastly, a winning grant proposal should contain a **budget** which clearly defines how every anticipated cost will be met and a plan for continuation of the project beyond the grant period. People prefer to fund projects which will eventually become self-funding as they move from start-up to maturity. Plans for this eventuality will increase your proposal's credibility and chances of success.

21.11 A FINAL WORD

Technical proposals have become an increasingly popular way for organizations to seek the information they need to make purchasing decisions. Competitive procurement markets insure that written proposals will become increasingly important as a method of gaining new contracts and business.

Successful proposals will come from organizations which prepare themselves for a team effort, and for a continued effort. Collaborative writing teams, good document design, and a competitive technical position can be coordinated to produce winning proposals.

Your tasks—and your opportunity—are to work effectively with your team, to pay attention to the process, and to contribute whatever you can through careful analysis of the clients' needs. Every opportunity you have to participate in the proposal preparation process will increase your ability to contribute to the next proposal.

TECHNICAL PROPOSAL EXERCISES

1. You are assigned to work on a project team which is preparing an external proposal for Laraki Enterprises, a small manufacturing company from Rabat, Morocco. This twenty-five person operation manufactures very fine quality cotton shirts which have proved a popular export to the marketplace.

The Moroccan embassy has circulated an eight hundred thousand dollar RFP/Q on behalf of this company. Sheila O'Brien, the head of the project team, isn't sure whether the Moroccans are just fishing for good ideas or looking for a competitive proposal. Gates and Associates would like very much to gain an entry point toward doing business with Morocco, so Sheila decides to do some preliminary investigation. *See Chapter 26, Communicating in an Organizational Environment,* for more details. She assigns you (or your team) to the following tasks:

(a) Prepare a one-page report describing Morocco, its present state of technical development, business climate, relations with the United States, and any opportunities for technical consulting which stand out during your research.

(b) Prepare a detailed outline which will provide the framework for the collaborative efforts of many people. You want to make this as detailed as possible, assuming that if you do not mention something in your outline, the other members of the team will not include it.

(c) Create a visual schedule for the proposal process, assuming a response time of thirty days. Include specific activities, stages, and deadlines for the complete process. Don't forget to include production and delivery in your work breakdown schedule.

(d) Design a storyboard for one step in the outline you created. Post the storyboard in the classroom and discuss with other students and groups.

2. Prepare an internal proposal for Gates and Associates which recommends a change in the way the company reviews RFP's and IFB's. (*See Chapter 26 for more information on Gates and Associates.*) Presently, these are analyzed and discussed at a weekly meeting of the Senior Officers. Promising opportunities are then directed to the technical departments for capability decisions.

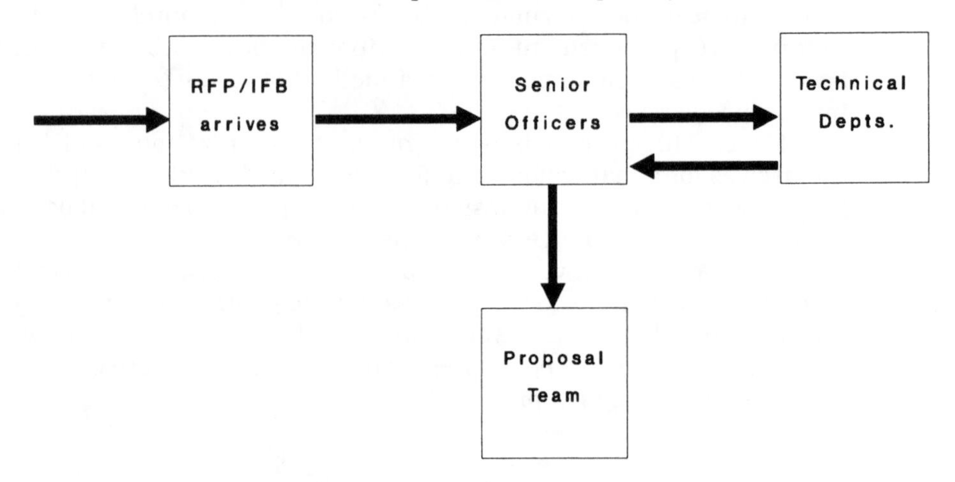

As Department Head in Research, you believe that this process slows our response to closely-timed requests. Days can be wasted before the technical capability review takes place. Your suggestion is simple. Incoming requests should receive a quick review by the appropriate technical departments before the Seniors meeting. You are sure this will save time and effort that could be better used. Your task is to write the memo.

3. Prepare an invitation to bid which will be circulated among minority suppliers and manufacturers. As the Federal contractor for many projects, Gates and Associates wants to assure the Federal Government that we are sincerely and actively searching for minority and women subcontractors. (*Consult Chapter 26 for more information on Gates and Associates.*)

   Mr. Gates is disturbed that we have only achieved a compliance rate of 4.1%. You have been asked to produce a two-page document which will encourage minority and women businesses to participate in these contracts.
4. The Lindbergh Grant Proposal is in many ways a typical grant proposal, offering to fund interesting and worthwhile projects which will contribute to a positive goal. See Fig. 21-8. Read this proposal request carefully and prepare a one-page memo to your instructor explaining why you would or would not be interested in pursuing this grant.
5. Prepare an internal proposal designed to persuade Sheila O'Brien to purchase a scanner for the Special Projects Investigation Team. You want the scanner as soon as possible so you can begin to add digitized images to your reports and proposals. Use a **problem** → **solution** strategy to put forward the benefits of your idea.
6. Find a Request For Proposal (RFP) in your local newspaper or in *The Federal Register.* Write a letter requesting a copy of the RFP. Prepare a compliance matrix (See Fig. 21-6) to demonstrate that your offer conforms to all specifications.

22

Revising, Editing, and Proofreading

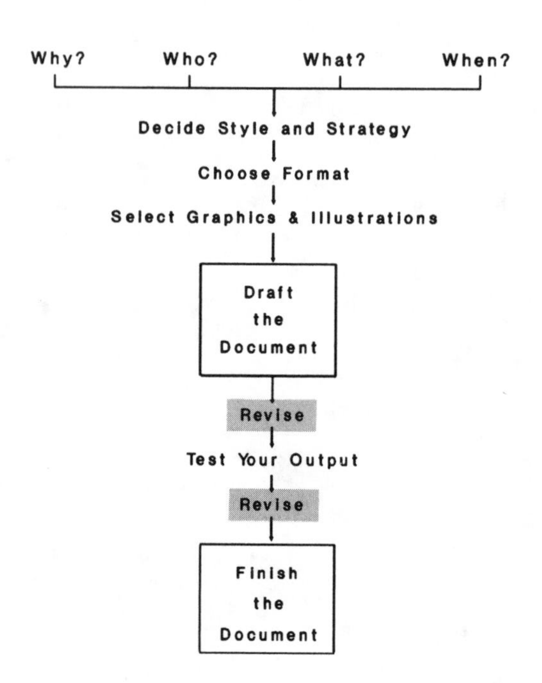

2.1 INTRODUCTION

Important documents need to be carefully checked before they are ready to be designated as complete. Omissions of key information and mechanical errors, for example, can create the impression that your work is flawed. As a result, you may not achieve your purpose.

Chapters 9 through 21 should give you specific guidelines on how to prepare a draft for a variety of technical documents. Once you write your draft, you need to work on revising, editing, and proofreading your document to ensure that you have improved the draft and eliminated any errors.

In this chapter, *revising* refers to the process of examining the content and design of your document; *editing* refers to the process of examining grammar, mechanics, and style; and *proofreading* refers to the process of eliminating typographical and other errors in your document. While we present revising, editing, and proofreading as separate processes, you may do all three simultaneously as you examine your draft, and distinctions between the three processes are frequently blurred. For example, can content (revising) be divorced from punctuation (editing)? Editors revise, edit, and proofread, and they do so without considering which process they are doing at each moment.

Look at your draft critically. Assume that you will be able to make improvements. Many people simply go through the motions, convincing themselves that the document is good enough. If you review your documents and seldom make changes, you may need to consider changing your perspective on revisions first.

Ideally, you should plan some time between the writing and the revising, editing, and proofreading. When you return to your document with a fresh eye, you can see ways of improving the manuscript you may have missed previously. You need to create some distance between yourself and your text. Examining your draft should be impersonal and objective: the goal is to improve the text, even if this means cutting out large sections of painfully-written material.

22.2 REVISING

The first step in revising your document is to return to your statement of purpose, the first step in the Document Design Process. Have you successfully fulfilled your goal? Will this document do what it was intended to do? The document must express your message before you proceed to edit for grammatical and typographical errors. After all, what good is a technically correct document that has nothing to say to the readers?

Experiment with the text. Revise the structure by moving blocks of information around. This is best done with a word processor; the ability to cut-and-paste rapidly frees you to experiment with your document, testing different arrangements of words, sentences, paragraphs, and sections. Refine your sentences until they convey your meaning precisely. If you have the least doubt about a sentence, then work on it until you have eliminated the doubt.

Revise for clarity. Technical documents must be understandable to the readers. Could someone unfamiliar with your subject comprehend your message? Sometimes you need a test audience to determine this. We refer to the process of finding a test audience as testing your output, which we discuss in depth in Chapter 23.

Revise for accuracy. Technical information must be conveyed accurately and precisely. A decimal that is out of place, an incorrect number, a misspelled word can cost your company money and can affect your standing within the company. Verify any factual statements with a trusted source. Make sure that you adhere to company procedures and policies. Above all, convey what you know to be true. If you are unsure, conduct more research and gather more information.

Revise any section of your document that contains discriminatory language. Categorizing individuals based on their association with an age group, a gender, a race, an ethnic group, or a sexual orientation is stereotyping. Avoid it. Stereotyping is inaccurate. It also destroys your credibility as an objective, thoughtful individual. Be aware that our language has an inherent gender-bias. That is, the dominant pronoun reference system until recently was male-oriented. If a person was unnamed, writers referred to this person with the pronouns *he, his,* and *him.* People who delivered the mail were mailmen, and it was assumed that scientists were male. Whenever possible, you should employ a neutral pronoun such as *they, their,* and *them.* This means that you should use plural nouns: The *technicians* who worked on the space shuttle did *their* work properly.

ETHICAL CONSIDERATIONS

Be aware that the words we choose and the styles we employ are a result of a subtle process of cultural conditioning. How you phrase your message can communicate a great deal more than just technical information; age, gender, race, ethnic, sexual and other cultural biases can destroy your attempts to communicate.

Make sure that the words you choose are accurate and appropriate. If you are unsure about a word, find another one, or consult a dictionary. Be aware that many words have both denotations and connotations; that is, they have an objective dictionary definition and they have implied meanings and associations for readers. For example, the word "skinny" has negative associations, and someone referred to as skinny could be hurt by the use of the word. Technical writing should be, as a general rule, objective.

Use jargon only with expert audiences. Jargon refers to words and phrases whose meaning is known only to those familiar with a subject. For a general audience, you must define your terms.

Support your main points. Provide enough details, explanations, and examples of your key statements. The only way to determine what constitutes *enough* support is to look carefully at your audience. How much do they know? To what degree must you explain your ideas for them? Always be specific. Define any terms your readers may not understand.

22.3 EDITING

No brief section on editing for grammatical correctness can express all of the rules of grammar nor all of the possible errors writers can make. This section suggests some of the ways you can improve your written documents through editing. For a more comprehensive look at the rules of grammar, consult one of the handbooks published by one of the major publishing houses. Such a reference work is an important tool for the technical writer.

Phrase your sentences so that each word helps convey your meaning. If a word does not add meaning, take it out. Use as few words as necessary. Avoid comma splices, fused sentences and sentence fragments.

Comma splices are sentences which combine two independent clauses (clauses that could stand alone as sentences) with a comma between them instead of a period or semi-colon.

Example

Jones stood on the bridge, he peered into the water.

Corrections

Jones stood on the bridge; he peered into the water.
Jones stood on the bridge. He peered into the water.
Jones stood on the bridge, peering into the water.
Jones stood on the bridge and peered into the water.

Fused or run-on sentences are sentences where punctuation has been omitted between independent clauses. Correct these errors as you do with the comma splice.

Example

R and D won the K-Sat contract as a result we will hire five engineers.

Corrections

R and D won the K-Sat contract. As a result we will hire five engineers.
R and D won the K-Sat contract; as a result we will hire five engineers.
Because R and D won the K-Sat contract, we will hire five engineers.
We will hire five engineers because R and D won the K-Sat contract.

Sentence fragments are strings of words and phrases masquerading as sentences. "*The train coming around the bend.*" is a sentence fragment because the statement is incomplete. There is no verb to describe the train. "*The train coming around the bend is going too fast.*" and "*The train is coming around the bend.*" are complete sentences.

For more information, consult a grammar handbook.

Edit for style. Make sure your style is appropriate for the task you are doing. Base your decisions on the level of formality of your document upon your analysis of the audience. For most internal communication within an organization, try for a style that is a mixture between professional and informal. *See Chapter 6, Choosing Styles and Strategy,* for a more detailed discussion of style.

Organize your paragraphs for your readers. Your paragraph structure tells your readers how to read your ideas. Each paragraph is a signal to your readers that they should see the paragraph text as a separate unit, a presentation of a particular topic. A new paragraph begins a new idea; there should be, however, a logical and easily-grasped sequence of development running through the paragraphs.

The length of the paragraphs should also reflect your assessment of your readers' needs. Let the nature of the document determine the length. For example, if you are writing a process analysis for a set of technical directions, then make your paragraphs very short. Your readers will not have the time to ponder long paragraphs as they attempt to follow your directions. In contrast, a paragraph in a technical journal describing abstract concepts may be much longer.

Always be aware, however, that most readers are comfortable with paragraphs of between three and seven sentences. A paragraph of more than seven sentences is a risk. You may lose your readers' attention. Reserve your one-sentence paragraphs for extremely important information. A one-sentence paragraph should be a signal to the readers that this idea is isolated for a reason.

Edit for voice. Generally, you should use the active voice rather than the passive voice. The active voice is more direct and forceful; the emphasis is on the action and the doer of the action. With the passive voice, an action has occurred.

Active Voice

I made the decision. (The emphasis is on *I*.)

Passive Voice

The decision *was made by me.* (The *decision* is emphasized.)

Correct your spelling. Although recent studies have shown little correlation between spelling ability and intelligence and between spelling ability and writing ability, your audience may react very negatively to spelling errors. An abundance of spelling errors is an indication that the writer did not take the proper care in completing a document. If you know that you are a poor speller, take every possible measure to correct your spelling. Most word processors have a spell checker feature that allows you to correct about 80-85% of your errors.

Make sure your punctuation is correct. We include punctuation in writing *for our readers.* Include punctuation that will help your readers understand what you are saying. If you do not have a clear reason for including a punctuation mark, leave it out.

When you edit, look for the types of errors you have made in the past. If you know that you have a tendency to make a particular kind of error, search the document for that error.

> *"A sentence should contain no unnecessary words, a paragraph no unnecessary sentences, for the same reason that a drawing should have no unnecessary lines and a machine no unnecessary parts."*
>
> Strunk & White
> *Elements of Style*

> Blot out, correct, insert, refine
> Enlarge, diminish, interline;
> Be mindful, when invention fails,
> To scratch your head, and bite your nails.
>
> Jonathan Swift, *On Poetry*

Edit for consistency. Check to see that your formats for your labels and captions for illustrations, headings, headers and footers, fonts, and symbols follow consistent patterns. Your document should have a uniform appearance.

22.4 REVISING AND EDITING WITH A WORD PROCESSOR

We encourage writers to use word processors because entering text into electronic memory encourages revising and editing. It is easy to move text around and to add sentences right up until the last moment. Adjusting to writing on a computer may not be easy, but for many writers it is beneficial.

Some features available with many word processors may help you with your writing. Three of the most common features are search and replace, spell checking, and a thesaurus.

Search and replace allows you to revise a word or phrase throughout the document quickly. For example, let's say you have used the phrase "the State of Massachusetts" throughout your document. While revising, you realize that you should say "the Commonwealth of Massachusetts." You can find all of the instances where you used State and substitute Commonwealth. If you choose to do an automatic search and replace, give specific commands, or you could change a sentence to read "The President gives The Commonwealth of the Nation Address in January of each year."

Spell checkers are particularly helpful for poor spellers. Most word processors contain a dictionary and will check your text against the dictionary. If a word is not contained in the dictionary, the program informs you. Often you are presented with a list of options, including entering a correction or selecting from a list of similar words.

Be aware: Spell checkers will not catch every error. If you meant to say *which* but wrote *witch*, the spell checker, unless it is extremely sophisticated, will overlook the error. In other words, even if you use a spell checker, you should carefully check the spelling of your words.

A thesaurus, or synonym checker, is helpful for long documents where you want to avoid repeating a word or phrase. Instead of physically looking up a word in a book, the thesaurus function provides you with a list of words with similar meanings.

A more comprehensive discussion of the kinds of computer software available for the technical professional is presented in *Chapter 25, Writing With Electronic Tools*. Some of the topics discussed in the chapter are grammar and style checkers, groupware, and outliners.

Note: Some writers are helped by seeing a print-out of their work. They will find errors on paper that they will not find on the computer screen. So, you may want to print a draft version of your document for revising, editing, and proofreading.

22.5 *PROOFREADING*

Proofreading refers to the practice of examining a manuscript to find typographical errors, or "keyboarding" errors. Many readers react negatively to such errors, for it implies a carelessness on the part of the writer. Be aware of the tendency that we have to overlook our own errors. If you made the mistake the first time, you are more likely to miss it when you proofread.

There are different ways to proofread. The techniques listed in the shaded box become more rigorous and more time-consuming as you proceed.

1. Read the manuscript looking for errors.
2. Read the manuscript aloud.
3. Read it aloud to someone who has a copy of the manuscript. Say aloud every word and every mark that appear on the page. For example, if you come to a period, say "period".
4. Read it backwards, word by word and number by number. Use this technique when your numbers must be correct. (You lose the flow of the sentence with this technique, but you gain attention to individual words.)

No system is foolproof, and you can always miss an error. For very important documents, find a careful, experienced proofreader to review your document.

22.6 *A FINAL WORD*

Let's say that you have a flaw in one of your tennis strokes. If you hit a tennis ball 1,000 times, you have extensive practice hitting a tennis ball incorrectly. You need someone, a tennis pro, for example, to tell you what you are doing incorrectly and offer advice on how to correct it. If you are able to correct the flaw, you can now gain extensive practice hitting the ball correctly and improve your game significantly.

The same is true of writing. You need the pro, your writing teacher or your editor, to inform you when you are doing things incorrectly and to provide you with ways to improve. If you ignore the feedback on your writing, you may continue to practice your mistakes. So pay careful attention to any comments and suggestions on improving writing in general and to any specific corrections and comments on your written work.

One other way of improving your document is discussed in Chapter 23, Testing Your Output. With some technical documents, you need to find a trial audience to read your work before you produce the final version.

REVISING, EDITING AND PROOFREADING EXERCISES

1. You are the supervisor of a group of ten employees. One of the persons who works in your project group, Roger Lewis, recently witnessed an accident near the loading dock. Asked to write an accident report, Roger has done his best, but you read it and quickly realize it is not good enough to submit to Human Resources. (Human Resources will need this document for potential insurance claims, Workman's Compensation filings, and legal actions.) Help Roger out by rewriting this memo so that it is less confusing, correctly organized, and more specific. Correct the paragraph structure and get rid of any irrelevant details.

October 24, 199
To: Human Resources
From: Roger Lewis
RE: *

Frank Wilson was injured while hanging a banner near the loading dock. He was unassisted while trying to complete this task. While he was high up on the ladder, he hurt his thumb when he banged it with his hammer. He was trying to hit a nail. By the way, the banner said "Welcome back, Harry." As he was wrapping his handkerchief around his thumb, he lost his balance and fell to the ground. The plant nurse, Phyllis Harry, arrived shortly after the accident becuase she was informed by Pete Stevens who saw the whole thing. Pete said that someone should have been holding the ladder. Phyllis asked me to call for an ambulance. The ambulance arrived at 9:54 a. m. They shouldn't have taken so long. The ambulance loaded Frank onto a stretcher. They took him to Mercy Hospital. Frank has a broken leg and and some other injuries. These type of injuries can be avoided. Why, my brother Hank was also hurt when he was hanging a picture at his house. He banged his thumb and fell off a ladder too.

*This line was left blank. You should supply a subject.

Please note: You can add some details that Roger Lewis has omitted, but do not change the overall situation.

2. Before he sends the following letter, Bob Curtain asks if he has the name of William Novack's presentation correct. When he shows you the letter, you realize that it cannot leave the company the way it is. Make the necessary improvements in the letter.

August 4, 1993

Dr. William Novac
General Dynamo
77 Sunset Lane
Smithtown, New York 11787

Dear Dr. Novack:

Thank you very much for being our guest speaker at our annual Appreciation Day ceremony. I'd like to apologize for the poor attendance by our employees to your speech, but there was a softball-game-and-keg-party going on at the same time. Everyone who bothered to attend found your talk on "Applying Quality Control Standards to the Entire Workplace" an interesting subject.

I didn't want to bring this up in front of the entire audience, but why didn't you use more humor? Your topic could have used some levity.

Anyway, thanks again.

Sincerely,

Bob Curtain

Bob Curtain

3. You have recently been promoted to Department Head of Quality Assurance. During the first week in your new position, you decide to review the lab notebooks of your personnel. When you look at the first notebook, you read the following:

June 21

Ran the XT-150s threw TP-41. All good. Problem with keyboard not solved. Escape sequences is wrong. F4 key not working. Why???

What is wrong with the notebook entry? What steps would you take to resolve the problems with this particular notebook? Write down what you would do.

4. Locate the last major report or research paper you submitted. Did the instructor make any corrections that you could have prevented? Make a list of the corrections and try to identify a pattern in the errors and omissions.

Now review the errors you committed in a series of five papers submitted for a course. Try to distinguish patterns in the errors you make. Make a list of the frequency of the errors.

5. Proofread your resume using the four techniques presented in the shaded box in Section 22.5. Keep track of how much time each technique takes. Which of the techniques would be most beneficial for a document such as a resume where you cannot afford to make a mistake?
6. This exercise requires the participation of small groups of three to five individuals. For the next writing task you have to complete, make enough copies for each individual in the group. Distribute the copies and let each one revise, edit, and proofread the document. Discuss each of the suggested improvements within the group and attempt to make each document error-free.

23

Testing Your Output

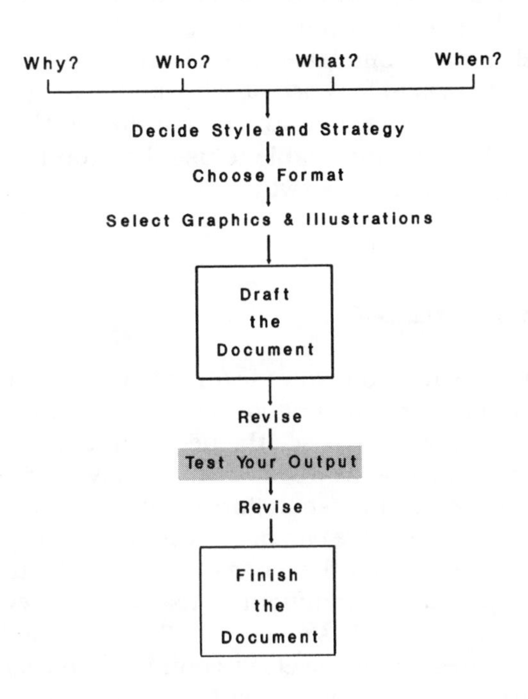

23.1 INTRODUCTION

Before introducing a new product, marketing people find a test group of consumers and elicit their responses: How do they like the product? What features did they like the most? What did they dislike? Would they buy the product if it cost *x* amount?

Before arguing an important case in court, many attorneys assemble a group of individuals and try out their arguments in front of this test jury. The attorneys ask the test jury which arguments were most influential, and they build their strategy based on this feedback.

Before starting to manufacture a new product or technology, engineers build a prototype and test it under a variety of conditions to see if it works. If it doesn't, they find out why it failed, make the necessary corrections, and run more tests.

The same principle should apply to your technical writing. The more important and the more complex the document, the more you need to get feedback from others; that is, the more you need to test your output. We send technical products through Quality Assurance and troubleshooting checks. We should do the same for written documents, and for verbal activities such as oral presentations.

The clearest example of this is a set of technical directions. Your directions may be poetically phrased, but if they do not tell someone clearly and accurately how to complete a process, they are useless. The best way to test your directions is to find people who qualify as typical users of the product or doers of the process, and let them follow the directions. Watch them closely, and if they are unable to use the product or complete the process, work with them to analyze why.

23.2 GETTING FEEDBACK

Sometimes we are too involved with or too familiar with the subject matter to see that our words do not convey the meaning we intend. When we read over our work, we do not see the need to change; we know what we meant to say, and when we re-read what we have written, we believe that our meaning is there. In these instances, it takes other people to say that the meaning is not clear. We can argue with them that they *should* have understood the meaning, but the truth is that they did not.

Many people find it difficult to ask others to evaluate their writing. The act of writing is viewed as so personally revealing that people find it intimidating to request someone else's comments. If this is true of you, you need to change your perspective on writing. What you need to realize is this:

> All writing could be improved.
> Good writers revise their work.
> Technical writing must be clear to the audience.

Once you make these assumptions, asking someone to test your output may be less intimidating.

Now you can find an appropriate audience and ask:

"How does this sound?"

"What do you think of this?"

"Can you suggest ways to improve this?"

"Do you understand the message? What is it?"

You have to decide who would qualify as an appropriate audience based on the communication task involved. In Section 17.6, Editing Technical Directions, we recommend requesting the assistance of a novice user, someone who relies completely on your written directions for what to do. In Section 20.6 of the chapter on User Manuals, we advise you to "field-test" your user manuals with novice users in order to cause failures which you can correct before the manual is finished. In Section 15.8.3, Rehearsal, in the chapter on Designing Presentations, we stressed the benefits of friendly faces in your test audience.

You may also rely on experts and professionals in fields other than your own to review what you have said. There will be times when you will need to have an attorney check through your work to minimize legal liability before the document is produced. Such documents as memos, phone records, minutes of meetings, and business letters can become evidence in a court of law, and companies are responsible for the work-related communications of their employees. Any document containing classified information needs to receive a security clearance before it leaves a secure area. You may also want to have a qualified expert check the accuracy of your information, particularly when you are outside of your field.

You do not want someone in your test audience who will say that your work is wonderful simply to please you and to boost your confidence. Neither do you want someone who would be hostile or overly critical of your work. Your audience need not make negative comments to provide valuable feedback. A simple comment such as "I don't understand this section." can improve your writing significantly. You want test readers who can and will articulate what they do not understand, what errors they picked up, and what suggestions they have to improve your work.

To get your audience involved in the process of testing your writing, have them do something based on what they have read. Ask them to identify the key point in the document, or give them a task that reflects the subject matter, and see if they can do it.

23.3 *LISTENING TO FEEDBACK*

The benefits of requesting feedback are lost if you will not listen to the feedback. Some writers are too stubborn to change what they have written, some writers too hurt by what they feel is a personal attack rather than a sincere attempt to help, and some writers are just too lazy to make revisions based on the comments of others. Whatever the case, these writers have missed the opportunity to improve their work before the intended audience sees it. After all, at this point the document has not reached the hands of the audience that truly matters.

Don't be defensive. Make sure you distinguish between your words and your self. Accept the comments as a sincere effort to improve your work which will have significant benefits for you.

Don't interfere. If you give your audience a document to read, don't tell them what is in it. Don't add your own commentary while peering over their shoulder as they read. If you are watching them follow your written instructions, let your words stand for themselves.

Don't apologize. You have, after all, asked them to help you improve your work. If they find errors, they have done their job, and therefore there is no reason to apologize to your test audience for your mistakes.

You may of course decide to discard some of the feedback you get. You may decide to trust your instincts since you are the one who discovered your purpose, analyzed your audience, researched the topic, and so forth. Even if you make no changes as a result of testing your output, you will benefit from the process of having your writing examined by others. You will gain an added confidence in the strength of your finished product.

23.4 *GIVING FEEDBACK*

When you work with other people on a writing project, or when you are the supervisor of people who write documents, you become responsible for what they write. If there are errors in a document, it does not matter who made them; the errors will reflect badly on you. Therefore, there will be times when you will have to give negative feedback to other writers.

First and foremost, the feedback you give should be accurate and precise. If you have only a vague feeling that a grammar rule has been broken, you need to check the rule before you say anything. Secondly, you need to be able to give feedback in a non-judgmental way. What this means is that your feedback should not injure the feelings of the writer. Stick to verifiable, descriptive feedback, such as "You use *clearly* six times in this paragraph." rather than "I don't like your repetition of *clearly* in this paragraph." A writer will be much more likely to hear what you have to say if you keep opinions and feelings out of your comments, and only refer to precise descriptive statements that can be verified in the text.

People can take criticism without being defensive, but they need to see that the criticism is a response to the words on the page, and not to them.

23.5 A FINAL WORD

If you look back at the Document Design Process, you will notice that we have included a second "revising" after testing your output. What we would like you to do in the Edit Stage is proofread and edit your draft, test your output, and then make revisions based on the comments and suggestions of your test audience. Be aware, however, that you need to proofread and edit your most recent revisions. In other words, you should do a second proofreading and a second spelling check right before you print your final version. Don't allow errors to appear in your work as a result of following a very positive, beneficial activity: testing your output.

TROUBLESHOOTING EXERCISES

1. Follow Exercise #2 at the end of Chapter 17, Giving Technical Directions. Write a memo to your instructor in which you explain what happened and what you learned when you watched someone attempt to fold the paper airplane from your instructions.
2. Do Exercise #7 at the end of Chapter 17, Giving Technical Directions.
3. Create a logo for Gates and Associates. Each individual in the class should do this. Transfer your logo to an overhead transparency. The entire class should view the logos and decide which three logos are the best designs. The class should discuss what criteria to use in the selection process, and should discuss why the three logos were selected. After the discussion, a second vote should be taken to decide which of the three logos is the best logo to represent Gates and Associates. (If you are at all surprised by the decision of the class, you have learned a valuable lesson about testing your output.)
4. See Exercise 1 in Chapter 22. Write a memo to Roger Lewis informing him of the changes you have made to his accident report. Make sure you explain why you have made the changes. Attach a copy of your changes.
5. This exercise can be part of a group project. Locate an example of an instance where failure to test a product properly had severe consequences for a company or organization. What was the situation, and what did the company or organization fail to do?

   Now find an instance where a company or organization failed to test written output. How could this situation have been prevented?

6. This exercise requires the participation of small groups of three to five individuals. For the next writing task you have to complete, make enough copies for each individual in the group. Distribute the copies and let each one revise, edit, and proofread the document. Discuss each of the suggested improvements within the group and attempt to make each document error-free. (This exercise also appears at the end of Chapter 22.)

24

Preparing the Finished Product

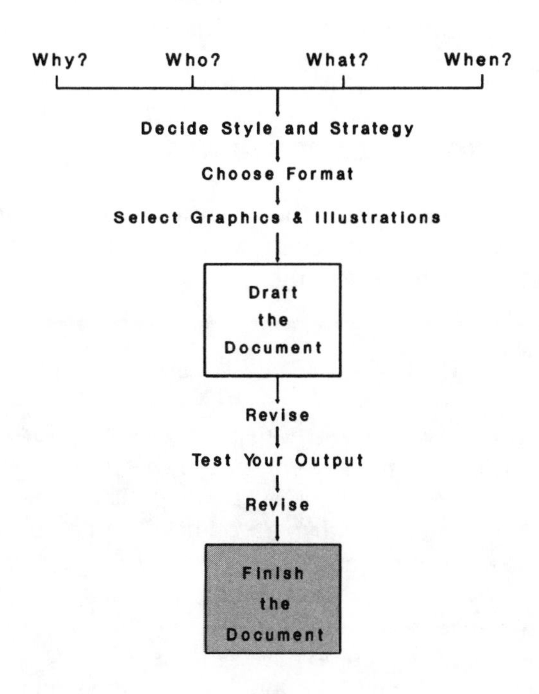

24.1 INTRODUCTION

This chapter concerns the last step in the process of producing your document. This is a very important step, for it determines what your audience sees of all of the work you have done. Remember, no matter how much time and effort you put into a document, your audience will see only the finished product. This final version must be clear and accurate to accomplish your purpose.

If you are a perfectionist, your writing is, seemingly, never done. It cannot be done because no writing can ever be perfect. This is why we need deadlines. *See Chapter 5, Scheduling Your Writing,* for more on the importance of deadlines. Ideally, when you schedule your writing, you allow time for each step in the Document Design Process, and this includes allowing time to complete your document.

You have already tested your output, and proofread and revised your work again. Now, in the Final Stage of the Document Design Process, you need to assemble all of the pieces of the completed document. If the document includes anything besides text, such as art work, photographs, tables, and charts, you need to incorporate these in their finished state. If the document is a collaborative effort, you need to get every contributor's final product and blend them together into a cohesive whole. Depending on your output source or sources, you need to allow for printing the final copy. Your writing may be done, but this very important final step cannot be overlooked.

24.2 HOW DO I KNOW WHEN IT'S DONE?

With anything you write, there comes a point where you decide to stop. While this will differ with every writing task you begin, there is one general rule that applies to all writing.

The Law of Diminishing Returns Applied to Writing

There is a point in the writing process where further effort will not produce any significant benefit for the writer. That is, you will not be able to influence the result of your writing by putting more time and effort into the piece of writing. For example, you have worked on a memo to your fellow employees for one hour. It will accomplish its purpose. You could spend another hour refining your message, but your efforts will not change the way your audience receives your message: they will do what you have asked them to do. Since you do have other tasks to do, you quickly check it one more time, print the final copy, sign the document, make a photocopy for your records, and then send it.

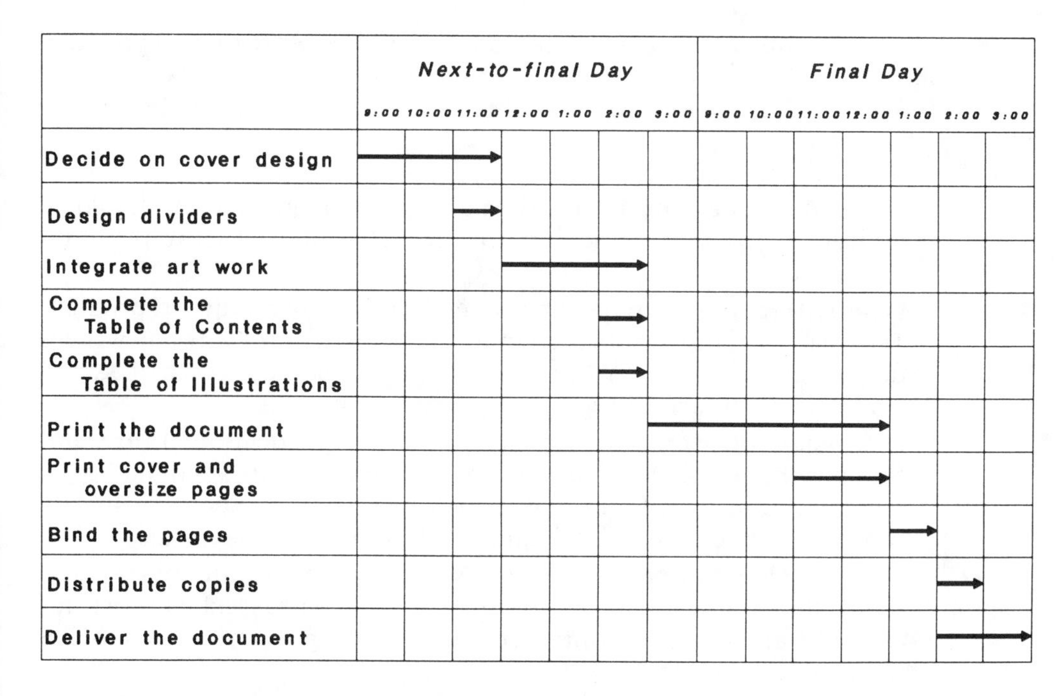

Figure 24-1 An example of the final stage of preparation of a document.

We would like to add one note of caution here: This is **not** a license to expend a minimum amount of energy on a project, but a guideline to follow when you are thinking about allocating your time. You can spend too much **or** too little time on a writing project, and both can have serious consequences. If you regularly find yourself running out of time and failing to produce the quality of work expected of you, there is one clear solution: plan your time better.

The ability to schedule your time efficiently is learned from experience. The more you write, the more you learn about how to budget your time so that you do have time in reserve at the end of the process. With a large project, you need to have some time at the end because you may not make your earlier deadlines. Some documents, such as proposal submissions, must be on time. You cannot afford to be late, so you need to include some extra time in your schedule.

24.3 HOW SHOULD THE FINISHED PRODUCT LOOK?

Once the writing is complete, you turn your attention to producing the document. You should consider how you will produce your document long before you actually finish writing. Although we are discussing it here at the end of the Document Design Process, an experienced writer would probably make this decision while selecting a format for the document. *See Chapter 7, Selecting a Format,* for some guidelines. Let's look at some of the options you have at your disposal:

- **pen and ink, or pencil:** Only the most informal documents should be written in pen or pencil, with the exception of graphics and illustrations.
- **typewriters:** A good typewriter can produce a good finished product. Typewritten manuscripts can smudge, however, and many typewriters do not have memories. You cannot save your document on a disk, and you cannot make major cut-and-paste revisions.
- **dot-matrix printer:** The standards for producing documents are continually changing. Dot-matrix output, once viewed as good quality, is now barely passable for formal reports and manuals. Some dot-matrix printers produce "near letter-quality" (NLQ) output that is acceptable.
- **letter-quality printer:** In between NLQ and laser output are some letter-quality printers. Printers in the Hewlett-Packard Deskjet series, for example, are not laser printers but produce output that is hard to distinguish from a laser's.
- **laser printer:** Although relatively costly, laser printers produce much of the formal written work produced by businesses today.
- **plotters:** Plotters are very useful for graphics and illustrations. Their major disadvantage is the time necessary to produce a finished product.
- **phototypesetters:** Very expensive and of extremely high quality, printing by phototypesetting is used extensively for very high quality output.

While you may not have access to many of the options discussed above, you should investigate all of the possibilities open to you. For example, there are many service bureaus and printshops which give you access to Linotype printers. You bring your work on a computer diskette, and they will produce the document using the most advanced equipment and techniques. Particularly if you intend on using many colors, you should consider using service bureaus. The higher the quality of the finished product, the higher the cost. (Many service bureaus also allow you to purchase time on their computers, usually loaded with advanced software, to produce your documents.)

The quality of the output should depend on the importance of the document. You should consider the expectations of the audience and the way you would like to be perceived by your audience. You need to be attuned to

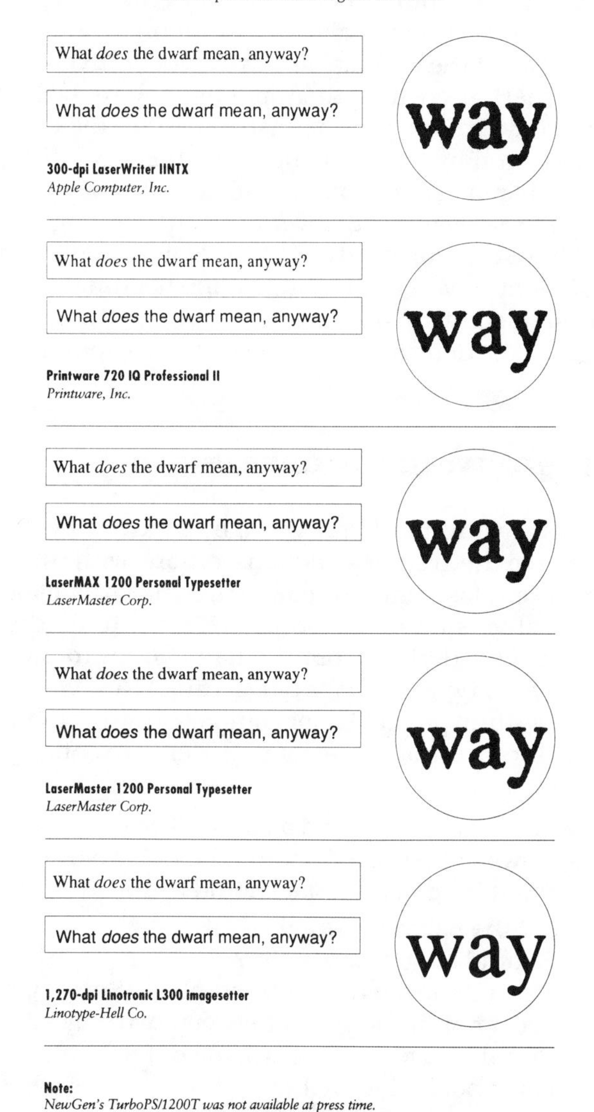
THE RIGHT TYPE
Samples are 10-point Times Roman and Italic, shown actual size and magnified 500%, and 10-point Helvetica Regular and Italic.
What does the dwarf mean, anyway?
What does the dwarf mean, anyway?
way
300-dpi LaserWriter IINTX
Apple Computer, Inc.
What does the dwarf mean, anyway?
What does the dwarf mean, anyway?
way
Printware 720 IQ Professional II
Printware, Inc.
What does the dwarf mean, anyway?
What does the dwarf mean, anyway?
way
LaserMAX 1200 Personal Typesetter
LaserMaster Corp.
What does the dwarf mean, anyway?
What does the dwarf mean, anyway?
way
LaserMaster 1200 Personal Typesetter
LaserMaster Corp.
What does the dwarf mean, anyway?
What does the dwarf mean, anyway?
way
1,270-dpi Linotronic L300 imagesetter
Linotype-Hell Co.
Note:
NewGen's TurboPS/1200T was not available at press time.
Printware and LaserMaster samples were supplied by vendors.

Figure 24-2 The essential difference between different types of output is the number of dots per inch (dpi) used to form letters. The higher the number, the better the quality of the output.
Reprinted with the permission of *Publish*.

the standards of the company you work for, and be aware of the relative costs of each type of output. Bear in mind that your document will be competing for attention with all of the other messages people receive in the Information Age. Your decision should reflect all of these factors. (Sometimes someone else will make the decision.)

While all of the options discussed above involve producing a hard copy (text on paper), there are other options that are being used much more frequently today. Information is distributed on diskette, by modem, by electronic mail, and by facsimile machine, thereby letting the users determine how they will read the document. Sometimes the users read the information off computer screens, and never print the message onto paper. As a writer, you still have control over the format of a document, but not the actual finished product. If you are writing online documentation in the form of help screens for a software product, for example, the actual appearance of your message will be determined by the user's computer system.

24.4 SOME LAST-MINUTE CONSIDERATIONS

Let's assume your final version will be a hard copy. You have already taken care to merge the graphics and illustrations with the text. You proofread the document one last time to make sure that it is clear and accurate. (This next-to-final version is sometimes referred to as the galleys or the galley proofs; the term derives from the name given to a tray used with printing presses to hold type ready for typesetting.) Now you are ready to print.

Your writing project is not complete until you have given some consideration to the question of packaging. Here are some questions to ask:

- Does the document need a cover? If yes,
  - How sophisticated should the cover be?
  - What kind of material should be used?
- Should the pages be attached? If yes,
  - Should I use a paper clip?
  - Should the document be stapled? (Some options are in the top left corner or in the gutter between two pages.)
  - Should I use binding? (Some options are spiral binding and hard cover book binding.)
  - Should I put in holes for a 3-ring binder or notebook?
  - Should I use some other way of attaching the pages?

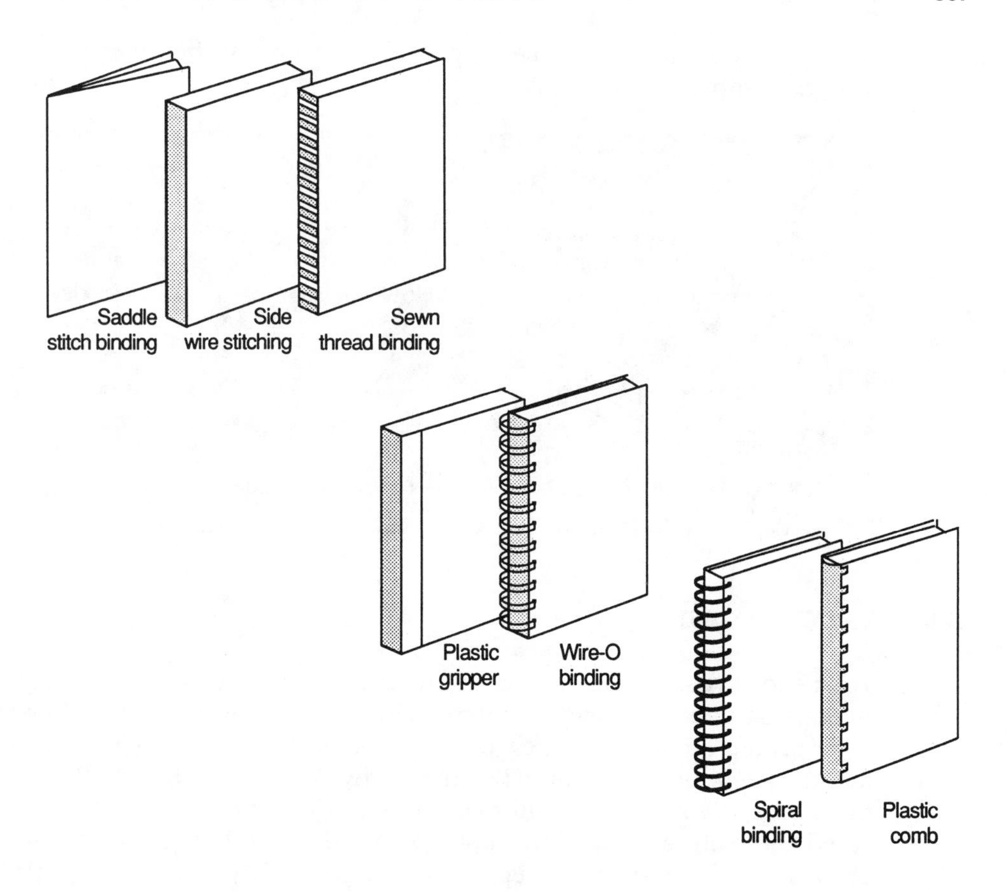

- Should the document be enclosed in an envelope? If yes,
  - What kind of envelope or folder?
  - How large should the envelope be?
  - What should be the format of the envelope?

As with printing options, these questions should be asked much earlier in the Document Design Process. To produce a very high quality document such as an annual report, decisions about the cover and the binding process need to be made months in advance. For most internal documents within a company, the decisions can be made a few days before the deadline.

Now you have the hard copy of the document. At the very last moment, you may want to go over this list.

A List of Very-Last-Minute Considerations

- Does your document require your signature? (For example, if a letter of transmittal accompanies a long report, have you signed the letter?)
- How many copies do you need? How are you going to make those copies? Do you need to consider the quality of the paper for the copies?
- How are you going to distribute this document? (For example, are you going to send it through a facsimile machine? By postal or other delivery service? Through the company mailroom? Are you going to hand-deliver it yourself?)
- Do you have a copy of the document for your files?

24.5 A FINAL WORD

You need to reserve time at the end of the process because *something could always go wrong.* Your dot-matrix printer could need a new ribbon, your laser printer could need a new toner cartridge, your copying machine could fail, the telephone lines could be affected by an electrical storm, the printing machines at your local printer could be out of service. The more sophisticated the technology you are using to produce your document, the more problems you could have. Prepare for this possibility by scheduling the production of the final copy at least one day before you absolutely need it.

COMPLETION EXERCISES

1. Make a list of all of the potential output sources you have available to you. Your list should include what is available at home, in your college, at your office, or within the community. (Be aware that there are businesses in most major cities where you can rent time on computers equipped with advanced software packages. You can produce your final output on very high quality printers.)
2. Find a technical document that you believe is of superior quality. Write an evaluation of the special features that make it a superior document. Speculate about how the document was produced.
3. Locate an assignment that you previously submitted for this course or for another course. Make a list of ways you could improve the final presentation of the assignment. How much time, effort, and money would be needed for each improvement?

4. You have completed the writing of a 75-page technical proposal to a Federal Government agency. (This is a very significant document; in fact, the fate of your company depends on the success of this proposal.) You need to produce 25 copies of the report, 5 copies for submission to the government agency and 20 copies for distribution to company employees, including the company president.

   What steps would you take to make sure that the document was successful? You have permission to "spare no expense" in terms of the cover, the printing, the graphics, the paper quality, and the binding. Write a memo to your instructor detailing your plans.

5. You need a very high grade in a course in your major taught by an instructor with a reputation as a very hard grader. The entire grade is based on a term paper to be submitted at the end of the semester.

   You have 10 weeks to complete the paper. Design a schedule that will allow you to submit the best possible paper you can. Focus on what you would do in the final week to improve your chances of receiving a high grade.

6. You have recently been appointed the Department Head in the Technical Editing Department at Gates and Associates. See Chapter 26 for details about Gates and Associates. Write a policy statement in which you detail a procedure to be followed in the final stage of preparation of all technical proposals and reports. Circulate the procedure to everyone in the company at and above the level of Department Head.

25

Writing with Electronic Tools

25.1 INTRODUCTION

The purpose of this chapter is to encourage you to learn more about computer-aided writing and presentation tools. Technological developments in the field of communication have been rapid and astonishing in recent years. The microcomputer has become a powerful and flexible device for technical communication.

Improvements in the technology of writing, drawing, and publishing have reduced the expense and production time of creating documents. New products are changing deadlines, costs, and expectations. Your documentation must keep up with industry standards.

Writing, particularly technical writing, has changed dramatically because of the personal computer. Today many writers use new methods, new skills, and new computer-aided writing resources. Effective technical

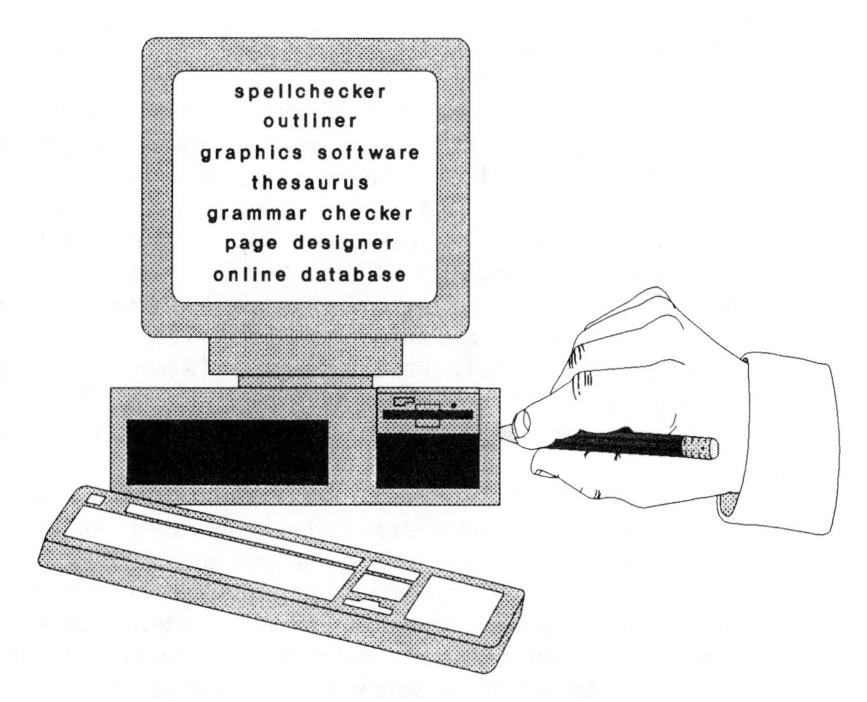

Figure 25-1 Some examples of electronic writing tools.

writers make full use of current electronic writing tools. These tools include word processors, outliners, spell checkers, style checkers, graphics applications, and online databases.

Even a few years ago most technical documents were drafted in longhand before they were typed. Rough draft versions of the illustrations were sketched by hand. Clerical tasks were often tedious and time-consuming. Today you can use a microcomputer to carry out a technical analysis; then to write, edit, and illustrate a report on the results of that analysis; and then to send the final document to a co-worker hundreds of miles away.

Documents created with electronic tools can read better and look neater. Readability and appearance can improve because you have the opportunity to write and rewrite, to organize and reorganize. You have the chance to polish your text. Electronic writing tools won't **make** you write better, but they will **allow** you to write better.

Facility with these developments is a highly-valued skill in the workplace. You should plan to learn how to apply the personal computer to your many technical communication tasks. By becoming familiar with the main features of the most popular commercial software products, you increase your value to employers and your own ability to perform your job well.

25.2 *HARDWARE AND SOFTWARE*

Some of the advantages of microcomputers include their portability and versatility. They can be used to collect, analyze, and report data from a variety of locations. Data and results can be transferred electronically to anywhere there is a telephone and modem. The personal computer is a multipurpose machine. You can use the same computer to accomplish hundreds of different tasks by changing the instruction programs.

A microcomputer system consists of two mutually dependent parts: hardware and software. **Hardware** is the physical equipment which makes up the computer. Generally this includes a keyboard, central processing unit, a monitor, storage and input devices, and a printer. These are the basic physical components of every computer system from mainframe to laptop.

This hardware will not do anything until you furnish it with data and instructions. These instructions are called programs or **software.** The term software refers to the fact that these programs are stored electronically and cannot be physically manipulated and touched like the hardware.

There are two general types of software. **Systems software** refers to programs that control the various computer components including input/output devices. **Applications software** is a set of programs that instructs the computer to carry out particular tasks and operations.

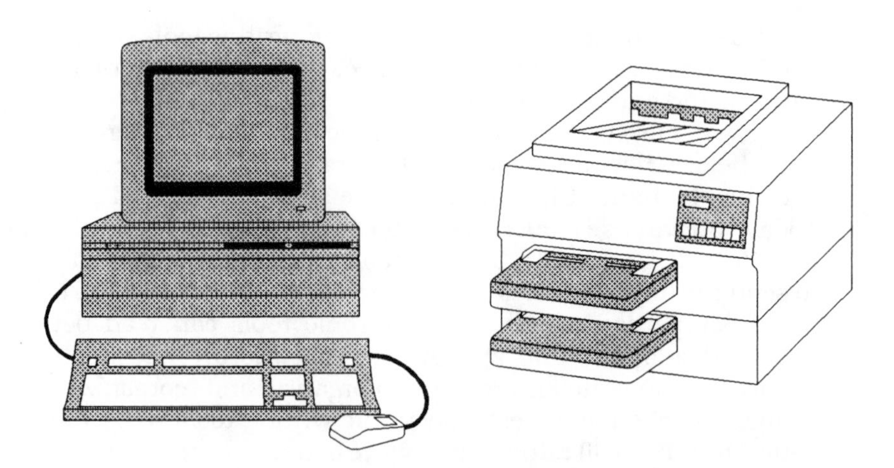

Figure 25-2 Hardware is physical equipment.

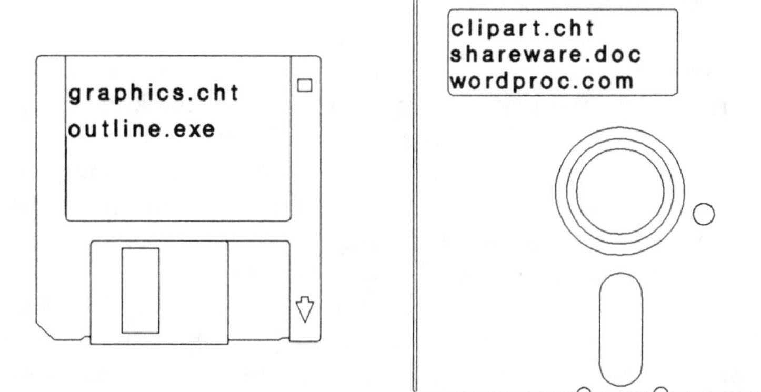

Figure 25-3 Software is electronic instructions and data.

ETHICAL CONSIDERATIONS

SOFTWARE AND INTELLECTUAL RIGHTS

Respect for intellectual labor and creativity is vital to academic discourse and enterprise. This principle applies to works of all authors and publishers in all media. It encompasses respect for the right to acknowledgment, right to privacy, and right to determine the form, manner, and terms of publication and distribution.

Because electronic information is volatile and easily reproduced, respect for the work and personal expression of others is especially critical in computer environments. Violations of authorial integrity, including plagiarism, invasion of privacy, unauthorized access, and trade secret and copyright violations, may be grounds for sanctions against members of the academic community.

Source: EDUCOM and ADAPSO. Reprinted with permission.

In the sections below we describe a variety of electronic writing tools: software applications which are designed to improve personal and group communications. We hope you will make the effort to investigate some of these and try them as ways to improve your own writing and presentations.

25.3 *WORD PROCESSORS*

During the past ten years personal computers have replaced dedicated word processors, using many different software packages with an astonishing array of features. Word processing software is the most common application used on microcomputers. While each of these packages has its own advocates, most of them offer similar features and advantages. What you should expect is the ability to display, store, retrieve, and manipulate your text in ways that are not possible with typewriters.

Word processors are particularly useful for any writing task which involves **repetition.** Writing tasks with a repetitive element—names, addresses, phrases, or summaries—can be automated to save your time and effort. Use the computer to free yourself from tedious and time-consuming retyping tasks.

Whether you are updating a proposal, making major revisions on a long report, changing a few words in a memo, or correcting typographical errors, **editing** is quick and easy with word processing programs. Correcting spelling, grammar, and sentence structure is fast and simple because you accomplish this electronically. You can move, delete, and insert text wherever you choose. You can see your changes on the screen as you make them.

Revisions include operations such as deleting and inserting words, moving paragraphs, and correcting misspellings. **Formatting** includes such operations as changing margins and line spacings, boldfacing, centering, and underlining. Both can be accomplished with just a few keystrokes.

Another advantage to word-processing is **appearance.** You can complete your document before you produce the paper copy. New printers provide crisp, sharp output. You can achieve a professional appearance without correction-fluid marks, erasures, or penciled-in changes. Remember, expectations have been elevated because of these new technologies.

Special features included in many word processors can save you time and effort while you polish your finished work. For example, full-featured word processors can automatically number footnotes and make certain that they appear properly formatted on the correct page. Other features allow you to quickly generate an index or table of contents by marking the text to be included from a document. More and more these programs allow you to bring graphic images into your document.

Many word processors allow you to bring graphic images into your document. Some allow you to wrap text around the figure. In this example, the figure box is placed at the left margin and the text is wrapped around the figure. The text is right justified.

Figure 25-4 An example of graphics within a text.

Other features common to word processors can save you an appreciable amount of time. The most fundamental time-saving technique is boilerplating—recycling information—by using small and large blocks of text over and over again. Text that you have reason to use more than once can be saved in a file and reloaded when necessary. Merge capabilities allow you to combine a single form document with a list of names, addresses, and personalized salutations or phrases. You can even record the format and print settings of documents for use again in similar documents.

Word processing is probably the most important application software for you to become familiar with. Many technical professionals are expected to prepare their own short reports, memos, and letters. Using a word processor can make these tasks simple and routine.

25.4 SPELL CHECKERS

New applications of computerized dictionary technology can change the way you write. For example, you can leave your concern for correct spelling aside until you are done creating your text. Then you can run the document through the spell checker and quickly repair your errors. Electronic proofers and computerized dictionary software allow you to find your mistakes and correct them. Most spell checkers offer alternate spellings and display misspelled words in context. Standard options allow you to accept a suggested spelling, type in another spelling, add the flagged word to a personal user dictionary, or continue on.

The quality control benefits of this tool cannot be exaggerated and it is widely available. Spelling checkers have gone from add-on, expensive stand-alone programs to standard features in most word processing packages. Your own spelling will improve from seeing the corrections on the screen. Even if you are an excellent speller, there is an insurance factor in having your work double-checked by the computer. Spelling mistakes will cost you credibility and may interfere with your message.

What you need to remember is that spell checkers will not recognize correctly spelled but incorrectly used words. Any speller checker is only as good as the contents of the dictionary on which it's based. Words like *dear* and *deer* can be spelled correctly but used incorrectly. The spelling checker will not help you. Some programs have homonym (sound-alike words) checkers and allow you to suggest alternatives.

According to some experts, the word *from* is the most commonly misspelled word in the English language, but since the usual mistake, *form*, is correctly spelled, this mistake will be overlooked. You should plan to proofread your writing even if you use the spell checker. Spelling mistakes will damage your writing or your presentations. Checking your spelling is a minimal, and required, step for technical documentation.

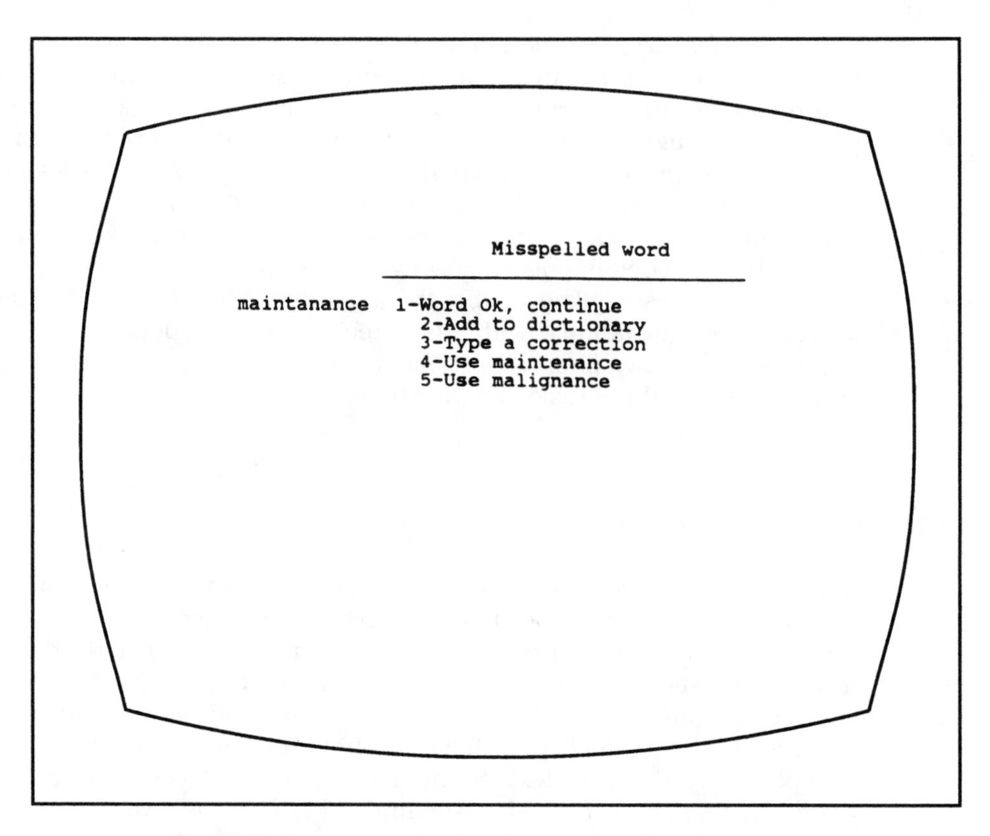

Figure 25-5 An example of an electronic spellchecker screen.

25.5 *ON-LINE THESAURUS*

Keeping a dictionary and thesaurus on your desk is one way to improve your spelling and vocabulary. Many word processing packages put both of these tools at your disposal. By providing you with a variety of alternative words, a computerized synonym finder can help you become more clear and precise. Some users find this an excellent way to avoid repetitious word use; others find it helps them choose the exact word they are looking for. Most of these programs allow you to substitute your synonym for the word you select to replace.

The biggest problem with this tool is that misuse of language can be as damaging to your writing as misspellings. Using a synonym incorrectly can confuse your readers. Some words are trite and overused. Other choices may be old-fashioned, likely to be misused, or inappropriate in a particular context. Language is tricky stuff, particularly when you are translating from one term you are not sure of to another you aren't sure of. Pragmatic word choice requires that you be familiar with the words you choose. Know what your words mean and what they suggest.

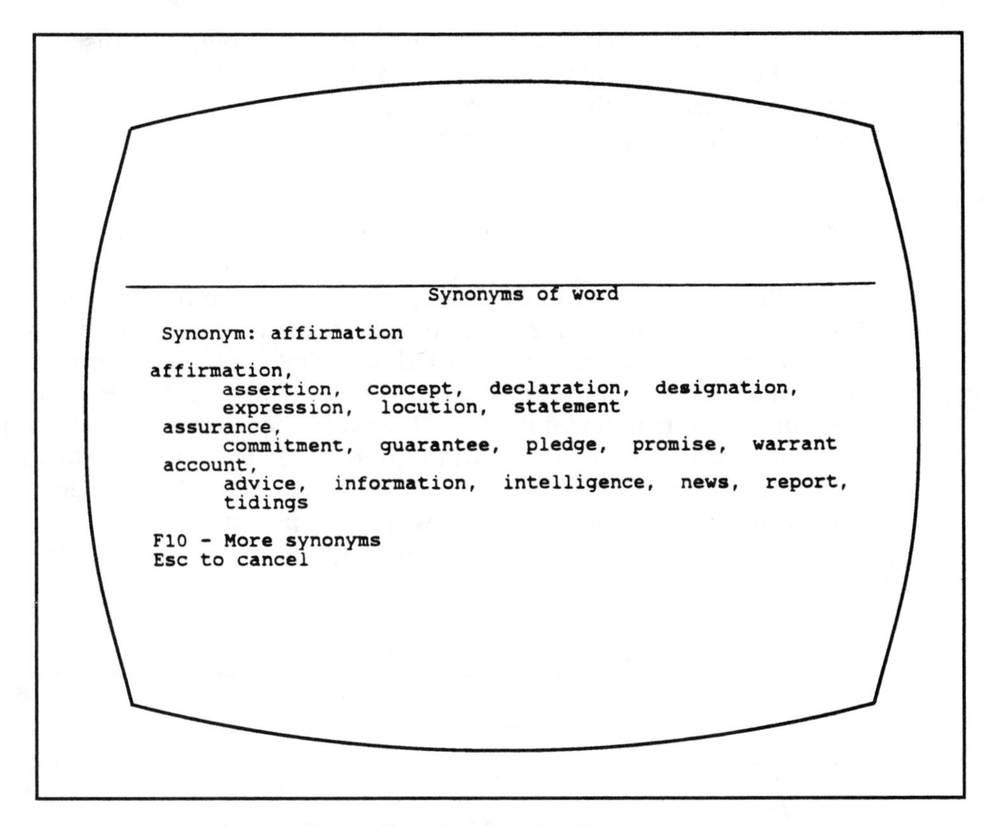

Figure 25-6 An example of an electronic thesaurus.

25.6 GRAMMAR AND STYLE CHECKERS

Automatic writing analyzers provide immediate, on-screen feedback about possible grammar and style errors. These programs use lists and algorithms to search for mistakes in usage, punctuation, grammar, and style. Some of these programs also provide a summary of your document, offering statistics about word, sentence, and paragraph length, and generating readability levels. Advanced grammar checkers use expert systems to parse the text and discover mistakes in agreement and usage.

One of the difficulties with these programs is their lack of flexibility in applying grammatical and stylistic rules. This software routinely rewards short words, short sentences and short paragraphs, even when they are inappropriate. Some of the suggestions these programs offer may be wrong and misleading. Some grammatical conventions such as split infinitives are really questions of style and author's approach. Good writing involves questions of organization, structure, audience, and intent: questions which cause problems for rule-based systems.

In general, we don't believe that grammar and style checkers will make significant differences in your writing. They may help you discover crude errors but careful proofreading will do the same. If you need a computer program to tell you that you are using too many long sentences, then we think you should spend more time looking at what you write.

25.7 OUTLINERS

An outliner is a drafting tool to help you organize your document. Your outline can help you decide if your material is connected, appropriate, and correctly located. Outlines can be sorted, shuffled, and shifted around before the first draft is even created. Stored outlines, called templates, can be used over and over for such standardized tasks as laboratory or progress reports.

Some word processors, and many writing strategies, are designed to use an outline approach. When you are working on multi-page documents, it is probably a good idea to begin with a simple outline, even if you use it only as a source for your headings and sub-headings. Outlines are easy to create with computer software.

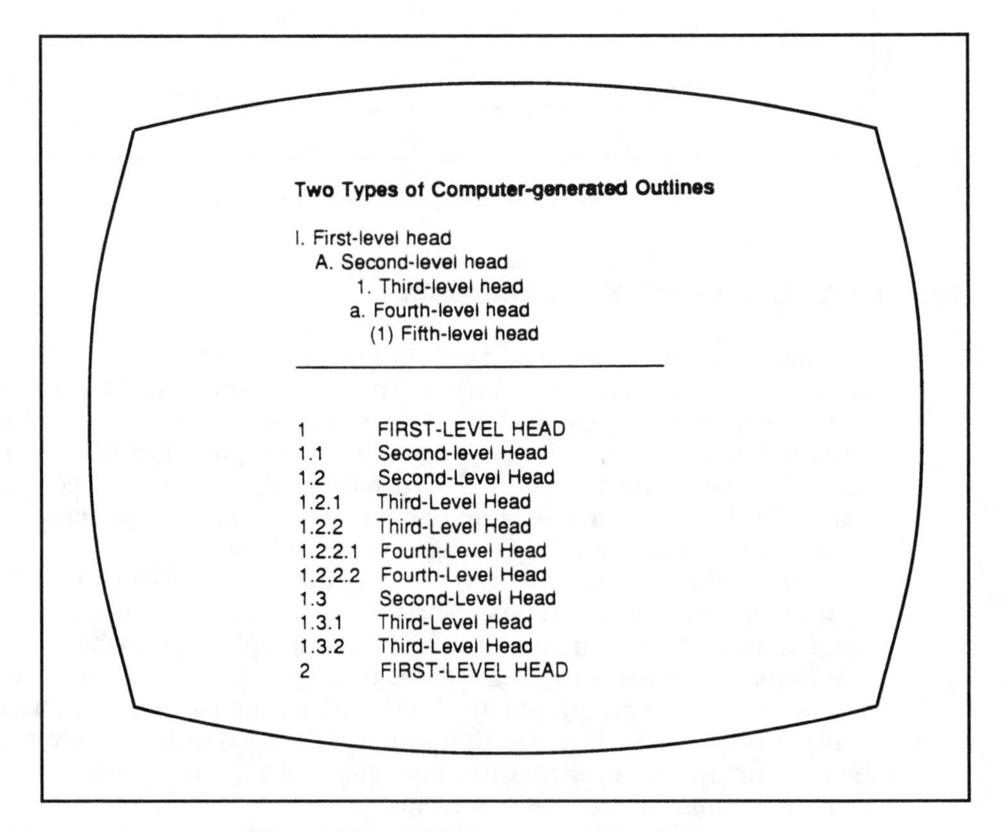

Figure 25-7 An example of an electronic outline.

The biggest advantage of an outline is that it allows you to shuffle information from place to place while you determine your final document design. An outline helps you rearrange a random series of notes to an ordered set of ideas grouped by topic and importance. As your ideas develop into a longer outline with more intricate headings and subheadings, you can test your alterations by viewing them immediately.

Current software allows you to cut-and-paste between different levels of headings. The outline can provide you with a detailed blueprint of your document plan. Many outliners allow you to collapse levels of headings and focus on the overall design. Even if you don't have a software outliner you can use almost any word processor for this function. The ability to delete and insert allows for easy changes.

25.8 GRAPHICS SOFTWARE

Graphic images are showing up more frequently in offices, shop floors, and technical training. Personal computers make it possible and affordable for you to provide illustrations to accompany your text or presentations. Desktop graphics, scanners, symbol libraries, and clip-art disks are easy to use and inexpensive. Output devices like laser printers, color plotters, and slide makers are widely available in larger companies and through service bureaus.

In the last few years, the quality of graphics generated on personal computers has moved from merely functional to boardroom appeal. More and more frequently, the responsibility for producing text charts and simple data charts falls directly on the person who will be using them. It is likely that you will be involved in creating at least some of the illustrations you will use.

Remember, it takes time to turn your ideas into symbols that work. Creating text charts and other simple graphics can help you think your ideas through, but careful consideration is essential. Restraint is the secret of good graphics. Don't get carried away with all the options in the program. Mechanical ability needs to be supported with design and layout skills. Keep your graphics simple and easy-to-understand. One advantage to doing it yourself is that you can make changes up to the last minute.

Another thing to remember is that good graphics can't save a poorly thought-out and badly organized presentation or document. Instead, the graphics will display the flaws in a way that draws full attention. Your images should support what you say, not compete with your message.

Text charts can reinforce your key points. They also double as a script for your presentation, keeping you relaxed, focused, and comfortable. Keep your charts consistent. Stay with the same facts, and remain with the same style throughout. Data-driven charts need to be kept simple or comparisons and significance will be lost. Don't clutter your images with unnecessary information.

Bullet Chart

- Use short phrases.
- Present your main ideas.
- Don't overload the list.
- Provide even spacing.

Figure 25-8 An illustration of a computer-designed text chart.

25.9 DESKTOP PUBLISHING

Desktop publishing software provides page design applications which can turn reports and memos into professional-looking documents that include graphics, captions, type in various fonts and sizes, and flowing columns of neatly proportioned text. This software is useful for flyers, newsletters, proposals, brochures, and other types of professional publications.

Desktop publishing software is part of a spectrum that extends from powerful graphics-capable word processors to extremely sophisticated page composition systems. There are entry-level products which can be used by beginners and full-fledged electronic publishing systems. These advanced systems allow you to design and produce complex, graphically-rich documents in an interactive mode. In simple words, you can see exactly what you are doing as you make changes in your document. These programs can provide input to a high-resolution phototypesetter, allowing for professional quality.

Don't begin with a package that is too sophisticated for what you need. A good word processing package with font and graphics support will allow you to merge text and graphics on the same page. This is enough for most simple documents.

25.10 PRESENTATION SOFTWARE

With currently available software you can use a personal computer to design and produce all four important presentation formats: paper, overheads, slides, and video. A polished presentation will generate interest and create understanding. Pictures provide a quick translation from numbers and ideas, and they help to focus your audience's attention on important information.

You can use paper images to create interesting handouts with a crisp, professional appearance. One advantage to paper presentations is that they can be delivered to persons who were unable to attend the meeting. They also make an effective supplement to your slides or overheads. Unlike projected images, a paper graph can be accompanied with as much explanation as you want to include. Use handouts to emphasize key points and serve as reminders to your audience.

You can include the output from graphics software in paper presentations or photocopy it onto transparent sheets for use with overhead projectors. Overheads are a convenient and inexpensive presentation tool, with projectors available in most companies and business locations. Some software provides predesigned layouts for bullet charts, title charts, and other common chart types. You enter your data and the program supplies large-scale lettering and punctuation symbols. Color overheads are more and more common as software capabilities improve and color printers and photocopiers become available.

Remember, the quality of your presentations will reflect the quality of your product or service. Hand-sketched graphs or dirty overheads will not help your credibility. Keep your overheads simple and to the point. They are used to amplify and echo your spoken words, not to replace them. Make sure that the text is big enough for everyone in the room to read. Overheads work best with fewer than forty people. With a larger audience you should use slides.

Color slides are a versatile and portable presentation format, suitable for large and small meetings. Slide projectors are standard equipment in many technical organizations. Graphics which you create on your personal computer can be captured on film and developed into color slides. Major graphics software packages usually offer a service bureau to provide slides. Since you design the visuals, the cost is cheaper than commercial preparation.

Some graphics packages allow you to simulate a slide show on the computer screen. Special projectors allow these images to be shown on a large screen. Typically this software lets you use special effects as you move from one image to another. Again, restraint will have a positive effect. Some screenshow features can distract your audience and cause them to wonder what your next effect will be instead of concentrating on your message. Keep it simple. You don't want your visual supports to compete with you.

Finally, you have the option of using one of the new animated presentation packages where you can create the illusion of motion within and between your graphic images. While animation offers an exciting visual element, it simply may not be worth the time and effort to create something so complex, at least for most presentations. Some persons are making video copies of their computer-generated graphics to show with a VCR. Animated video is developing quite rapidly and may become common in the next few years.

Whether you choose paper, transparency, film, or screenshow, the personal computer can help you maintain an image of quality and professionalism. Remember though, a poorly prepared, inadequate presentation will not be saved by flashy graphics. Instead, you will call attention to the flaws.

25.11 ONLINE INFORMATION SERVICES

Online databases are one of the new information technologies which make it possible for you to access technical and business data electronically. For example, you can search the *Readers' Guide to Periodical Literature*, check for recent stock market quotes, or make airline and hotel reservations.

Information-retrieval services such as BRS and Dialog provide commercial access to hundreds of databases. These encyclopedic enterprises attempt to provide a gateway to a comprehensive collection of information files, with an emphasis on business, technical, and professional interests. Narrowly focused databases can provide information on such specialized subjects as industrial chemicals or biomedical engineering.

You should prepare before you use these services. Practice with local bulletin board systems to master the basics of logging on, downloading and uploading, and moving around the system before you experience the connect-time charges of commercial online services. Become familiar with the commands and overall structure of the service before you log on. Online commands will become familiar and routine if you use them regularly.

25.12 GROUPWARE

Groupware is computer software designed to support task-oriented project teams in business and technical activities. Users located in different parts of a building, or different parts of the world, can work together almost as closely as if they were sitting at the same desk. There are new group writing, group communications, and group productivity tools. Electronic meeting rooms use same-time/same-place groupware designed to improve face-to-face meetings. All of these programs are designed to help people work together in more productive and effective ways.

Portable computers and laptops allow for telecommuters, persons who work from a distance, to participate in file sharing from many different locations. Workgroup software makes the editing process easier because only a single document is accessed by the group. This software allows you to keep track of changes made and also numbers the revision cycles for easy reference.

25.13 BULLETIN BOARD SYSTEMS

Bulletin board systems allow users to call a central file server where they can leave messages, transfer files, and carry on real-time conversations. You need a personal computer and a modem to access the systems. These boards offer electronic mail, interactive forums, public domain and shareware programs, and other services.

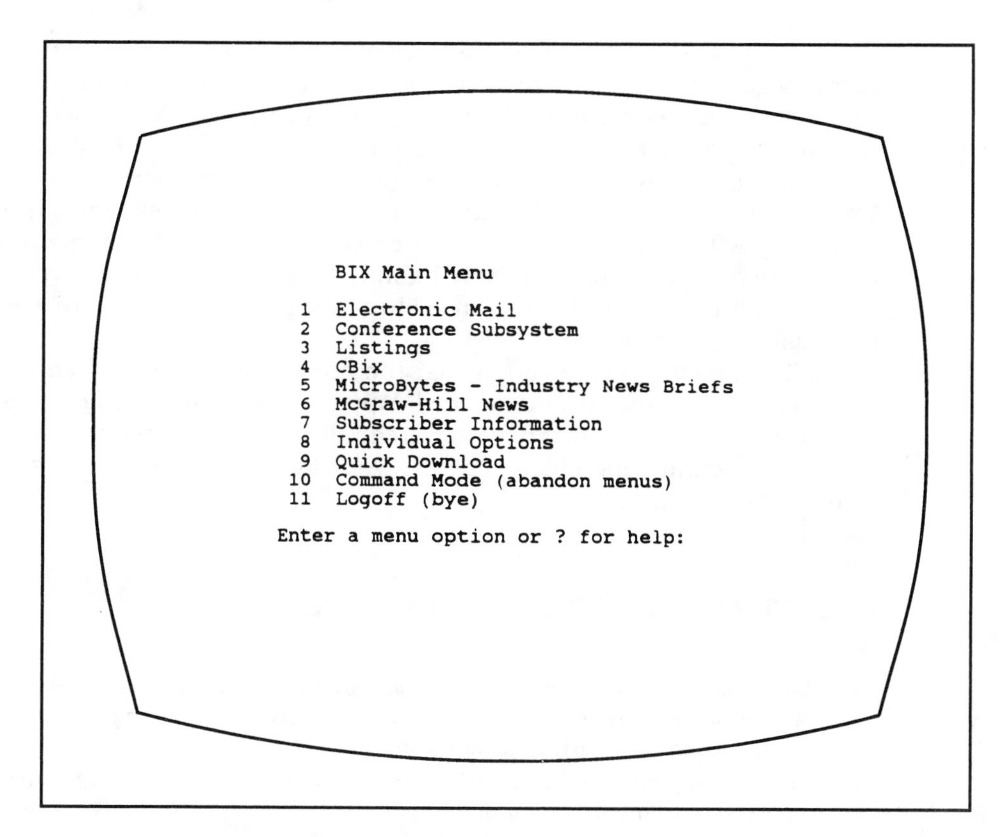

BIX Main Menu
1 Electronic Mail
2 Conference Subsystem
3 Listings
4 CBix
5 MicroBytes - Industry News Briefs
6 McGraw-Hill News
7 Subscriber Information
8 Individual Options
9 Quick Download
10 Command Mode (abandon menus)
11 Logoff (bye)
Enter a menu option or ? for help:

Figure 25-9 An example of an electronic bulletin board menu.

Some companies have found electronic bulletin boards to be a practical and inexpensive way to provide up-dated technical information. Customers can leave questions and problems in the mail system and expect to receive an expert answer within a day or two. Technical forums allow customers to make suggestions for future up-dates and to share problems and solutions they have encountered. Remote field-site employees can send a message or leave a request at any time convenient to them.

25.14 A FINAL WORD

A whole variety of computer-aided communication tools have elevated standards for everything from the appearance of correspondence to the visual quality of technical reports and presentations. We want to encourage you to learn to take advantage of the full range of these technologies.

At the same time, we want you to realize that many of these software programs have very steep learning curves. This means you will have to expend a good deal of time and effort before you become productive with them. Begin with applications which you can use in your current situation, whether you are going to school or working, or both. Software which can help you do your job is easier to practice and learn on a daily basis.

Some people will tell you that writing with a computer will save time. We don't claim this. Certainly you can accomplish some tasks more quickly and easily with a computer, but your capacity to do *more* may make your task more complex. Your increased ability to accomplish high level tasks such as editing, reorganization, and illustrating may lead to more work as you push for perfection.

What we can assure you is that the professional appearance of your work will be recognized and noticed. Appearance communicates design, organization, care, importance and credibility to your audience. You can make your documents look as good as your ideas by learning to use electronic writing tools.

ELECTRONIC TOOLS EXERCISES

1. Make a list of the software programs you are familiar with. Write down the key features of these programs and your level of expertise with each program. Bring your list to class for further discussion.
2. The following data shows the increase in online databases available for electronic distribution of information:

Year	# of Databases
1980	400
1986	2901
1987	3369
1988	3699
1989	4062
1990	4465

Source: *Directory of Online Databases*

What conclusions can you draw from the data? With a personal computer, design a sample graph which illustrates your results. If your library has the *Directory of Online Databases*, find more up-to-date statistics.

3. Small groups of students or participants should work together on this exercise. Each group should create lists of ten words you think will stump the electronic thesaurus. What generalizations can you draw after testing your lists? What kinds of failures did you find? Were there any surprises? Compare your lists with the lists produced by other groups.

4. Create a three-level outline for a long report on the topic of computer access for the disabled. Show your outline to other persons in your group and ask for suggestions. Revise your first outline by cutting-and-pasting. Then compare the two outlines.
5. If you have access to a personal computer with a modem, try logging on to a BBS (bulletin board service) in your local area. Bring some printscreens to class to show others what the system offered.
6. Mr. Gates called Sheila O'Brien this morning and asked her whether Gates and Associates was using "shareware" programs. A friend of Mr. Gates showed him some very useful software which was available at very low cost. Mr. Gates wants a short report from your Special Projects group. "*Four* times he said short," Sheila tells you as she hands you the assignment. "He just wants to know the advantages and disadvantages." See Chapter 26 for more information about Gates and Associates.
7. See Exercise 6 before completing this exercise. Mr. Gates called back this morning. He was impressed with your short report. He wants to make a decision today, and would like a memo from you describing your sources of information for the report. He wants to be sure that it is reliable information.
8. Gates and Associates has been invited to make a presentation to the Ministry of Development in Surabaja, Indonesia. What computer equipment will we be able to use in the hotels there? What is the availability of audiovisual equipment? Will we encounter compatibility problems? Check all of this out. Report to Sheila O'Brien as soon as possible. For more information about Gates and Associates, read Chapter 26.
9. Choose a document you have designed, and print it out in three different fonts. How does the communication change with the change in type style?
10. Choose three diffferent short documents and put them through a grammar or style checker. Write a short report to your instructor summarizing the response of the software program. Add a discussion section to your report: Do you think it is worth the time and expense to use this program? Do you think this software would be useful to you or to your organization?

26

Communicating in an Organizational Environment

26.1 INTRODUCTION

Many of the exercises and many of the examples in this text involve a small consulting firm named Gates and Associates. By becoming acquainted with this company and the way the company operates, you gain a **context*** for your writing. You will not be asked to consider an abstract situation that could happen to *any* company, but you can understand and appreciate the circumstances of a particular situation.

You will be assigned to various tasks within the company. Your assigned role defines who you are within the company. Your task defines what you must do. The different situations you will face will require different communication solutions. Imagine yourself as an employee of Gates and Associates rather than as a student. You will still be writing to earn a grade, but your writing tasks will fulfill a specific purpose within the com-

*Context refers to the specific details of a communication *situation*. For example, if you were to think of a problem you were having with a parent, you would need to consider the context of that communication situation. You would need to think about previous conversations with that parent, about an entire history of events leading to the present situation. That history and the details of the current problem form the context of the situation.

pany. You will gain valuable experience doing some of the kinds of writing you will do during your career.

26.2 *A BRIEF HISTORY OF GATES AND ASSOCIATES*

Gates and Associates was founded by F. Robert Gates in 1979. Gates had worked for the Department of Transportation for 15 years after graduating from Carnegie Institute of Technology in Pittsburgh with a Masters degree in Mechanical Engineering in 1959. During his tenure in a variety of positions for the Federal government, Gates gained an appreciation for how the government operates. For the period 1972-1974 he served on a panel that reviewed applications for government grants for studies of technology.

In 1974 he was hired by the Bolton Group, a 200-member privately-funded research institute that analyzes technological disasters and catastrophes. Here Gates gained considerable knowledge of how the Federal government responds in an emergency, and how to quickly and clearly identify the causes of a disaster and solutions to remedy the situation.

Cost Breakdown of the Consulting Fees

for Gates and Associates

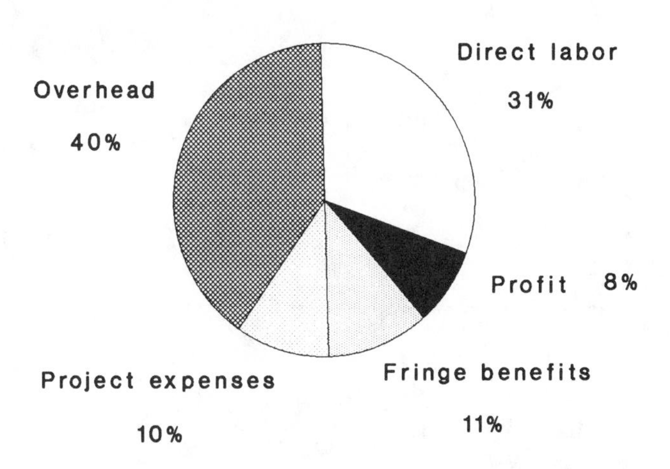

In 1978 Gates began to do outside consulting, with the approval of the Bolton Group. He was able to utilize the experience he had gained during his years with the Federal government. The number of clients expanded rapidly, and in 1980 Gates left his position with the Bolton Group and devoted himself full-time to his company.

He began hiring assistants and a clerical staff, rented offices on North Capitol St. in Washington, D. C., and began developing the company into a

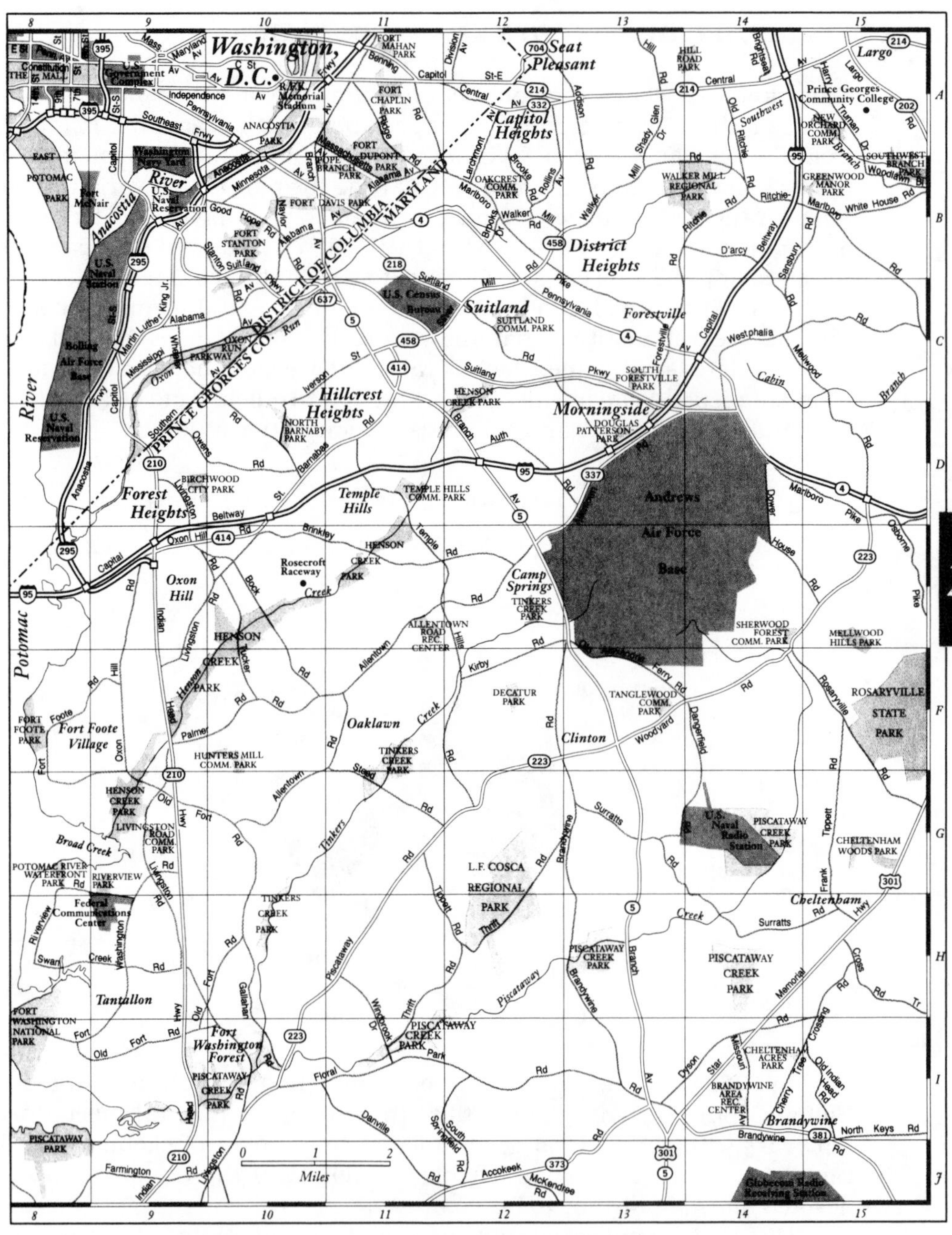

Figure 26-1 A street map of Washington, D.C.

Reprinted with the permission of C & P Telephone Yellow Pages.

29

Figure 26-2
Logo design reprinted with permission of Joseph Mercurio

widely known and highly regarded consulting firm. The major client remained the Federal Government, but many of the Fortune 500 companies, the largest companies in the United States, would utilize the services of Gates and Associates.

Currently, the company has a staff of 85 full-time employees beyond the clerical staff. You are one of them. Your job description entails the following:

- investigate crises and disasters when they occur.
- travel when necessary to the site of a disaster.
- prepare short and long reports for clients.
- research a variety of situations.
- make presentations.
- make recommendations.
- attend regular meetings of the project team.

26.3 THE CEO

F. Robert Gates is an aggressive, creative leader. He is competitive, driven by the desire to encourage the growth of the firm he started. He works long hours, and he demands that his employees do the same. Gates wants his people to be smart and imaginative, so he pays his people well, and he listens to what they have to say.

F. Robert Gates is a strong family man. He has been married for 22 years to the former Barbara Hurst with whom he has three children: Kristin (18), Thomas (16), and Rebecca (13). He has strong ties to the community in which he lives, which is Bethesda, Maryland, and a strong sense of morality, of following an ethical code. He is an active participant in the Rotary Club.

Gates' leisure time activities include playing golf, fishing for trout, and reading 19th century novels.

F. Robert Gates is a visual man, one who likes to "see" ideas on paper. He wants graphs and charts, and is not swayed as much by arguments and reasons as by statistics. He is very fond of saying "Show me the numbers."

In other words, if you want to make an impression on Gates, you have to be prepared to present information visually. You need to conduct your research so as to get accurate statistical information, and you need to be able to design clear, readable tables, charts, and graphs with the information you gain from your research.

26.4 YOUR POSITION IN THE COMPANY HIERARCHY

Gates and Associates is arranged as a hierarchy, with F. Robert Gates at the top of the pyramid. He has final decision-making authority on any major company operation. Gates is quite good at delegating authority, however, and he likes to give important assignments to new and relatively inexperienced employees. Gates is a firm believer that someone matures under pressure, and he wants to see how his new employees react under stress. It is possible that you will receive a phone call from Gates asking you to complete a difficult assignment quickly.

If you look at Fig. 26-3, you will see where you are placed within the structure of the company. Since the firm is small, advancement means taking a big leap in terms of salary and benefits. It also means that you know most of the people in the firm on a first-name basis.

You are in the Research Department, a member of one of the Special Projects Investigation Teams that conduct investigations into the causes of and solutions for large-scale accidents and disasters. For example, a recent project of your team was to investigate the derailing of a train carrying toxic chemicals. Gates and Associates was hired by the Federal Office of Technology Assessment to carry out a thorough investigation of the incident. The details of this situation are outlined in Section 26.8.

The specific nature of this work necessitates that you be able to react to situations quickly and accurately. You may be called upon to travel to another part of the country at a moment's notice. Your problem-solving ability is your greatest asset, because you rarely see the same situation twice. Although you are sometimes witness to gruesome scenes, the job is exciting and challenging.

Although the company frequently operates in a very informal manner, there are also some strict rules of conduct that must be followed in emergencies. You are required to write down all of your observations, and you must file daily reports with the office in Washington. You must maintain your professionalism at all times.

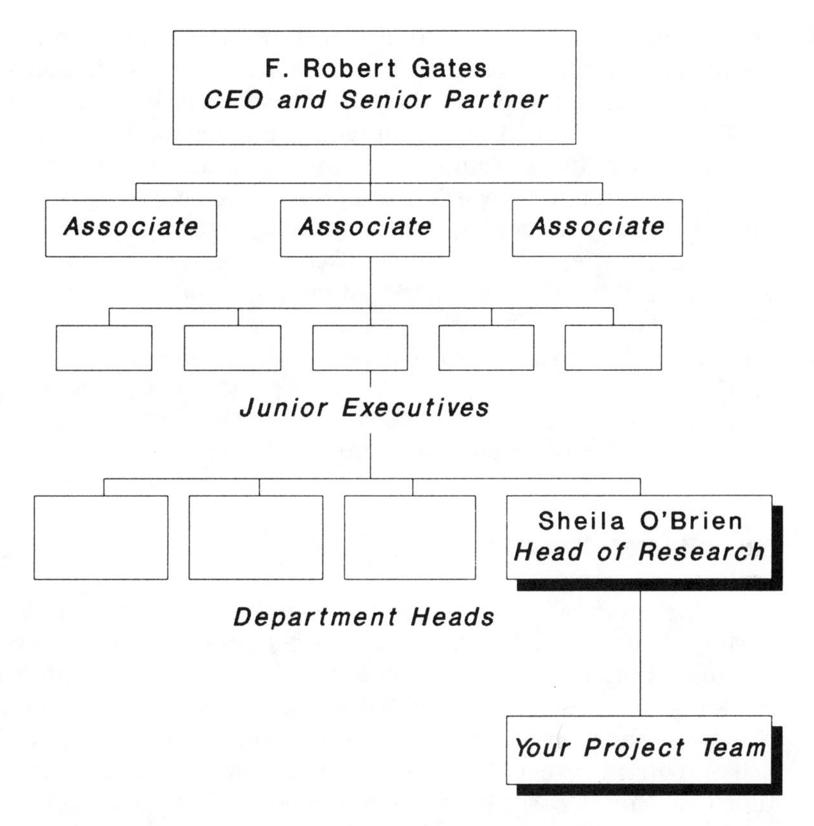

Figure 26-3 The organizational structure of Gates and Associates.

26.5 YOUR SUPERVISOR

Sheila O'Brien has been with Gates and Associates for five years. After graduation from Syracuse University in Syracuse, New York with a Bachelor's Degree in Journalism, Sheila worked for three years as an investigative reporter with *Newsday*, a highly respected daily newspaper based in Long Island, New York.

She came to the attention of F. Robert Gates when she interviewed him for a series of articles on Washington area consulting firms. Soon after, Sheila took a job with *The Washington Post*. After reading a number of her investigative articles, Gates hired her as an investigator with the Special Projects Investigation Team. Within three years, Sheila was promoted to Department Head. She oversees four teams, combining them when appropriate for particularly large tasks.

In contrast to F. Robert Gates, your supervisor is an "overview" person. She wants to be given a list of reasons why an action should be taken. Her years as a reporter have enabled her to size up a situation quickly. She trusts that you have all the necessary details, and that you can provide her with a synopsis of a situation. She will, however, occasionally test her people by making them produce all of their notes on a particular situation.

When presenting information to her, think in terms of being concise and to the point. As a matter of fact, it has been said of her that she only reads the first two paragraphs of any document she receives. (This is not true; she reads every word. She prefers writing that is concise and to the point.) You should consider summarizing material for her, and organizing material in order of importance. If you have a complicated message, you should consider sending a series of memos to her.

Sheila O'Brien is a very competent individual and is viewed with the highest respect by F. Robert Gates and the Senior Associates in the firm. It has been rumored that she will soon be promoted to Junior Partner.

Sheila is engaged to be married to Anthony Jones, a Public Defender in the Capitol court system in January of next year.

26.6 *YOUR PROJECT TEAM*

Your Special Projects Investigation Team consists of four individuals with a variety of backgrounds. In major emergency situations, all four members of the team may be sent to the location. When this happens, your ability to work in concert with other individuals in a pressure situation is crucial.

The other three members of your team, George Kershaw, Pamela Russell and Jorge Ortega, have all been with Gates and Associates for a longer time than you. George Kershaw has been with the Special Projects team for eight years, while Pamela and Jorge were hired four years ago. All are single and all are under 35. The team enjoys an *esprit de corps*, a close sense of friendship born of the pressure situations and rapid response time that are such integral parts of the job.

The following two sections are descriptions of two recent projects undertaken by your team. In both situations, your team was expected to fly to the site of a disaster, assess the situation, learn about specific technological problems, and make recommendations so that future problems could be minimized.

26.7 *CASE 1: THE COLLAPSE OF THE TRENT RIVER BRIDGE*

The Trent River Bridge, an eighty-foot-long suspension bridge across the Trent River, collapsed six days ago. Since the disaster occurred during rush hour, there were a significant number of vehicles on the bridge. Eight people were killed and four were injured.

The disaster attracted nationwide attention, and the United States Federal Government has hired Gates and Associates to investigate the causes of the disaster and to make recommendations to insure that such catastrophes do not occur again.

The investigation conducted by all of the professionals at the scene reveals that it is metal fatigue and corrosion that are responsible for the accident. The Special Projects Investigation Team concludes that the deterioration of one pin in the pin-and-hanger assembly was the cause of the collapse of the bridge. Each assembly contained two pins: an upper pin attached to the shore-side portion of the bridge, and a lower pin attached to the suspended span. Two steel hangers connected each set of pins. It was the lower pin where the corrosion occurred, and one of the hangers gradually separated from the pin. After the weight of one corner of the bridge shifted to the other hanger, a fatigue crack developed and the bridge plunged into the water.

Since the pins were capped, visual inspection of the pin would not have revealed the problem. The Trent River Bridge's pin-and-hanger assembly is a common construction technique, and there are probably many more bridges throughout the United States with similar problems. Indeed, a 1987 Federal Highway Administration survey found that 42 percent (250,000) of the highway bridges in the United States were "deficient". The estimate for repairs or replacement necessary for all of the bridges was $51,400,000,000.

The problems are manifold. There is not enough money at the Federal, State, and local levels of government to make the required repairs. Budget cutbacks have limited the number of inspectors, and the inspectors that remain are overworked. Very little money is spent on basic research. (Construction industries in the United States spend less than 0.4 percent of revenues on basic research.) Many of the bridges were built long before modern heavy equipment and 18-wheeler trucks frequented America's highways. Traffic flow has steadily increased, and many bridges were designed before rush hour traffic became a fact of life.

There are also some technological gains that may help solve the problem. Portable x-ray equipment, pulse-echo scanners, pachometers, radar, and a gel electrode technique can help diagnose potential failures. Sensors and strain gauges can be embedded in new construction to monitor structural stability.

After five days at the disaster site, the Special Projects Investigation Team returns to Washington to conduct further research, decide upon recommendations, and write up a report.

26.8 CASE 2: THE TOXIC TRAIN CRASH

A freight train carrying toxic chemicals derailed two weeks ago in a suburban location outside of Sacramento, California. When firefighters arrived at the scene of the crash, they were unable to find out the exact nature of the chemicals. Neither the Chief Engineer on the train nor any of his co-workers were able to provide any data on the chemicals or the company that was shipping the chemicals.

The problem is this: the firefighters are forced to guess which procedure should be used to fight the fire and control the clouds of toxic gas that may result from their efforts to put out the fire. You can contain some toxic fires with water, but pouring water on some chemicals can make a fire much, much more deadly. If they know the contents of the train, the firefighters know how to proceed. As it is, sometimes they are forced to choose a course of action that causes more problems.

When the Special Projects Investigation Team looks into similar disasters, they discover that the lack of adequate information about toxic chemicals is fairly common during incidents of this nature. They further discover that the number of firefighters who contract specific forms of cancer is significantly higher than the rest of the population.

Gates and Associates is hired by the U. S. Federal Government's Office of Technology Assessment, in conjunction with the Department of Transportation, to develop a plan to solve the current problem. What is needed is a way to identify quickly and accurately the specific toxic chemicals that are being transported by rail or by road in the event of an emergency. The information must be conveyed as quickly as possible to the firefighters so that they can take the most appropriate action.

ORGANIZATIONAL ENVIRONMENT EXERCISES

1. You have been asked by your supervisor to write a report investigating the possibility of moving the offices of Gates and Associates out of Washington, D. C. and into a new location within a 20-mile radius of your current location. You are to choose an appropriate site somewhere near where you currently reside and make a recommendation in a formal report.

What you need to know

physical dimensions of the office space you will need
- 10,000 square feet

number of offices required
- 45 plus 2 large offices of at least 50 feet by 40 feet

number of parking spaces
- 30

(a) Locate an atlas that contains information on the area in question, the 20-mile radius surrounding the city or town where you attend school. Then locate a street map of the major roads in the area. Consult an almanac for any other useful information. Now write a preliminary report based on the information you have found. This report should be addressed to Sheila O'Brien, your immediate supervisor at Gates and Associates.

(b) Select three sites that seem suitable for your office. At a library locate recent copies of a local newspaper, and find out about current rental and purchasing prices in the locations you have selected. Write a short report that compares the three locales in terms of price, location, mileage from the nearest business center, and proximity to restaurants and other conveniences.

(c) Obtain information on the transit system in your town. Do any subway lines run near the location? What about commuter railroads? Sheila O'Brien has asked you to research this information. Write a memo in response to her request.

(d) Access to airports is very important for the employees of Gates and Associates. Sheila O'Brien has asked you to look into travel times to airports from each of the three locations you have chosen as options. Send her a memo that details the travel times during both regular and rush hours. (Assume that a car can average only 15 mph in rush hour traffic.)

(e) Now write a final report in which you will recommend one of the locations as the proposed site for the new offices of Gates and Associates. Send the report to Sheila O'Brien and to F. Robert Gates. Make sure you stress the benefits of the move to your chosen site.

2. This exercise is based on the scenario presented in Section 26.7. The exercises deal with the investigation into the collapse of the Trent River Bridge.

(a) Locate a street map of the area in which you currently live. Choose a bridge in a downtown or residential section for the site of the Trent River Bridge. Include a photocopy of this map with the location of the Trent River Bridge clearly designated. (You can use "white-out" to hide the actual name of the bridge.) Now write a preliminary memorandum to Sheila O'Brien in which you describe the location of the Bridge. Briefly inform her of the disaster. Refer to the map in your description.

(b) Conduct an investigation into the causes of metal fatigue and corrosion in general. What are the potential costs? (In 1988 the National Bureau of Standards estimated the annual cost of corrosion damage at $175 billion.) What methods are used to diagnose potential problems? Write an informational memo to the other members of your project team conveying what you find.

(c) Now narrow your focus to bridges. What problems are unique to the construction of bridges? How widespread are the problems? Write a short report to the other members of your project team to outline the situation.

(d) This exercise requires that other individuals assume roles within the Special Project Investigation Team. Four people should be designated as the team. Work together to arrive at potential solutions to the very large and potentially dangerous situation. F. Robert Gates would like you to prepare an oral presentation of your findings and recommendations four weeks from today. Representatives of the National Bureau of Standards and the State Department of Highways will be present.

3. This exercise is based on the scenario detailed in Section 26.8.

(a) Form a team of three to five members. Conduct research into the problem created by the lack of information about toxic chemicals being transported across the United States. Find information about incidents that involved toxic chemicals and what occurred during these incidents. Write a report to Sheila O'Brien in which you provide her with the details of your investigation.

(b) Now develop three possible solutions to the problem and write up these solutions in a memo to Sheila O'Brien. Make sure you describe the solution in detail and explain the benefits of the solution. (Hints: One possible way of solving this problem is to have identification required on some location on the train or truck; another is to have a computer network that firefighters can access. If you use these hints, work out the details in much more depth.)

(c) Sheila O'Brien responds to your two memos as follows: "You've done a great job. Now write a report to the Office of Technology Assessment and the Department of Transportation." Your report should include sections on the situation, previous incidents, statistics on the frequency of toxic accidents within the United States in the most recent calendar year, and possible solutions to the problem. Provide charts, tables and graphs when necessary.

(d) Your team should decide on which of the solutions you would recommend. Write a separate memo report to the Office of Technology Assessment in which you offer your recommendation and your reasons for selecting this solution. (Consider such factors as cost, ease of implementation, and effectiveness when making your decision.)

(e) Pamela Russell of your Project Team needs more information about the Hazardous Material Training Center in Pueblo, Colorado run by the Association of American Railroads. She is busy on another project. Send her whatever material you can locate.

(f) Create a flowchart that details the steps to follow upon the arrival at the scene of a train fire. (Here are some possibilities to get you started: search for leaks, diagram derailments, check bills of lading.) See Chapter 8, Including Graphics and Illustrations, for an example of a flowchart and a set of flowchart symbols.

(g) Write a short, informational report to Sheila O'Brien concerning the high incidence of cancer among firefighters. Try to locate sources which provide some statistics on the situation. A copy of the report should be sent to each of the other members of your team.

4. Your Special Project team is debating whether to put the following graphic on each of the reports filed by the team. What is your opinion? Why? Write a memo to Sheila O'Brien explaining your response.

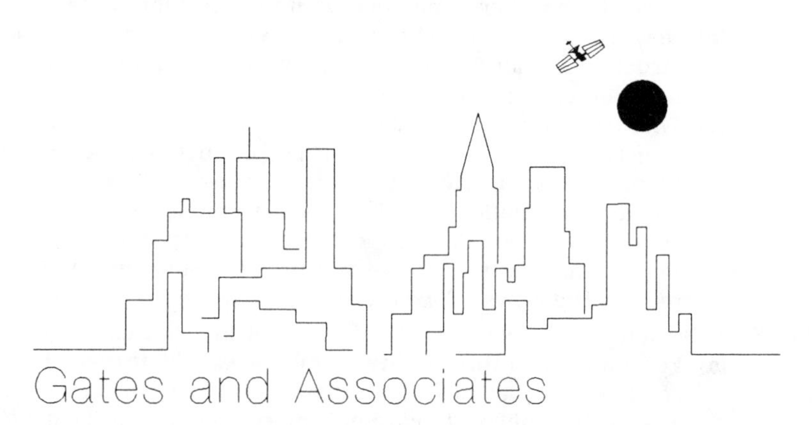

27

Communicating Across Cultures

27.1 INTRODUCTION

The purpose of this chapter is to describe some techniques and strategies which can be useful in designing technical communications for a global audience. Technical English is used to communicate important information to persons all around the world, yet very few technical writers or technical managers recognize the difficulties in communicating across cultures, even in the same language.

Many technical writers, for example, use local idioms and culture-bound references in an attempt to make their documents user-friendly and more accessible. To international readers whose command of English is not strong, these attempts to be friendly can make technical documents confusing and hard to understand.

Increasingly, technical documents will need to be designed for a world audience whose English is global rather than local. English is the primary language of technology in such diverse places as Nigeria, Singapore, Canada, and Guyana. Technical communicators need to consider these distinct cultures and their different varieties of English, and design documentation for a world audience.

In this chapter you will learn how to:

- avoid cultural stereotypes and cliches.
- read and decipher cultural cues.
- refine cross-cultural communication strategies.
- address issues from several cultural perspectives.

27.2 DEFINITION

Cross-cultural communication involves the sending and receiving of messages within the context of different cultures. As cultural differences increase or decrease, the difficulty of these tasks increases or decreases. In other words, it may be relatively simple for you to design your message for a Canadian audience, but it will be substantially more difficult to design the same message for your technical counterpart in a remote country like Zambia.

If your audience is similar to you, then you can refer to common experience, use language that you both understand, and communicate freely because you can be confident about how your message will be received. When your audience is from another culture, particularly one that is not familiar to you, then you will be less certain about how to design a message that is clear and effective.

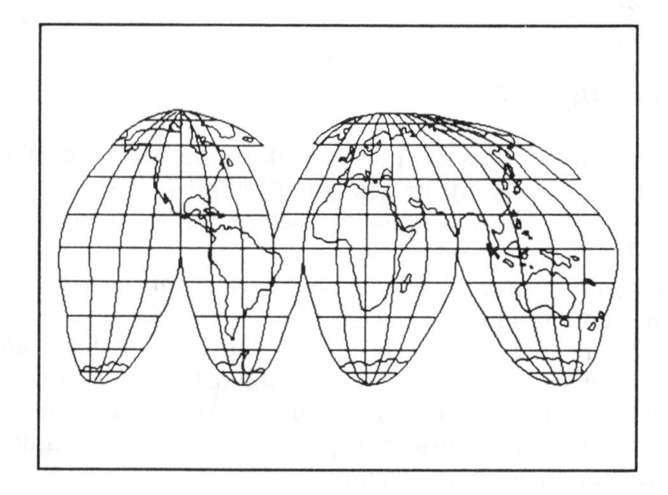

Culture influences people's self-identities, values, and patterns of language. It affects the ways we communicate and the ways we understand. Persons from other regions or different countries may communicate differently than you do, and may respond differently to what you write or say.

Your own culture seems normal and correct to you from the inside, and the behavior of persons in other cultures will often seem strange to you. Whatever puzzlement you might feel, however, you want to design your international communications with particular care and respect.

You will need to recognize, learn about, and appreciate other cultures if you want to be successful with the task of international communication. Foreign technical firms and agencies will frequently have different practices and perceptions than U.S. companies. Your abilities to understand and interact with different groups and cultures will have a significant effect on your career.

INTERNATIONAL RESEARCH RESOURCES

* World Trade Centers Association
* U.S. Department of Commerce/ Country Desk Officers
* International Trade Directors of the National Association of State Development Agencies
* Directors of Small Business Development Centers
* Small Business Administration Field Offices

27.3 *CULTURAL CUES*

Technical communications always take place in a context of human cultures: national, regional, ethnic, and even a culture within a community. Culture is a complex set of social attitudes and behaviors which make up the distinctive way of life of a people.

A culture reveals itself in many different ways: language, social organization, economic and political systems, technology, education, the role of sexes and age groups, religions, values, and attitudes. All of these dimensions blend together in a set of expectations about what is appropriate, acceptable, and natural behavior.

When we encounter members of different cultures and communicate with them, we need to be aware of alternative ways of behaving and thinking. Technology is a global enterprise and you will very likely find yourself working with persons from many different cultures.

Every culture offers certain cues or clues about its beliefs and attitudes. Styles of food, clothing, and architecture can give you useful insights. Body language will keep you informed about what is going on. Attentive observation will alert you to obvious norms and there are many books and training opportunities to prepare you for interacting with people through formal correspondence.

> There are a number of books that can help you understand other cultures and gain specific information about customs and business practices. Among them are:
>
> *The Do's and Taboos of International Trade: A Small Business Primer* by Roger E. Axtell. Published by John Wiley & Sons.
>
> *Do's and Taboos Around the World* by Roger E. Axtell. Published by John Wiley & Sons.
>
> *The Economist Business Traveller's Guides.* Published by Prentice Hall.

27.4 THE FOUR P's OF INTERNATIONAL PROTOCOL

Punctuality is one business behavior that varies from country to country and culture to culture. Here in the United States, punctuality is considered very important. You are expected to arrive on time and not to keep people waiting. In other cultures, however, time is perceived much more fluidly. A one o'clock meeting schedule may in fact indicate that the meeting will take place sometime in the mid-afternoon.

Patience is often a difficult quality for Americans to develop. We are an action-oriented, results-focused culture which likes decisive, short-term outcomes, quickly achieved and easily measured. There are many other cultures, however, the Japanese, for example, who do business very slowly and carefully and whose planning and strategies are aimed at long-term results.

Presents or small gifts are viewed differently by different cultures. Strict ethical codes often set clear limits on what presents you may or may not offer, but you should try to find out what attitudes are shared by the persons you are dealing with. An inexpensive gift, carefully and thoughtfully selected, can provide evidence of your interest and respect.

Preparation is always the key to success when you are dealing with persons from different cultures. Genuine curiosity will be perceived as a compliment to almost everyone. Indications that you have prepared ahead by reading books and articles, viewing films, and talking with foreign visitors will show your interest in and respect for the other person's culture. The fact that you are familiar with a literary figure, or an athlete's latest achievement, or local art history, can demonstrate your regard and gain you the good will which is essential to successful professional relationships.

> **BUSINESS CARDS ARE IMPORTANT!**
>
> In many parts of the world your business card is evidence that you are an important person. Rank and profession are taken very seriously in many countries. In addition, people you deal with will be able to understand your name and position more easily if they read it.
>
> Some simple suggestions:
>
> * Always include your company name and your position.
>
> * Include titles such as Research Associate, Division Manager, or Marketing Director.
>
> * No abbreviations. These may not be understood.
>
> * If English is not the primary language, have your cards printed in the local language on the reverse side.
>
> * Bring plenty of business cards and distribute them liberally.

27.5 *CULTURAL STEREOTYPES*

Something you do not want to burden yourself with are cultural stereotypes—ready-made images which are frequently wrong and incorrect. People from England are cold and formal, Italians demonstrative, and the French rude. Every nation has its stock adjectives, but they simply do not convey the complexity of various national cultures.

As easy as it is to generalize about different cultures, it is particularly important that we remember cultures are made up of individual persons who exhibit the whole range and variety of human behaviors. There are boisterous Britons, shy Italians, and polite French. Simple judgments, based on biased perceptions, will not lead to more effective communication.

27.6 COMMUNICATION STRATEGIES

Designing technical communications for international audiences requires special sensitivity to culture and anticipated formats. Business correspondence, telephone discussions, and face-to-face conversations should receive careful planning and consideration.

27.6.1 Writing

- **Keep sentences short and simple.** This is always a good rule for technical documents, but it is particularly important for an international audience. Accuracy and clarity are your two most important objectives. A simple, direct, professional style is helpful.
- **Avoid slang, local references, and humor.** Culture-bound references and local idioms do not translate. A reference to a *dog-and-pony show* can cause long and fruitless searches through English dictionaries and leave confusion behind.

- **Repeat key ideas.** This is an engineering principle: redundancy, repetition of critical elements. One way to make your main ideas clear is to repeat them frequently. This will make them visible and unmistakable, even for someone whose command of English is very weak.
- **Be polite but not ornate.** Even if you receive an elaborate and polished document filled with rhetorical flourishes, it is appropriate to respond simply and directly. People need to use what you are writing, not appreciate it for its fine prose. International communication should be simple and direct.

- **Proofread for possible misinterpretations.** Obviously you will want to check your document for mechanical errors. Bad writing causes most miscommunications. However, you also want to check your document for possible cultural confusions. Be sure your audience will know what you are referring to and what you intend to say. Again, simple and direct writing is most effective.

27.6.2 Telephoning

- **Speak carefully and slowly.** Most persons for whom English is a second or unfamiliar language complain that we speak too quickly. Native speakers of a language tend to blend words and sentences into a rapid porridge of sound. A slow, deliberate pace can help communication.
- **Repeat key ideas.** Repetition can focus your listener's attention on your main points. However confusing a conversation may become, the reoccurrence of key ideas demonstrates their importance. There is nothing wrong with repeating your central message. Your objective is communication.

- **Ask for ways to put it in writing: FAX, TELEX, E-MAIL.** A telephone connection and two modems will allow you to send written confirmation of your message almost instantly. TELEX and FAX facilities allow written messages and graphics to be transmitted from point to point throughout most of the world. Written confirmation will support your verbal message and ensure clarity and accuracy.
- **Know what time it is where you are calling.** This is a simple courtesy. Use maps or reference books to know the local time before you make your call. You do not want to call someone in the middle of the night, and you show respect for the listener's situation by being aware of this information.

- **Confirm your conversation in writing.** Written documents function as the memory of the technical and business world. A memo or a letter, written to confirm the details of a telephone conversation, will frequently establish the historical record of what was agreed upon. Particularly when dealing with persons whose English is problematic, it is important that you confirm key details in writing.

27.6.3 Conversing

- **Speak carefully and slowly.** This is the same as with telephone conversations. Your main objective is to be understood, not to sound profound. Pronounce your words carefully and speak deliberately. You do not need to speak more loudly with foreign listeners; you do need to speak deliberately.
- **Summarize: review what you are saying.** Summaries which assemble, review, and pull information together can be very useful to your listeners. You should pause to review key information and ideas, and to make sure that your audience is following what you are saying.

- **Echo what the other person is saying by repeating it.** You can assist the speaker in the conversation by repeating key ideas and information. Letting the speaker know you follow what is being said is courteous and helpful to the progress of the communication.
- **Paraphrase: express your key points in different ways.** Don't use the same words to say the same things over and over. If your audience

is confused or hesitant about asking questions, you can help them by repeating key ideas in different ways. You can refer to an input device, for example, as a keyboard, or a digitizer, or a mouse. Choose concrete words and examples whenever possible.

27.7 GLOBAL ENGLISH

The English language has come to dominate technical and professional communications at the international level. More than half of the world's scientific and technological literature is published in English. Worldwide organizations such as UNESCO, WHO, and even OPEC conduct all of their business in English. One fifth of the world's population has English as their first, second, or "official" language.

Consider what this means. For one thing, there are as many varieties of English as there are cultures which use the language. Persons in Hong Kong or Kenya may understand different things from the same expression. You must keep two questions in mind:

- Will your document be clear to persons for whom English is a foreign language?
- How will your message translate?

You want to avoid bad writing, culture-bound references and colloquial expressions. *April Fools!* may mean something to most people in the United States, but a technician in Kuwait will not get the reference. You also want to avoid a compressed style and inconsistent use of terms. Your objective is to communicate accurately and clearly. This means special care for your language.

Problems to Avoid
with global English

- Bad writing
- Inconsistency
- Culture-bound references
- Local expressions
- Compressed style

27.8 A FINAL WORD

International commerce is critical to the economy of the United States, and to the future of many technology enterprises. We need to export our technical products and services around the world. Technical development creates global interdependence and increased international communication. Your ability to communicate directly and accurately with overseas clients will be an important asset.

Individuals who can handle the complex task of cross-cultural communication will have a significant advantage in their careers. Perhaps the single most effective step that you can take to improve your skills in this area is to learn another language. The ability to speak and write someone else's language is invaluable. If you cannot learn another language, then take the trouble to learn a few words. You should always learn how to say thank you in the language of persons with whom you are dealing. Common expressions, including a toast or a folk saying, show that you have approached the culture with respect and that you are interested in the language of your audience.

CROSS-CULTURAL COMMUNICATIONS EXERCISES

1. Go out to lunch or dinner at an international restaurant which serves foods from another culture. Be brave and experiment; try some different dishes. Give a short oral report to your class or your group telling them about your experience. What did you learn about the culture from its food?
2. Imagine that you have been asked to entertain a visiting civil engineer from Indonesia. You are planning to take him to a baseball game. Make a list of the terms you will need to explain to your guest, terms such as *walking the batter, sacrificing,* or *stealing second base.* You need a lot of vocabulary to understand a baseball game.
3. One way to explore a culture, long distance, is to read its literature. Novels, plays, poetry, essays, biographies and histories, all reflect the inner and more personal views of a culture. Choose a culture you are interested in learning about, then use your library to discover some literature from that country or area. Write a one-page memo to your instructor describing what you learn.
4. Mr. Gates has been invited to visit Ibadan, Nigeria, where he will participate in a five-day seminar sponsored by the United States Information Service. He wants you to provide the following information as soon as possible. Use a memo format, but design this for use. He will take this with him and refer to it on a regular basis.

    - a list of national holidays, with dates and explanations.
    - data on local measurement systems, voltages, and any special requirements.

- information on climate and preferred business dress.
- customary office hours and methods of payment.

Include any other information you think would be useful to Mr. Gates. He is scheduled to be in Ibadan during the last week of March.

5. Use your library to discover at least three magazines published in other countries. Skim them quickly, with special attention for the advertisements. What do you notice about different styles and strategies, document designs, and cultural conventions? Discuss your findings with your group.
6. Design a graphic which will illustrate the relative importance of punctuality in three or four different countries from at least three different continents.
7. Sheila O'Brien assigns you to write a five-page report on the implications of the United States–Canada free trade agreement. Mr. Gates wants to know how this will affect our ability to bid on Canadian projects. He also wants to know what the rules are for Canadian competitors operating in the United States. Sheila explains that she chose you "because you write so well," but this is little solace. You know Mr. Gates will be reading this carefully.

Appendix A Metric Conversion

From the end of the Fiscal Year 1992 (September 30, 1992), the metric system is the system of weights and measures for the entire Federal government procurement process. The Omnibus Trade and Competitiveness Act of 1988 has already mandated the metric system as the preferred system for U.S. trade and commerce. Use the following table to convert to the metric system:

To convert	*To*	*Multiply by*
feet	meters	0.3048
inches	millimeters	25.400
inches	centimeters	2.5400
inches	meters	0.0254
miles	kilometers	1.6093
yards	meters	0.9144
acres	hectares	0.4047
square feet	square meters	0.0929
square miles	square kilometers	2.5900
square yards	square meters	0.8361
cubic feet	cubic meters	0.0283
cubic yards	cubic meters	0.7646
bushels (U.S.)	hectoliters	0.3524
gallons (U.S.)	liters	3.7853
ounces	grams	28.3495
pecks	liters	8.8096
pints (dry)	liters	0.5506
pints (liquid)	liters	0.4732
pounds	kilograms	0.4536
quarts (dry)	liters	1.1012
quarts (liquid)	liters	0.9463
tons (long)	metric tons	1.0160
tons (short)	metric tons	0.9072

LINEAR MEASURE	
10 millimeters (mm)	= 1 centimeter (cm)
10 centimeters	= 1 decimeter (dm)
10 decimeters	= 1 meter (m)
10 meters	= 1 dekameter (dam)
10 dekameters	= 1 hectometer (hm)
10 hectometers	= 1 kilometer (km)

SQUARE MEASURE	
100 square millimeters (mm^2)	= 1 sq centimeter (cm^2)
10,000 square centimeters	= 1 sq meter (m^2)
100 square meters	= 1 are (a)
100 ares	= 1 hectare (ha)
100 hectares	= 1 sq kilometer (km^2)

CUBIC MEASURE	
1000 cubic millimeters (mm^3)	= 1 cu centimeter (cm^3)
1000 cubic centimeters	= 1 cu decimeter (dm^3)
1000 cubic decimeters	= 1 cu meter (m^3)

VOLUME MEASURE	
10 milliliters (ml)	= 1 centiliter (cl)
10 centiliters	= 1 deciliter (dl)
10 deciliters	= 1 liter (l)
10 liters	= 1 dekaliter (dal)
10 dekaliters	= 1 hectoliter (hl)
10 hectoliters	= 1 kiloliter (kl)

WEIGHT	
10 milligrams (mg)	= 1 centigram (cg)
10 centigrams	= 1 decigram (dg)
10 decigrams	= 1 gram (g)
10 grams	= 1 dekagram (dag)
10 dekagrams	= 1 hectogram (hg)
10 hectograms	= 1 kilogram (kg)
1000 kilograms	= 1 metric ton (MT)

To convert Fahrenheit temperatures to Celsius, subtract 32 degrees, multiply by 5, and divide by nine. Thus, the normal body temperature of 98.6 degrees Fahrenheit can be converted as follows:

$$98.6°\text{F} - 32 = 66.6 \times 5 = 333 \div 9 = 37°\text{C}$$

	Fahrenheit	**Celsius**
Freezing point of water	32°	0°
Boiling point of water	212°	100°
Absolute zero	−459.6°	−273.1°

Appendix B Abbreviations

When you are communicating to a general audience, write out scientific and technical words and phrases unless the words would be well known to your readers. Never assume that your readers will undertand an abbreviation simply because you know what it means. There will be occasions, however, when you can be reasonably certain that your audience will understand your abbreviations. If you are writing to your state senators, you can be reasonably assured that they will know that OMB stand for the Office of Management and Budget.

Write out the words to be abbreviated when you use them for the first time and include the abbreviation in parentheses directly after the words. For example, ***the Accreditation Board for Engineering and Technology (ABET) regularly evaluates the programs of institutions that give degrees to technical students.*** Whenever you use an abbreviation, make sure to employ standard abbreviations by checking the abbreviation in a dictionary. Never guess at an abbreviation.

Use abbreviations over the full name when the abbreviation is more familiar to readers. For example, use NASA over the National Aeronautics and Space Administration and PVC piping over polyvinyl chloride piping.

Abbeviate organizational names only if the organization does. For example, use IBM and FBI because the organizations do. In general, respect the way an organization presents the name. For example, use the ampersand with AT&T.

Omit internal punctuation in most abbreviations, particularly acronyms (abbreviations formed from the initial letters of words and said as a single word) and initialisms (abbreviations formed from the initial letters of words but pronounced letter by letter). The final period is used, to avoid confusion, when the abbreviation spells out a whole word. For example, use *in.* because *in* is a word.

Use a period at the end of an abbreviation when the last letter is lower case.

Titles:	Mr., Mrs., Dr., Rev.
Addresses:	Ave., Rd., St., Blvd., Hwy.
References:	Ch., Fig., no.
Businesses:	Co., Corp., Inc., Ltd.
Days, Months:	Sat., Sun., Feb., Oct.

Do not use an apostrophe to form the plural of an abbreviation. Use the apostrophe to form the possessive of abbreviations. For example, use *RBIs* to represent the baseball term *runs batted in* and *the CIA's decision* to convey *the decision of the CIA.*

Avoid abbreviating weights and measurements, state names, and Latin abbreviations in the text of a document when employing professional and formal styles. Include the abbreviations in tables, charts, and illustrations.

Latin Abbreviations

e.g. = for example	i.e. = that is
p. = page	no. = number
pp. = pages	ed. = edition
etc. = et cetera	Fig. = figure

A.D. = anno domini (employing the Christian calendar)
B.C. = Before Christ (before the beginning of the Christian calendar)
A.M. or a.m. or AM = ante meridiem (from midnight to noon)
P.M. or p.m. or PM = post meridiem (from noon to midnight)

State Abbreviations

Alabama	AL	Montana	MT
Alaska	AK	Nebraska	NE
Arizona	AZ	Nevada	NV
Arkansas	AR	New Hampshire	NH
California	CA	New Jersey	NJ
Colorado	CO	New Mexico	NM
Connecticut	CT	New York	NY
Delaware	DE	North Carolina	NC
District of Columbia	DC	North Dakota	ND
Florida	FL	Ohio	OH
Georgia	GA	Oklahoma	OK
Guam	GU	Oregon	OR
Hawaii	HI	Pennsylvania	PA
Idaho	ID	Puerto Rico	PR
Illinois	IL	Rhode Island	RI
Indiana	IN	South Carolina	SC
Iowa	IA	South Dakota	SD
Kansas	KS	Tennessee	TN
Kentucky	KY	Texas	TX
Louisiana	LA	Utah	UT
Maine	ME	Vermont	VT
Maryland	MD	Virginia	VA
Massachusetts	MA	Virgin Islands	VI
Michigan	MI	Washington	WA
Minnesota	MN	West Virginia	WV
Mississippi	MS	Wisconsin	WI
Missouri	MO	Wyoming	WY

Appendix C Proofreading symbols

(delete mark)	Delete
(delete and close up mark)	Deleete and close up
⁀	Cl ose up
#	Insertspace
stet	Let it stand (i.e., the crossed out material above the dots)
¶	Begin a new paragraph
no ¶	Run two paragraphs together
(sp)	Spell out (e.g., 20 ft.)
tr	Trnaspose
lc	Lowercase a Capital letter
cap	capitalize a lowercase letter
o/	Correct an errer
^	Caret (indicates the point at which a marginal addition is to be inserted)
⊙	Period
(comma mark)	Comma
:/	Colon
;/	Semicolon
(apostrophe mark)	Apostrophe or single quotation mark
“/”	Quotation marks
?/	Question mark
!/	Exclamation point
=	Hyphen
(dash mark)	Dash
(/)	Parentheses
[/]	Brackets

Index